# LATENT HEAT-BASED THERMAL ENERGY STORAGE SYSTEMS

*Materials, Applications, and the Energy Market*

# LATENT HEAT-BASED THERMAL ENERGY STORAGE SYSTEMS

*Materials, Applications, and the Energy Market*

*Edited by*
**Amritanshu Shukla, PhD**
**Atul Sharma, PhD**
**Pascal Henry Biwolé, PhD**

APPLE
ACADEMIC
PRESS

Apple Academic Press Inc.
4164 Lakeshore Road
Burlington ON L7L 1A4, Canada

Apple Academic Press Inc.
1265 Goldenrod Circle NE
Palm Bay, Florida 32905, USA

First issued in paperback 2021

---

### Library and Archives Canada Cataloguing in Publication

Title: Latent heat-based thermal energy storage systems : materials, applications, and the energy market / edited by Amritanshu Shukla, PhD, Atul Sharma, PhD, Pascal Henry Biwolé, PhD.

Names: Shukla, Amritanshu, editor. | Sharma, Atul (Professor of environmental studies), editor. | Biwolé, Pascal Henry, editor.

Description: Includes bibliographical references and index.

Identifiers: Canadiana (print) 2020025538X | Canadiana (ebook) 20200255665 | ISBN 9781771888585 (hardcover) | ISBN 9780429328640 (ebook)

Subjects: LCSH: Heat storage.

Classification: LCC TJ260 .L38 2021 | DDC 621.402/8—dc23

---

### Library of Congress Cataloging-in-Publication Data

Names: Shukla, Amritanshu, editor. | Sharma, Atul (Professor of environmental studies) editor. | Biwolé, Pascal Henry, editor.

Title: Latent heat-based thermal energy storage systems : materials, applications, and the energy market / edited by Amritanshu Shukla, PhD, Atul Sharma, PhD, Pascal Henry Biwolé, PhD.

Description: Burlington, ON, Canada ; Palm Bay, Florida, USA : Apple Academic Press Inc., [2021] | Includes bibliographical references and index. | Summary: "In light of increasing human-induced global climate change, there is a greater need for clean energy resources and zero carbon projects. This new volume, Latent heat-based thermal energy storage systems: Materials, applications, and the energy market, offers up-to-date coverage of the fundamentals as well as recent advancements in energy efficient thermal energy storage materials, their characterization, and technological applications. Thermal energy storage (TES) systems offer very high-energy savings for many of our day-to-day applications and could be a strong component for enhancing the usage of renewable/clean energy-based devices. Because of its beneficial environmental impact, this technology has received wide attention in recent past, and dedicated research efforts have led to the development of novel materials, as well as to innovative applications in very many fields, ranging from buildings to textile, healthcare to agriculture, space to automobiles, etc. This book offers a valuable and informed systematic treatment of latent heat-based thermal energy storage systems, covering current energy research and important developmental work. The book provides valuable reference material for researchers in academia as well as in industry, enabling them to get an overall research update of latent heat-based thermal energy storage technology toward the ultimate goal of producing clean energy. Key features: Gives a comprehensive view of latent heat-based thermal energy storage systems, from the fundamentals to applications, while also addressing market issues and societal challenges. Includes research updates on the subject, covering novel materials as well as applications. Provides a balanced view between theoretical aspects and feasibility with chapters written by experts from academia as well as from industry. Covers the needs of faculty and students with appropriate details to serve as a professional reference"-- Provided by publisher.

Identifiers: LCCN 2020023466 (print) | LCCN 2020023467 (ebook) | ISBN 9781771888585 (hardcover) | ISBN 9780429328640 (ebook)

Subjects: LCSH: Heat storage devices--Materials. | Heat--Transmission. | Heat-transfer media. | Change of state (Physics)

Classification: LCC TJ260 .L39 2021 (print) | LCC TJ260 (ebook) | DDC 621.402/8--dc23

LC record available at https://lccn.loc.gov/2020023466

LC ebook record available at https://lccn.loc.gov/2020023467

---

# About the Editors

 Amritanshu Shukla, PhD, is currently working as an Associate Professor in Physics at the Rajiv Gandhi Institute of Petroleum Technology (RGIPT). He is also serving as Head of the Division of Basic Sciences and Humanities and Chief Vigilance Officer of the institute.

His research interests include theoretical physics, nuclear physics, and physics of renewable energy resources, including thermal energy storage materials. He has published more than 100 research papers in various international journals as well as in international and national conference proceedings. He has also written five international books, including *Energy Security & Sustainability: Present and Future* (CRC Press/Taylor & Francis), *Sustainability through Energy Efficient Buildings* (CRC Press/Taylor & Francis), and *Low Carbon Energy Supply -Trends, Technology, Management* (Springer Nature). He has delivered invited talks at various national and international institutes and is involved with several national as well as international projects and active research collaborations in India and abroad (such as with the University of Lund, Sweden; Kunshan University, Taiwan; Université Clermont Auvergne, France; Virginia Tech University, USA) on the topics of his research interests.

Formerly he has worked at some of the international research institutes, namely the Institute of Physics Bhubaneswar (Department of Atomic Energy, Govt. of India); Physical Research Laboratory Ahmedabad (Department of Space, Govt. of India); University of Rome/Gran Sasso National Laboratory, Italy; and University of North Carolina Chapel Hill, USA. He was awarded prestigious young scientist award in 2004 to attend a meeting with Nobel Laureates held at Lindau, Germany.

Atul Sharma, PhD, is currently Associate Professor at the Rajiv Gandhi Institute of Petroleum Technology (RGIPT), Jais, Amethi, Uttar Pradesh, India. Dr. Sharma has recently published an edited book, *Low Carbon Energy Supply: Trends, Technology, Management* (Springer Nature Singapore). He has also published several research papers in various international journals and conferences. He also holds several patents related to the PCM technology in the Taiwan region. He is working on the development and application of phase change materials, green building, solar water heating systems, solar air heating systems, solar drying systems, etc. Dr. Sharma is conducting research at the Non-Conventional Energy Laboratory, RGIPT, and is currently engaged with the Council of Science & Technology, a U.P.-sponsored project at his lab. In addition, he has served as an editorial board member and as a reviewer for many national and international journals, project reports, and book chapters.

He formerly worked as a Scientific Officer at the Regional Testing Centre Cum Technical Backup Unit for Solar Thermal Devices at the School of Energy and Environmental Studies, Devi Ahilya University, Indore, funded by the Ministry of Non-Conventional Energy Sources of the Government of India. He also worked as a Research Assistant at the Solar Thermal Research Center, New and Renewable Energy Research Department at the Korea Institute of Energy Research, Daejeon, South Korea, and as a Visiting Professor at the Department of Mechanical Engineering, Kun Shan University, Tainan, Taiwan.

Pascal Henry Biwolé, PhD, is a Professor at the Université Clermont Auvergne and Research Associate at MINES ParisTech Graduate School, Center for Processes, Renewable Energies and Energy Systems (PERSEE), Paris, France. His research interests are the modeling of heat and mass transfers in building innovative envelopes and systems, the modeling of phase change materials and aerogel-based insulating materials for energy applications, and the development of 3D particle tracking velocimetry for airflow study in buildings. His teaching interests are related to thermal energy storage, heat and mass transfers, fluid dynamics, and environmental building design and modeling. Dr. Biwolé graduated from the Special Military Academy of St-Cyr Coëtquidan, Guer, France. He holds a PhD degree in Civil Engineering from the National Institute of Applied Sciences of Lyon, France, and a research directorship habilitation degree from the University of Nice Sophia Antipolis, Nice, France. As secondary academic interest, he also holds a master's degree in Contemporary History from Paris-Sorbonne University. He currently heads the University Institute of Technology of Allier, France.

# Contents

# Contributors

**Patrick Achard**
MINES ParisTech, PSL Research University PERSEE–Center for Processes,
Renewable Energies and Energy Systems, CS 10207, F-06904 Sophia Antipolis, France

**Abhishek Anand**
Non-Conventional Energy Laboratory, Rajiv Gandhi Institute of Petroleum Technology,
Jais, Amethi, India

**Rohit Bansal**
Department of Management Studies, Rajiv Gandhi Institute of Petroleum Technology,
Raebareli, Uttar Pradesh, India

**Debajyoti Bose**
School of Engineering, University of Petroleum and Energy Studies, Energy Acres,
Bidholi, Dehradun 248007, Uttarakhand, India

**Pascal Henry Biwolé**
Université Clermont Auvergne, CNRS, SIGMA Clermont, Institut Pascal,
F-63000 Clermont–Ferrand, France
MINES Paris Tech, PSL Research University, PERSEE-Center for Processes,
Renewable Energies and Energy Systems, CS 10207, 06 904 Sophia Antipolis, France

**Abhay Kumar Choubey**
Division of Sciences & Humanities, Rajiv Gandhi Institute of Petroleum Technology,
Jais, Amethi 229304, India

**G. Raam Dheep**
Solar Thermal Energy Laboratory, Department of Green Energy Technology,
Pondicherry Central University, Pondicherry 605014, India

**Farouk Fardoun**
Lebanese University, Centre de Modélisation, Ecole Doctorale des Sciences et Technologie, Beirut,
Lebanon University Institute of Technology, Department GIM, Lebanese University, Saida, Lebanon

**Lucia Ianniciello**
MINES ParisTech, PSL Research University PERSEE–Center for Processes,
Renewable Energies and Energy Systems, CS 10207, F-06904 Sophia Antipolis, France

**Rachit Jaiswal**
Department of Management Studies, Rajiv Gandhi Institute of Petroleum Technology, Raebareli,
Uttar Pradesh, India

**Karunesh Kant**
Non-Conventional Energy Laboratory, Rajiv Gandhi Institute of Petroleum Technology,
Jais, Amethi 229304, India
Department of Mechanical Engineering, Eindhoven University of Technology,
5600 MB-Eindhoven, Netherlands

**Saurabh Mishra**
Rajiv Gandhi Institute of Petroleum Technology, Jais, Amethi, Uttar Pradesh, India

**M. B. Pereira**
Centre for Mechanical Technology and Automation (TEMA–UA),
Department of Mechanical Engineering, University of Aveiro, Aveiro, Portugal

**Atul Sharma**
Rajiv Gandhi Institute of Petroleum Technology (RGIPT), Jais, Amethi, Uttar Pradesh, India

**Madhu Sharma**
School of Engineering,University of Petroleum and Energy Studies, Energy Acres, Bidholi,
Dehradun 248007, Uttarakhand, India

**Amritanshu Shukla**
Non-Conventional energy laboratory, Rajiv Gandhi Institute of Petroleum Technology,
Jais, Amethi 229304, India
Department of Mechanical Engineering, Eindhoven University of Technology,
5600 MB-Eindhoven, Netherlands

**Shailendra Singh**
Non-Conventional energy laboratory, Rajiv Gandhi Institute of Petroleum Technology,
Jais, Amethi 229304, India

**Manoj K. Singh**
Department of Physics under School of Engineering and Technology,
Central University of Haryana, Haryana 123031, India

**Farah Souayfane**
Université Cote d'Azur, J.A. Dieudonné Laboratory, UMR CNRS 7351, 06108 Nice, France
Lebanese University, Centre de Modélisation, Ecole Doctorale des Sciences et Technologie,
Beirut, Lebanon

**Antonio C. M. Sousa**
Centre for Mechanical Technology and Automation (TEMA–UA),
Department of Mechanical Engineering, University of Aveiro, Aveiro, Portugal

**A. Sreekumar**
Solar Thermal Energy Laboratory, Department of Green Energy Technology,
Pondicherry Central University, Pondicherry 605014, India

**Vartika Srivastava**
Division of Sciences & Humanities, Rajiv Gandhi Institute of Petroleum Technology,
Jais, Amethi 229304, India

**L. Syam Sundar**
Centre for Mechanical Technology and Automation (TEMA–UA),
Department of Mechanical Engineering, University of Aveiro, Aveiro, Portugal

**Vikas**
Department of Management Studies, Rajiv Gandhi Institute of Petroleum Technology,
Raebareli, Uttar Pradesh, India

# Abbreviations

| | |
|---|---|
| AC | air conditioning |
| APCM | advanced phase change material |
| ASHP | air source heat pump |
| ASTM | American Society for Testing and Materials |
| BiMEP | Biscay Marine Energy Platform |
| BTMS | battery thermal management systems |
| CAES | compressed air energy storage |
| CLHSM | composite latent heat storage materials |
| CSP | concentrating solar power plant |
| DCNS | Direction des Construction Navales Services |
| DI | discomfort index |
| DNES | Department of Nonconventional Energy Sources |
| DOD | depth of discharge |
| DSC | differential scanning calorimeter |
| DTA | differential thermal analysis |
| EDF | Electricité de France |
| ETSC | evacuated tube solar collector |
| FTC | frequency of thermal comfort |
| HTF | heat transfer flow |
| IRENA | International Renewable Energy Agency |
| LPG | liquid petroleum gas |
| LHS | latent heat storage |
| LHTES | latent heat thermal storage |
| MNRE | Ministry of New and Renewable Energy |
| NZEB | net zero energy building |
| ORC | organic cycle |
| OTEC | ocean thermal energy conversion |
| OWC | oscillating water column |
| PCM | phase change material |
| PCM-SAHP | PCM-based solar air source heat pump |
| PHPS | hydro power storage |
| PSC | parabolic solar cooker |
| PV | photovoltaic |

| | |
|---|---|
| SEC | specific energy consumption |
| SEM | scanning electron microscope |
| SEGS | solar electricity generation systems |
| SHS | sensible heat storage |
| SMER | specific moisture extraction rate |
| SOC | state of charge |
| SSPCM | shape-stabilized phase-change material |
| SWHS | solar water heating systems |
| TES | thermal energy storage |
| TCS | thermo-chemical storage |
| WETS | Wave Energy Test Center |

# Preface

If one thinks of direct consequences of climate change as a kind of disorder our planet might be undergoing, it is the increasing visibility of extreme weather events. Prior to human-induced global warming, climate changes have been relatively slow over the course of hundreds to thousands of years. In the early start of the year 2019, the midwest part of USA, parts of Europe, and places in Himalayan belt were already experiencing sub-zero record temperatures, whereas places in Australia were adversely hit by extreme heat waves. The saga of climate change and its crisis is a topic to be debated/discussed, not only by the members of the scientific community but also by those involved in government, policymaking, technology, and so on, due to its large impact on mankind. Clean energy resources and zero carbon projects are being widely explored in every possible sector of energy usage.

In this connection, the present book offers up-to-date latest coverage of the fundamentals as well as recent advancements in energy efficient thermal energy storage materials, their characterization, and their technological applications. The thermal energy storage (TES) systems offer very high energy savings to many of our day-to-day applications and could be a strong component of enhancing the usage of renewable/clean energy based devices. Due to high environmental impact, this technology has received wide attention in the recent past, and dedicated research efforts have led to the development of novel materials as well as innovative applications in many fields, ranging from buildings to textile, healthcare to agriculture, space to automobiles, etc. A systematic treatment of latent heat-based thermal energy storage systems is an important aspect of current energy research and developmental work, which has been covered in this book.

We hope readers find this book interesting and that succeeds in serving as a graduate text, and reference material for researchers in academia as well as in industry, enabling them to get an overall research update of latent heat-based thermal energy storage technology. Although the ultimate goal of producing 100% clean energy still remains elusive, increasing attention and efforts in this direction are commendable and should give pleasure to all.

We would like to convey our appreciation to all the contributors as well as our academic collaborators, who have been always a source of inspiration to work with in this field of research. Our special thanks are also due to Dr. A. K. Haghi and his all team members from Apple Academic Press for their kind support and great effort in bringing the book to fruition. Last but not the least, editors are sincerely thankful to their family members, whose unending support and affection has always been a driving force in making any professional assignment a success!

**—Amritanshu Shukla**
**Atul Sharma**
**Pascal Henry Biwolé**

# CHAPTER 1

# Thermal Energy Storage Using Phase Change Materials: An Overview

ABHISHEK ANAND*, AMRITANSHU SHUKLA, and ATUL SHARMA

*Non-Conventional Energy Laboratory,*
*Rajiv Gandhi Institute of Petroleum Technology, Jais, Amethi, India*

*Corresponding author. E-mail: pre17001@rgipt.ac.in*

## ABSTRACT

Thermal energy storage has been reported to be an attractive option for renewable-based technology. There are several ways, i.e., sensible and latent by which this can be stored and utilized but the most proficient and efficacious way of doing it is through latent heat storage. It has the advantages, i.e., the energy is stored at the constant temperature, a practically large amount of energy is stored, minimum volume change during the phase transition, etc. This provides an attractive option for the phase change materials (PCMs) to be used in several applications, i.e., solar photovoltaic system, buildings, solar water heating system, solar greenhouse, solar cookers, solar air heating, etc. This chapter discusses the various energy storage systems with a focus on thermal energy storage. The thermal energy storage with phase change materials has been discussed in detail. Finally, the application of PCMs in various systems has been provided.

## 1.1 INTRODUCTION

The thermal energy storage (TES) is not a new concept, and it has been used for centuries and has always been one of the most critical components of energy storage systems. The TES can also be defined as the temporary storage of thermal energy at high or low temperatures. It provides a reservoir of energy to adjust the mismatch between energy demand and energy supply and to meet the energy needs at all times. Generally, it is used as a bridge to cross the gap between supply and demand and also plays an important role in energy conservation.[1-2] It leads to saving of premium fuels and makes the system more cost effective by reducing the wastage of energy and capital cost. For example, storage would improve the performance of a power generation plant by load leveling and higher efficiency would lead to energy conservation and lesser generation cost.

The storage of thermal energy in the form of sensible and latent heat has become an important aspect of energy management with the emphasis on the efficient use and conservation of the waste heat and solar energy in the industry and buildings. The use of phase change material (PCM) in a latent heat storage system is an effective way of storing thermal energy and has the advantages of high energy storage density and the isothermal nature of the storage process.

In the last few decades, many researchers and technologists developed and concluded different aspects to efficiently use PCMs in various applications such as heat pumps, agricultural greenhouse, solar cooker, solar water heating systems, solar air heating systems, waste heat recovery system, and thermal control in buildings. In this chapter, different methods of TES are first described with respect to their basic characteristics. This chapter will help to find the suitable PCM for various purposes, suitable heat exchangers with ways to enhance the heat transfer, and it will also help to provide a variety of designs to store thermal energy using PCMs for different applications. The different forms of energy that can be stored include mechanical, electrical, and thermal energy, which are described in the subsequent sections.[3-4]

### 1.1.1  MECHANICAL ENERGY STORAGE

Gravitational energy storage or pumped hydro power storage (PHPS), compressed air energy storage (CAES), and Flywheels are the basic mechanical energy storage systems. The PHPS and CAES technologies can be used for large-scale utility energy storage while flywheels are more suitable for intermediate storage. CAES technology also stores low-cost off-peak energy, in the form of compressed air in an underground reservoir.

### 1.1.2 ELECTRICAL STORAGE

Electrical energy storage systems are expected to work for load leveling, fluctuation smoothing, uninterruptible power supply generated by a wind turbine, or photovoltaic plants. Energy storage through batteries is an option for storing the electrical energy. The most common type of storage batteries are the lead acid and Ni-Cd.

### 1.1.3 THERMOCHEMICAL ENERGY STORAGE

Thermochemical storage is a new concept being considered for solar thermal applications. This storage technique relies on the energy absorbed and released in breaking and reforming molecular bonds in a completely reversible chemical reaction. In this case, the heat stored depends on the amount of storage material, the endothermic heat of reaction, and the extent of conversion.

$$Q = a_r.m.\Delta h_r \qquad (1.1)$$

where, $a_r$ = fraction reacted, $m$ = amount of storage material (kg) and $\Delta h_r$ = endothermic heat of reaction.

### 1.1.4 THERMAL ENERGY STORAGE (TES)

TES systems remove heat from or add heat to a storage medium for use at another time. Energy may be charged, stored, and discharged daily, weekly, annually, or in seasonal or rapid batch-process cycles. High-temperature storage is typically associated with solar energy or high-temperature heating, and cool storage with air-conditioning, refrigeration, or cryogenic-temperature processes.

Among these storage methods, TES, which has been considered an advanced energy technology method, is of great importance in a wide variety of energy applications.[5] TES can be stored as a change in internal energy of a material as sensible heat, latent heat, and thermochemical or combination of these.[6,7] An overview of the major techniques of TES is shown in Figure 1.1.[1,4,8]

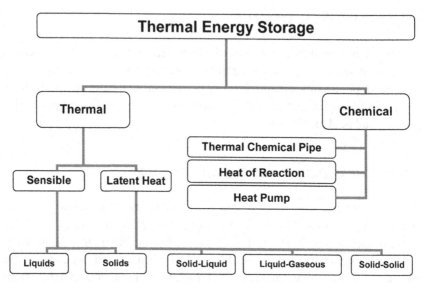

**FFIGURE 1.1**   Different types of thermal energy storage.

### 1.1.4.1   SENSIBLE HEAT STORAGE

Sensible heat storage (SHS) systems (e.g., rock beds, water tanks, aquifers) commonly use rocks or water as the storage medium and are available for short- and long-term storage. Thermal energy is stored by raising the temperature of a solid or liquid. SHS system utilizes the heat capacity and the change in temperature of the material during the process of charging and discharging. Water appears to be the best SHS liquid available because it is inexpensive and has a high specific heat. However, above 100°C, oils, molten salts, liquid metals, etc., are used. The amount of heat stored depends on the specific heat of the medium, the temperature change, and the amount of storage material.[4,9,10]

$$Q = \int_{T_i}^{T_f} m.Cp.dT \tag{1.2}$$

$$= m\, C_{ap}\, (T_f - T_i) \tag{1.3}$$

where, $Q$ = quantity of heat stored (J), $T_i$ = initial temperature (°C), $T_f$ = final temperature (°C), $m$ = mass of heat storage medium (kg), $C_p$ = specific heat (J/kg K), and $C_{ap}$ = average specific heat between $T_i$ and $T_f$ (J/kg K).

## 1.1.4.2 LATENT HEAT STORAGE

The selection of the TES systems mainly depends on the storage period required (i.e., diurnal or seasonal), economic viability, operating conditions, etc. In practice, many research and development activities related to energy have been concentrated on efficient energy use and energy savings, leading to energy conservation. In this regard, TES systems appear to be one of the most attractive thermal applications. Limited performance evaluation studies on the energy analysis of TES systems have already been undertaken.[11–15]

Latent heat storage (LHS) systems, which store energy in PCM (e.g., salt hydrates and organic substances), are used for short-term storage. The heat is stored when the material changes phase from a solid to a liquid. Materials used for this purpose are called PCMs. The storage capacity of the LHS system with a PCM medium[4,9,10] is given by

$$Q = \int_{T_i}^{T_m} m.Cp.dT + m.a_m.\Delta h_m + \int_{T_m}^{T_f} m.Cp.dT \qquad (1.4)$$

$$Q = m.\left[ C_{sp}.(T_m - T_i) + a_m.\Delta h_m + C_{lp}.(T_f - T_m) \right] \qquad (1.5)$$

where, $m$ = amount of storage material (kg), $a_m$ = fraction melted, $\Delta h_m$ = heat of fusion per unit mass (J/kg), $T_i$ = initial temperature (°C), $T_f$ = final temperature (°C), $T_m$ = melting temperature (°C), $C_{sp}$ = average specific heat between $T_i$ and $T_m$ (kJ/kg K), and $C_{lp}$ = average specific heat between $T_m$ and $T_f$ (J/kg K).

Among aforementioned thermal heat storage techniques, latent heat TES is particularly attractive due to its ability to provide high energy storage density and its characteristics to store heat at a constant temperature corresponding to the phase transition temperature of PCM. Phase change can be in the following form: solid–solid, solid–liquid, solid–gas, liquid–gas, and vice versa.

In solid–solid transitions, heat is stored as the material is transformed from one crystalline to another. These transitions generally have small latent heat and small volume changes than solid–liquid transitions. Solid–solid PCM offer the advantages of less-stringent container requirements and greater design flexibility.[14]

Solid–gas and liquid–gas transition through have the high latent heat of phase transition but their large volume changes on phase transition are associated with the containment problems and rule out their potential utility in TES systems. Large changes in volume make the system complex and impractical.[15] Solid–liquid transformations have comparatively smaller latent heat than liquid–gas. However, these transformations involve only a small change (of the order of 10% or less) in volume. Solid–liquid transitions have proved to be economically attractive for use in TES systems. PCMs cannot be used as a heat transfer medium. A separate heat transfer medium must be employed with a heat exchanger in between to transfer energy from the source to the PCM and from PCM to the load. The heat exchanger to be used has to be designed specially, in view of the low thermal diffusivity of PCMs in general.

Any LHS system, therefore, possesses at least the following three components:

i) A suitable PCM with its melting point in the desired temperature range
ii) A suitable heat exchange surface, and
iii) A suitable container compatible with the PCM

A wide range of technical options available for storing low-temperature TES is shown in Figure 1.2.[4,14]

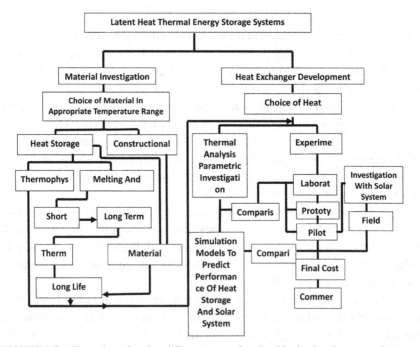

**FIGURE 1.2** Flow chart showing different stages involved in the development of a latent heat thermal energy storage system.

## 1.2 LATENT HEAT STORAGE MATERIALS

The latent heat storage materials are commonly known as PCMs. The thermal energy transfer occurs when a material changes from solid to liquid, or liquid to solid. This is called a change in state, or "phase." Initially, these solid–liquid PCMs perform like conventional storage materials, their temperature rises as they absorb heat. Unlike conventional (sensible) storage materials, PCM absorbs and releases heat at a nearly constant temperature. A large number of PCMs are known to melt with a heat of fusion in any required range. However, for their employment as PCMs, these materials must exhibit certain desirable thermodynamic, kinetic, and chemical properties, which are as given in Table 1.1.[4,8,16] Moreover, economic considerations and easy availability of these materials have to be kept in mind.

**TABLE 1.1** Main Desirable Properties of PCMs.

| | |
|---|---|
| Thermal properties | • Suitable phase-transition temperature |
| | • High latent heat of transition |
| | • High thermal conductivity in both liquid and solid phases |
| | • Good heat transfer |
| Physical properties | • Favorable phase equilibrium |
| | • High density |
| | • Small volume change |
| | • Low vapor pressure |
| Kinetic properties | • No supercooling |
| | • Sufficient crystallization rate |
| Chemical properties | • Long-term chemical stability |
| | • Compatibility with materials of construction |
| | • No toxicity |
| | • No fire hazard |
| Economic properties | • Abundant |
| | • Available |
| | • Cost effective |

### 1.2.1 CLASSIFICATION OF PCMs

In 1983, Abhat[15] gave a useful classification of the substances used for TES. A large number of PCMs (organic, inorganic, and eutectic) are available in any required temperature range.[4,9,16–21] A classification of PCMs is given in Figure 1.3. In practice, several PCMs are known, such as paraffin's, fatty acids, organic and inorganic salt hydrates, and organic and inorganic eutectic compounds. A comparison of the advantages and disadvantages of organic and inorganic materials is shown in Table 1.2.[4,8,19]

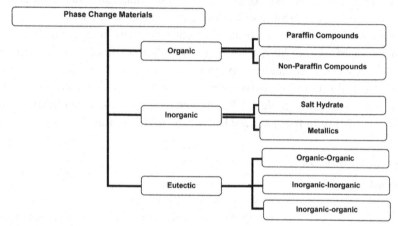

**FIGURE 1.3**   Classification of PCMs. Reprinted with permission from Ref. 4. © 2009 Elsevier.

**TABLE 1.2**   Comparison of Organic and Inorganic Materials for Heat Storage.

| Classification | Advantages | Disadvantages |
|---|---|---|
| Organics | • Chemical and thermal stability, suffer little or no supercooling, noncorrosives, nontoxic, high heat of fusion, low vapor pressure | • Low thermal conductivity, high changes in volumes on phase change, inflammability, lower phase change enthalpy |
| Inorganic | • High heat of fusion, good thermal conductivity, cheap, nonflammable, small volume changes on melting | • Lack of thermal stability, supercooling, corrosion, phase decomposition and suffer from loss of hydrate |
| Eutectics | • Sharp melting temperature<br><br>• High volumetric thermal storage density | • Lack of currently available test data of thermophysical properties |

### 1.2.1.1   ORGANIC PCMs

Organic materials are further described as paraffin and nonparaffin. Organic materials include congruent melting, which means they melt and freeze repeatedly without phase segregation and consequent degradation of their latent heat of fusion; self-nucleation means they crystallize with little or no supercooling and usually are noncorrosiveness.

### 1.2.1.1.1 Paraffins

Paraffin wax consists of a mixture of mostly straight chain n-alkanes CH3-(CH2)-CH3. The crystallization of the (CH3)–chain release a large amount of latent heat. Both the melting point and latent heat of fusion increase with chain length. Paraffin qualifies as the heat of fusion storage materials due to their availability in a large temperature range. Some selected paraffins are shown in Table 1.3 along with their melting-point and latent heat of fusion.[4,8]

TABLE 1.3 Physical Properties of Some Paraffins.

| Paraffin | Freezing point/range (°C) | Heat of fusion (kJ/kg) |
|---|---|---|
| 6106[a] | 42–44 | 189 |
| P116[b] | 45–48 | 210 |
| 5838 | 48–50 | 189 |
| 6035 | 58–60 | 189 |
| 6403 | 62–64 | 189 |
| 6499 | 66–68 | 189 |

[a]Manufacturer of technical Grade Paraffins 6106, 5838, 6035, 6403 and 6499: Ter Hell Paraffin Hamburg, FRG.
[b]Manufacturer of Paraffins P116: Sun Company, USA.

### 1.2.1.1.2 Nonparaffins

The nonparaffin organic are the most numerous of the PCMs with highly varied properties. Each of these materials will have its own properties unlike the paraffin's, which have very similar properties. This is the largest category of candidate's materials for phase change storage. Buddhi and Sawhney[8] and Abhat et al.[22] have conducted an extensive survey of organic materials and identified a number of esters, fatty acids, alcohols, and glycols suitable for energy storage. These materials are flammable and should not be exposed to excessively high temperature, flames, or oxidizing agents. Fatty acids also show reproducible melting and freezing behavior and freeze with no supercooling.[23-26] Their major drawback, however, is their cost, which is 2–2.5 times greater than that of technical grade paraffin. They are also mild corrosive. Some fatty acids show interest to low-temperature TES applications and are tabulated in Table 1.4.

**TABLE 1.4**   Melting Point and Latent Heat of Fusion: Fatty Acids.

| Material | Melting point (°C) | Latent heat (kJ/ kg) |
| --- | --- | --- |
| Acetic acid | 16.7 | 184 |
| Polyethylene glycol 600 | 20–25 | 146 |
| Capric acid | 36 | 152 |
| Eladic acid | 47 | 218 |
| Lauric acid | 49 | 178 |
| Pentadecanoic acid | 52.5 | 178 |
| Myristic acid | 58 | 199 |
| Palmatic acid | 55 | 163 |
| Stearic acid | 69.4 | 199 |
| Acetamide | 81 | 241 |

## 1.2.1.2   INORGANIC PHASE CHANGE MATERIALS

Inorganic materials are further classified as salt hydrate and metallics. These PCMs do not supercool appreciably and their heats of fusion do not degrade with cycling.

### 1.2.1.2.1   Salt Hydrates

Salt hydrates are the most important group of PCMs, which have been extensively studied for their use in TES systems. They are not very corrosive, are compatible with plastics, and are only slightly toxic. Many salt hydrates are sufficiently inexpensive for the use in storage.[27]

Salt hydrates may be regarded as alloys of inorganic salts and water forming a typical crystalline solid of general formula AB. $nH_2O$. The solid–liquid transformation of salt hydrates is actually a dehydration of hydration of the salt, although this process resembles melting or freezing thermodynamically. A salt hydrate usually melts to either to a salt hydrate with fewer moles of water, that is,

$$AB.nH_2O \text{-----------------} AB.mH_2O + (n-m)H_2O \text{-----------------}$$

or to its anhydrous form

$$AB.nH_2O \text{-----------------} AB + n\,H_2O. \text{-----------------}$$

At the fusion temperature, the rate of nucleation is generally very low. To achieve a reasonable rate of nucleation, the solution has to be suspended and hence energy instead of being discharged at fusion temperature is discharged at a much lower temperature. A list of salt hydrates is given in Table 1.5.

**TABLE 1.5**  Melting Point and Latent Heat of Fusion: Salt Hydrates.

| Material | Melting point (°C) | Latent heat (kJ/ kg) |
|---|---|---|
| $CaCl_2.12\,H_2O$ | 29.8 | 174 |
| $LiNO_3.3\,H_2O$ | 30 | 189 |
| $Na_2CO_3.10\,H_2O$ | 32.0 | 267 |
| $Na_2SO_4.10\,H_2O$ | 32.4 | 241 |
| $LiBr_2.2\,H_2O$ | 34 | 124 |
| $Zn(NO_3)_2.6\,H_2O$ | 36.1 | 134 |
| $Mn(NO_3)_2.4\,H_2O$ | 37.1 | 115 |
| $Na_2HPO_4.12\,H_2O$ | 40.0 | 279 |
| $K_2HPO_4.7\,H_2O$ | 45.0 | 145 |
| $Mg(NO_3).4\,H_2O$ | 47.0 | 142 |
| $Fe(NO_3)_3.9\,H_2O$ | 47 | 155 |
| $K_2HPO_4.3\,H_2O$ | 48 | 99 |
| $Na_2S_2O_3.5\,H_2O$ | 48.5 | 210 |
| $Ni(NO_3)_2.6\,H_2O$ | 57.0 | 169 |
| $CH_3COONa.3\,H_2O$ | 58.0 | 265 |
| $NaAl(SO_4)_2.10\,H_2O$ | 61.0 | 181 |
| $Ba(OH)_2.8H_2O$ | 78 | 265 |
| $Mg(NO_3)_2.6H_2O$ | 89.9 | 167 |
| $MgCl_2.6H_2O$ | 117 | 167 |

Generally, three types of the behavior of the melted salts can be identi-fied: congruent, incongruent, and semicongruent melting.

i)   Congruent melting occurs when the anhydrous salt is completely soluble in its water of hydration at the melting temperature.

ii)  Semicongruent melting occurs when the liquid and solid phases in equilibrium during a phase transition is of different melting composition because of conversion of the hydrate to a lower hydrated material through loss of water.

iii) Incongruent melting occurs when the salt is not entirely soluble in its water of hydration at the melting point. Thus, the resulting solution is supersaturated at the melting temperature. Some of the ways in which incongruent melting can be minimized are as follows:

- Mechanical stirring.
- Encapsulation of PCMs to reduce separation.
- By adding of the thickening agents, which prevent setting of solid salts by holding it in suspension.
- By use of excess water so that melted crystals do not produce a supersaturated solution.
- By modifying the chemical composition of the system and making incongruent melting congruent.

### 1.2.1.2.2   Metallics

This category includes the low melting metals and metal eutectics. These metallics have not yet been seriously considered for PCM technology because of weight penalties. However, when the volume is a consideration, they are likely candidates because of the high heat of fusion per unit volume. They have high thermal conductivities, so fillers with added weight penalties are not required. The use of metallics poses a number of unusual engineering problems. A major difference between the metallics and other PCMs is their high thermal conductivity.

### 1.2.1.3   EUTECTICS

A eutectic is a minimum-melting composition of two or more components, each of which melts and freezes congruently forming a mixture of the component crystals during crystallization.[28] Eutectic nearly always melts and freezes without segregation since they freeze to an intimate mixture of crystals, leaving little opportunity for the components to separate. On melting, both components liquefy simultaneously, again with separation unlikely.

## 1.3 MEASUREMENT TECHNIQUES OF LATENT HEAT OF FUSION AND MELTING TEMPERATURE

Generally, two techniques (i) differential thermal analysis (DTA) and (ii) differential scanning calorimeter (DSC) are used to measure the basic properties such as latent heat and melting temperature of the PCM.[29]

## 1.4 APPLICATIONS OF PCMs

PCMs are able to store and release thermal energy under controlled circumstances. Nowadays, PCMs have a lot of applications and are widely used in TES systems. Some of the different applications found in the literature are given in the following points:

- Thermal storage of solar energy
- Passive storage in bioclimatic building/architecture (HDPE)
- Cooling: use of off-peak rates and reduction of installed power, ice bank
- Heating and sanitary hot water: using off-peak rate and adapting unloading curves
- Safety: temperature maintenance in rooms with computers or electrical appliances
- Thermal protection of food: transport, hotel trade, ice-cream, etc.
- Food agro-industry, wine, milk products (absorbing peaks in demand), greenhouses
- Thermal protection of electronic devices (integrated into the appliance
- Medical applications: transport of blood, operating tables, hot–cold therapies
- Cooling of engines (electric and combustion)
- Thermal comfort in vehicles
- Softening of exothermic temperature peaks in chemical reactions
- Spacecraft thermal systems
- Solar power plants

Many companies (Climator, Cristopia, EPS Ltd., Mitsubishi Chemical Corporation, Rubitherm GmbH TEAP, Pluss Polymer) are engaged in the development of PCMs for several applications.[30-36]

Due to phase change in the process of exploitation, PCMs are usually encapsulated in various containers. The investigation of chemical compatibility of low-temperature PCMs and design structure materials have shown that stainless steel, polypropylene, and polyolefin can be used in most cases as suitable container materials. Many companies use polyolefin and polypropylene to encapsulate its PCM products. The storage materials are packed in the container in the form of spherical balls, rectangular, or cylindrical bars and they are hermetically sealed. Here, Figure 1.4 shows some of the commercially manufactured phase change storage products.[21]

**TEAP Polyolefine spherical capsule**

**TEAP Polypropylene flat panel**

**EPS Ltd stainless ball capsule**

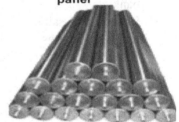

**EPS Ltd module beam**

**FIGURE 1.4**   Some of the commercially manufactured phase change storage products. Reprinted with permission from Ref. 21. © 2007 Elsevier.

### 1.4.1   SOLAR WATER HEATING SYSTEM

Integrating solar energy collection system into the building shell and mechanical systems may reduce the cost of the solar energy systems as well improve the efficiency of the collection. The solar water heater is getting popularity since they are relatively inexpensive and simple to fabricate and maintain. Bhargava[37] utilized the PCM for a solar water heater and concluded that the efficiency of the system and the outlet water temperature during the evening hours increases with the increase in the thermal conductivity of the solid–liquid phases of the materials. Hasan[38] has investigated some fatty acids as PCMs for domestic water heating. They recommended that myristic acid, palmitic acid, and stearic acid, with melting temperature between 50°C and 70°C are the most promising PCMs for water heating.

Kurklu et al.[39] designed and developed a new type of solar collector with PCM. The solar collector, which exhibited a net solar aperture area of 1.44 m², consisted of two adjoining sections one filled with water and the other with a PCM (paraffin wax) with a melting and freezing range of about 45–50°C. The PCM functioned both as an energy storage material for the stabilization, theoretically, of the water temperature and as an insulation material due to its low thermal conductivity value. The results of the study indicated that the water temperature exceeded 55°C during a typical day of high solar radiation and it was kept over 30°C during the whole night.

Suat et al.[40] presented a conventional open-loop passive solar water-heating system combined with sodium thiosulfate pentahydrate (Fig. 1.5). Long and Zhu[41] developed an air source heat pump water heater with PCM for TES to take advantage of off-peak electrical energy. Shukla et al.[42] investigated and analyzed TES incorporating with and without PCM for use in solar water heaters.

Mohsen et al.[43] conducted a study on a compact solar water heater, in order to evaluate the performance of the heater and to determine the optimal depth of the storage tank of depths of 5, 10, and 15 cm with single and double glazing. The 10 cm depth of the tank is optimum which can supply hot water for 24 h. The rise of water temperature is slightly higher in the case of single glazing than the double glazed system, while the double glazed system is more effective in retaining higher temperatures during night hours. Kumar and Rosen[44] developed an integrated solar water heater with a corrugated absorber surface and observed that the modified water heater does not increase the system cost appreciably over a plane surface, as most parts of the design of water heater are identical, except in the shape of the absorptive surface.

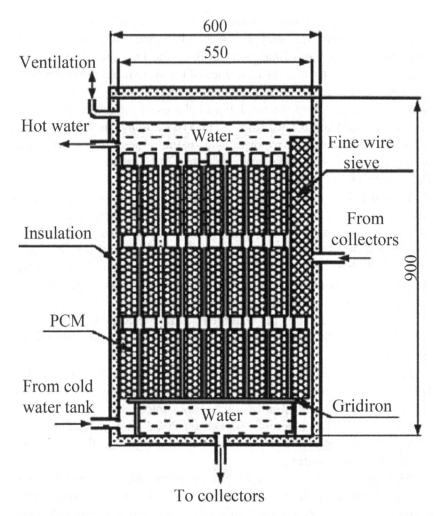

**FIGURE 1.5** Detailed cross-sectional view of the cylindrical heat storage tank combined with PCM. Reprinted with permission from Ref. 40. © 2005 Elsevier.

## 1.4.2 *SOLAR AIR HEATING SYSTEMS*

A solar air heating system consists of a number of individual solar air heating modules that can be arranged end-to-end or side-by-side to provide any desired heat or temperature output under the design conditions. Since the availability of solar energy is usually not coincident with the demand, heat collected from solar radiation is also stored in a TES unit, which is used to store heat during the day so that stored heat can be supplied at night or when there is no sunshine. During times of sunshine and when heating is required, air is passed through the collector and subsequently into the home.

Morrison, Abdel Khalik, and Jurinak in their different studies[45,46] evaluated the performance of air-based solar heating systems utilizing phase change energy storage unit. It was also found that air-based system utilizing sodium sulfate decahydrate as a storage medium requires roughly one-fourth the storage volume of a pebble bed and one-half the storage volume of a water tank. Ghoneim and Klein[47] compared theoretically the performance of phase change and sensible heat storage for air and water-based solar heating systems. Fath[48] developed a simple solar air heater integrated with TES system. Heat storage materials were filled in the tubes and set as an absorber of the collector. Paraffin wax and $Na_2SO_4.10H_2O$ are used as PCM. With paraffin wax, the outlet air temperature was maintained at 5°C above the ambient temperature for about 16 h (almost 4 h after sunset). He recommended that $Na_2SO_4.10H_2O$ (melting point 32°C) should be used for lower ambient temperature and solar irradiance.

University of South Australia (UniSA)[49] has developed a roof-integrated solar air heating/storage system, which uses existing corrugated iron roof sheets as a solar collector for heating air. Adequate amounts of fresh air are introduced when the solar heating system is delivering heat into the home Enibe[50] presented design, construction, and performance evaluation of a passive solar powered air heating system consisting of a single-glazed flat plate solar collector integrated with a PCM heat storage system. The PCM is prepared in modules, with the modules equispaced across the absorber plate. The spaces between the module pairs serve as the air heating channels, the channels being connected to the common air inlet and discharge headers at the ambient temperature range of 19–41°C.

Arkar[51] designed a solar-assisted ventilation system based on PCM storage. He recommended that paraffin-spherical encapsulations provide a homogeneous porous effect in a ventilation duct, which can improve the thermal conductivity of paraffin. RT20 paraffin was used as a PCM in this study. Hammou and Lacroix[52] have investigated a hybrid TES system for managing simultaneously the storage of heat from solar and electric energy. They have proposed a mathematical model to examine the effect of various storage materials and the operating conditions on the thermal behavior of the system.

Zhou et al.[53] investigated numerically the performance of a hybrid heating-system combined with thermal storage by shape-stabilized phase-change material (SSPCM) plates. A direct gain passive-solar house in Beijing was considered: it includes SSPCM plates as inner linings of walls and the ceiling. The results indicated the thermal-storage effect of SSPCM plates, which improves the indoor thermal comfort level and saves about 47% of normal-and-peak-hour energy use and 12% of total energy consumption in winter. This hybrid heating system can level the electrical load for power plants and would provide significant economic benefits in areas where night and day electricity tariff policy is used.

Diaconu et al.[54] experimentally investigated a low-temperature heat storage system through the PCM based slurry for solar air conditioning applications. It was found that the values of the heat transfer coefficient for the PCM slurry were higher than for water, under identical temperature conditions inside the phase change interval.

### 1.4.3 SOLAR COOKERS

A solar cooker is a device that utilizes the solar energy to cook the food in a daytime. But its use is limited due to no sunlight in the evening and night time. That's why solar cooker cannot cook the food in the late evening. That drawback can be solved by the storage unit associated within a solar cooker. Therefore, in this section, an attempt has been taken to summarize the investigation of the solar cooking system incorporating with PCMs.

Domanski et al.[55] have studied the use of a PCM ($Mg(NO_3)_2.6H_2O$) as a storage medium for a box type solar cooker designed to cook the food in the late evening hours and/or during the non-sunshine hours. Buddhi and Sahoo[56] filled commercial grade stearic acid below the absorbing plate of the box-type solar cooker. Sharma et al.[57] developed a PCM storage unit with acetamide for a box type solar cooker to cook the food in the late evening. They recommended that the melting temperature of a PCM should be between 105°C and 110°C for evening cooking. Later Buddhi et al.[58] developed a storage unit with acetanilide for a box type solar cooker to store a larger quantity of heat through a PCM, so that they used three reflectors to get more input solar radiation (Fig. 1.6). Sharma et al.[59] also used erythritol as a latent heat storage material for the solar cooker based on an evacuated tube solar collector.

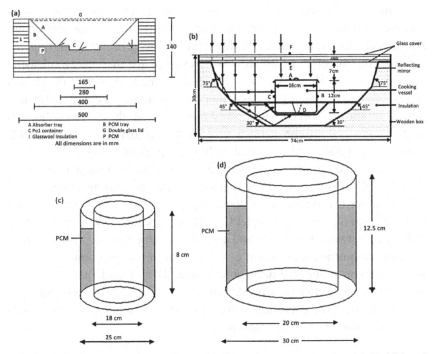

**FIGURE 1.6** Box-type solar cookers with latent heat storage units: (a) Buddhi and Sahoo,[56] (b) Domanski et al.,[55] (c) Sharma et al.,[57] and (d) Buddhi et al.[58]
(a) Reprinted with permission from Ref. 56. © 1997 Elsevier. (b) Reprinted with permission from Ref. 55. © 1995 Elsevier. (c) and (d) Reprinted with permission from Ref. 58. © 2003 Elsevier.

### 1.4.4  SOLAR GREEN HOUSE

The greenhouse that needs additional heat at night to ensure an inside air temperature above 20°C, and applies a new type of shape-stabilized PCM to maintain the temperature inside the greenhouse. The most effective greenhouses require control of temperature, humidity, solar irradiance, and internal gas composition with rational consumption of energy. On this basis, studies are classified in three groups according to the PCMs used, namely; calcium chloride hexahydrate ($CaCl_2.6H_2O$), sodium sulfate decahydrate ($Na_2SO_4.10H_2O$), and others (polyethylene glycol PEG, and paraffins).

Kern and Aldrich[60] employed 1650 kg of $CaCl_2.6H_2O$ in aerosol cans each weighing 0.74 kg was used to investigate energy storage possibilities both inside and outside a 36 m²-ground area greenhouse covered with tedlar-coated fiberglass. While the energy storage unit inside the greenhouse collected warm air from the ridge of the greenhouse during the daytime, the direction of air flow was reversed for the energy-releasing process at night. Boulard et al.[61] tested a greenhouse with a PCM heat storage system containing a quasi-eutectic mixture with a classical lettuce-tomato rotation. Experiments showed that in the south of France that this heat storage system can keep a greenhouse temperature roughly 10°C higher than outside during typical nights of March and April.

Hüseyin and Aydın[62] had conducted an experimental study to evaluate the thermal performance of five types of ten pieces solar air collector and PCMs, under a wide range of operating conditions. The results showed that the solar air collectors and PCM system created 6–9°C temperature difference between the inside and outside the greenhouse. Xinxia et al.[63] had done a contrast experiment of temperature between compound PCM wall greenhouse and ordinary greenhouse. They found that PCM wall greenhouse had better heat preservation than an ordinary greenhouse in the low-temperature condition as well as it improved indoor heat environment and reduced environmental pollution.

### 1.4.5  BUILDINGS

The application of PCMs in a building can have two different goals. First, using natural heat that is solar energy for heating or night cold for cooling. Second, using man-made heat or cold sources. In any case, storage of heat or cold is necessary to match availability and demand with respect to time and also with respect to power. Basically, three different ways to use PCMs for heating and cooling of buildings are:

i)   PCMs in building walls,
ii)  PCMs in other building components other than walls, and
iii) PCMs in heat and cold storage units.

Last one decade showed many developments in the use of PCMs in buildings. Researchers had proposed many new technological aspects in the use of PCMs in the heating and cooling of buildings.

Gutherz and Schiler[64] developed a space heating system that incorporates a PCM located in the ceiling. Sun reflectors were used to direct the solar energy entering via the windows on to the PCM. The main advantage of the system was that it allowed a large area to be dedicated to heat storage without the need for large volumes of storage medium that would be required with sensible heat storage. Nagano et al.[65] presented a floor air conditioning system with PCM in buildings. Granulated PCM was made of foamed waste glass beads and a mixture of paraffin. The PCM packed bed of 3 cm thickness was installed under the floorboard with multiple small holes. The change in room temperature and the amount of stored heat were measured and results showed the possibilities of cooling load shifting by using packed granulated PCM.

Takeshi Kondo[66] studied the effects of the peak cut control of the air conditioning system using PCM for ceiling board in the office buildings. The author concluded that the PCM ceiling system is effective for the peak-cut control. Stirith[67] developed a cheap TES system using paraffin wax as a TES medium, with an ordinary air/water heat exchanger to store and utilize the heat or cold. Xu et al.[68] studied the thermal performance of a room using a floor with shape –stabilized PCM. The experiments showed that the mean indoor temperature of a room with the PCM floor was about 2°C higher than that of the room without PCM floor and the indoor temperature swing range was narrowed greatly.

Huang et al.[69] experimentally evaluated PCM for thermal management of photovoltaic devices. The two PCMs used are RT 25 and GR 40. By experiment, it is shown that using RT 25 with internal fins, the temperature rise of the PV/PCM systems can be reduced by more than 30°C when compared with the datum of a single flat aluminum plate during phase-change. Hasan et al.[70] evaluated PCM for thermal regulation enhancement of building integrated photo-voltaic. Five PCMs were selected for evaluation with $t_m \approx$ 21–29°C and heat of fusion between 140 and 213 kJ/kg. They evaluated the performance of each PCM in four different PV/PCM systems. The experiments showed that the thermal regulation performance of a PCM depends on the thermal mass of PCM and thermal conductivity of both PCM and overall PV/PCM system.

Pasupathy et al.[71] discussed the major challenges and fact on the concept of green buildings. They also discussed the method of energy efficient charging and discharging, the effect of phase change temperature, insulation and geographical location of PCMs in buildings.

Haghshenaskashani and Pasdarshahri[72] presented a two-dimensional model for simulation and analysis of PCMs in order to minimize energy consumption in buildings. The effect of using PCM in integrated brick was investigated numerically. The experiment showed that by utilizing PCM in brick the maximum inlet flux may reduce about 32.8% depending upon PCM quantity. It also showed that to obtain maximum performance the PCM should be located near outdoor.

Zhou et al.[73] numerically investigated the effect of SSPCM plates (Fig. 1.7) combined with night ventilation in summer. These plates are applied as inner linings of walls and the ceiling of a building without air-conditioning. Effecting factors such as the thermo-physical properties of the SSPCM, its thickness, and air change per hour at both night-time and daytime are investigated. The results show that SSPCM plates could decrease the daily maximum temperature by up to 2°C due to cool storage at night.

Raj and Velraj[74] investigated the thermal performance of an eutectic PCM (48% $CaCl_2$ + 4.3% NaCl +0.4% KCl + 47.3% $H_2O$) for thermal management in a residential building and found it promising. Castell et al.[75] had done an experimental study of using PCM for passive cooling. Commercial PCM's RT 27 and SP 25 A8 was used during the study. The experiment showed that the PCM can reduce the peak temperature up to 1°C and in summer, the electrical consumption was reduced in the PCM cubicles about 15%. Thus, about 1–1.5 kg/year/$m^2$ of $CO_2$ emissions was saved in PCM cubicles. Bogdan and Diaconu[76] proposed a PCM-enhanced wall system in which the effect of occupancy pattern and ventilation on the energy savings potential of the wall system is studied and found that PCM melting point is influenced by the occupancy pattern as well as the ventilation and its pattern reduces the relative value of the energy savings. The PCM with melting point ≈19°C had the highest potential for energy savings. Karlessi et al.[77] investigated the performance of organic PCMs when incorporated in coatings for buildings and urban fabric. The thermal characteristics of these coatings were compared to infrared reflective and common coatings of the same color. The results demonstrated that during the experimental period surface temperatures of PCM coatings were lower than simple cool color matching common and simple cool coatings. PCM coatings can be up to 8°C cooler than common coatings of the same color presenting a surface temperature reduction of 12%.

**FIGURE 1.7**   The photos of the shape-stabilized PCM. Reprinted with permission from Ref. 73. © 2009 Elsevier.


## 1.5   CONCLUSION

This book chapter presents the TES state of the art of PCMs and its applications, that is, solar water heating systems, solar air heating systems, solar cooking, solar greenhouse, space heating, and cooling application for buildings. PCMs based on TES technologies are very beneficial for the humans as well as for the energy conservation. This chapter presents the review in this particular field, with the main focus being on the assessment of the thermal properties of various PCMs. It is obvious that these research results will help to diversify applications of PCMs in general with due to the respect of TES point of view. A very important criterion concerning TES with PCMs is the necessary discharge power of the storage. For short-term storage of thermal energy, the discharge power normally has to be relatively high in comparison to the storage volume. This requires certain conditions concerning PCM module sizes or the thermal conductivity of the PCM material respectively.

## ACKNOWLEDGMENT

Authors are highly thankful to Council of Science and Technology, UP (Reference No. CST 3012-dt.26-12-2016) for providing the research grants to carry out the work at the institute.

## KEYWORDS

- **thermal energy storage**
- **phase change materials**
- **photovoltaic**
- **solar air heating system**
- **solar greenhouse**
- **solar water heating system**
- **solar cookers**

## REFERENCES

1. Garg, H. P.; Mullick, S. C.; Bhargava, A. K. *Solar Thermal Energy Storage*; D. Reidel Publishing Co.: Holland, 1985.
2. Buddhi, D. Energy Conservation through Thermal Energy Storage. An AICTE Project, 2000.
3. Khartchenko, N. V. *Advanced Energy Systems*; Institute of Energy Engg. & Technology University: Berlin, 1997.
4. Sharma, A.; Tyagi, V. V.; Chen, C. R.; Buddhi, D. Review on Thermal Energy Storage with Phase Change Materials and Applications. *Renew. Sustain. Energy Rev.* **2009,** *13*, 318–345.
5. Dincer, I.; Dost, S. A. Perspective on Thermal Energy Storage Systems for Solar Energy Applications. *Internat. J. Energy Res.* **1996,** *20*, 547–557.
6. Baylin, F. *Low Temperature Thermal Energy Storage: A State of the Art Survey*; Report No. SERI/RR/-54-164, Golden, Colorado, USA, 1979.
7. Dincer, I.; Dost, S.; Li, X. Performance Analyses of Sensible Heat Storage Systems for Thermal Applications. *Internat. J. Energy Res.* **1997,** *21*, 1157–1171.
8. Buddhi, D.; Sawhney, R. L. Proceeding on Thermal Energy Storage and Energy Conversion, School of Energy and Environmental Studies; Devi Ahilya University, Indore, India, 1994.
9. Lane, G. A. *Solar Heat Storage—Latent Heat Materials*, Vol I; CRC Press, Inc.: BoCaRaton, Florida, 1983.

10. Dincer, I.; Dost, S.; Li, X. *Thermal Energy Storage Systems and Energy Savings*, Proceedings TIEES-96, First Trabzon Int. Energy and Environment Symposium, Karadeniz Technical University, Trabzon, Turkey, July 29–31, 1996; Ayhan, T., Dincer, I., Olgun, H., Dost, S., Cuhadaroglu, B., Eds.; *1*, 373–379.

11. Rosen, M. A.; Hooper, F. C.; Barbaris, L. N. Energy Analysis for the Evaluation of the Performance of Closed Thermal Energy Storage Systems. *ASME J. Solar Energy Engineer.* **1988**, *110*, 255–261.

12. Krane, R. J.; Krane, M. J. M. The Optimum Design of Stratified Thermal Energy Storage Systems—Part I: Development of the Basic Analytical Model. *ASME J. Energy Res. Technol.* **1992**, *114*, 197–203.

13. Rosen, M. A. Appropriate Thermodynamic Performance Measures for Closed Systems for Thermal Energy Storage. *ASME J. Solar Energy Engineer.* **1992**, *114*, 100–105.

14. Pillai, K. K.; Brinkwarth, B. J. The Storage of Low Grade Thermal Energy Using Phase Change Materials. *Appl. Energy* **1976**, *2*, 205–216.

15. Abhat, A. Low Temperature Latent Heat Thermal Energy Storage. In *Thermal Energy Storage*; Beghi, C., Ed.; D. Reidel Publication Co.: Dordrect, Holland, 1981.

16. Abhat, A. Low Temperature Latent Heat Thermal Energy Storage: Heat Storage Materials. *Solar Energy* **1981**, *30*, 313–332.

17. Lane, G. A. Solar Heat Storage: Latent Heat Material. In *Technology*, Vol. II; CRC Press: Florida, 1986.

18. Dincer, I.; Rosen, M. A. *Thermal Energy Storage: Systems and Applications*; John Wiley & Sons: Chichester, England, 2002.

19. Zalba, B.; Marín, J. M.; Cabeza, L. F.; Mehling, H. Review on Thermal Energy Storage with Phase Change: Materials, Heat Transfer Analysis and Applications. *Appl. Therm. Engineer.* **2003**, *23*, 251–283.

20. Farid, M. M.; Khudhair, A. M.; Razack, S. A. K.; Said, A. H. A Review on Phase Change Energy Storage: Materials and Applications. *Energy Convers. Manage.*, **2004**, *45*, 1597–1615.

21. Kenisarin, M.; Mahkamov, K. Solar Energy Storage Using Phase Change Materials. *Renew. Sustain. Energy Rev.* **2007**, *11*, 1913–1965.

22. Abhat, A., et al. *Development of a Modular Heat Exchanger with an Integrated Latent Heat Storage*. Report No. BMFT FBT 81-050, Germany Ministry of Science and Technology Bonn, 1981.

23. Sharma, A.; Sharma, S. D.; Buddhi, D. Accelerated Thermal Cycle Test of Acetamide, Stearic Acid and Paraffin Wax for Solar Thermal Latent Heat Storage Applications. *Energy Conver. Manage.* **2002**, *43*, 1923–1930.

24. Lane, G. A.; Glew, D. N. *Heat of Fusion System for Solar Energy Storage*. In Proceedings of the Workshop on Solar Energy Storage Subsystems for the Heating and Cooling of Buildings, Charlothensville, Virginia, 43–55, 1975.

25. Herrick, S.; Golibersuch, D. C. *Quantitative Behavior of a New Latent Heat Storage Device for Solar Heating/Cooling Systems*. General International Solar Energy Society Conference, Denver Co.: Valley Forge, PA, 1978.

26. Sarı, A.; Kaygusuz, K. Some Fatty Acids Used for Latent Heat Storage: Thermal Stability and Corrosion of Metals with Respect to Thermal Cycling. *Renew. Energy* **2003**, *28*, 939–948.

27. Lane, G. A., et al. *Macro-Encapsulation of PCM*. Report No. ORO/5117-8, Dow Chemical Company, Midland, Michigan, 152, 1978.
28. Lane, G. A. Hand Book of Thermal Design. In *Phase Change Thermal Storage Materials*; Guyer, C., Ed.; McGraw Hill Book Co.: USA, 1981.
29. Buddhi, D.; Sawhney, R. L.; Seghal, P. N.; Bansal, N. K. A Simplification of the Differential Thermal Analysis Method to Determine the Latent Heat of Fusion of Phase Change Materials. *J. Phys. D: Appl. Phys.* **1987,** *20*, 1601–1605.
30. http://www.cristopia.com
31. http://www.pcm-solutions.com/hvac.html
32. http://pcmenergy.com/products.htm
33. http://www.teappcm.com/products.htm
34. http://www.pcmproducts.net
35. http://www.pluss.co.in
36. http://www.epsltd.co.uk
37. Bhargava, A. K. Solar Water Heater Based on Phase Changing Material. *Appl. Energy* **1983,** *14*, 197–209.
38. Hasan, A. Thermal Energy Storage System with Stearic Acid as Phase Change Material. *Energy Conver. Manage.* **1994,** *35*, 843–856.
39. Kurklu, A.; Ozmerzi, A.; Bilgin, S. Thermal Performance of a Water-Phase Change Material Solar Collector. *Renew. Energy* **2002,** *26*, 391–399.
40. Canbazoglu, S.; Sahinaslan, A.; Ekmekyapar, A.; Gokhan, A. Y.; Akarsu, F. Enhancement of Solar Thermal Energy Storage Performance Using Sodium Thiosulfate Pentahydrate of a Conventional Solar Water-Heating System. *Energy Build.* **2005,** *37*, 235–242.
41. Long, J. Y.; Zhu, D. S. Numerical and Experimental Study on Heat Pump Water Heater with PCM for Thermal Storage. *Energy Build.* **2008,** *40*, 666–672.
42. Shukla, A.; Buddhi, D.; Sawhney, R. L. Solar Water Heaters with Phase Change Material Thermal Energy Storage Medium: A Review. *Renew. Sustain. Energy Rev.* **2009,** *13*, 2119–2125.
43. Mohsen, M. S.; Al-Ghandoor, A.; Hinti, I. A. Thermal Analysis of Compact Solar Water Heater Under Local Climatic Conditions. *Internat. Commu. Heat Mass Transf.* **2009,** *36*, 962–968.
44. Kumar, R.; Rosen, M. A. Thermal Performance of Integrated Collector Storage Solar Water Heater with Corrugated Absorber Surface. *Appl. Therm. Engineer.* **2010,** *30*, 1764–1768.
45. Morrison, D. J.; Abdel Khalik, S. I. Effects of Phase Change Energy Storage on the Performance of Air-Based and Liquid-Based Solar Heating Systems. *Solar Energy* **1978,** *20*, 57–67.
46. Jurinak, J. J.; Adbel Khalik, S. I. On the Performance of Air-Based Solar Heating Systems Utilizing Phase Change Energy Storage. *Solar Energy* **1979,** *24*, 503–522.
47. Ghoneim, A. A.; Klein, S. A. The Effect of Phase Change Material Properties on the Performance of Solar Air-Based Heating Systems. *Solar Energy* **1989,** *42*, 441–447.
48. Fath, H. E. S. Thermal Performance of a Simple Design Solar Air Heater with Built-In Thermal Energy Storage System. *Energy Conver. Manage.* **1995,** *36*, 989–997.
49. Saman, W. Y.; Belusko, M. *Roof Integrated Unglazed Transpired Solar Air Heater*. In Proceedings of the 1997 Australian and New Zealand Solar Energy Society, Paper 66, Canberra, Australia; Lee, T., Ed., 1997.

50. Enibe, S. O. Performance of a Natural Circulation Solar Air Heating System with Phase Change Material Energy Storage. *Renew. Energy* **2002,** *27,* 69–86.

51. Arkar, C.; Medved, S. Influence of Accuracy of Thermal Property Data of a Phase Change Material on the Result of a Numerical Model of a Packed Bed Latent Heat Storage with Spheres. *Thermochimica Acta* **2005,** *438,* 192–201.

52. Hammou, Z. A.; Lacroix, M. A Hybrid Thermal Energy Storage System for Managing Simultaneously Solar and Electric Energy. *Energy Conver. Manage.* **2006,** *47,* 273–288.

53. Zhou, G.; Zhang, Y.; Zhang, Q.; Lin, K.; Di, H. Performance of a Hybrid Heating System with Thermal Storage Using Shape-Stabilized Phase-Change Material Plates. *Appl. Energy* **2007,** *84,* 1068–1077.

54. Diaconu, B. M.; Varga, S.; Oliveira, A. C. Experimental Assessment of Heat Storage Properties and Heat Transfer Characteristics of a Phase Change Material Slurry for Air Conditioning Applications. *Appl. Energy* **2010,** *87,* 620–628.

55. Domanski, R.; El-Sebaii, A. A.; Jaworski, M. Cooking During Off Sunshine Hours Using PCMs as Storage Media. *Energy* **1995,** *20* (7), 607–616.

56. Buddhi, D.; Sahoo, L. K. Cooker with Latent Heat Storage Design and Experimental Testing. *Energy Conver. Manage.* **1997,** *38,* 493–498.

57. Sharma, S. D.; Buddhi, D.; Sawhney, R. L.; Sharma, A. Design, Development and Performance Evaluation of a Latent Heat Unit for Evening Cooking in a Solar Cooker. *Energy Conver. Manage.* **2000,** *41,* 1497–1508.

58. Buddhi, D.; Sharma, S. D.; Sharma, A. Thermal Performance Evaluation of a Latent Heat Storage Unit for Late Evening Cooking in a Solar Cooker having Three Reflectors. *Energy Conver. Manage.* **2003,** *44,* 809–817.

59. Sharma, S. D.; Iwata, T.; Kitano, H.; Sagara, K. Thermal Performance of a Solar Cooker Based on an Evacuated Tube Solar Collector with a PCM Storage Unit. *Solar Energy* **2005,** *78,* 416–426.

60. Kern, M.; Aldrich, R. A. *Phase Change Energy Storage in a Greenhouse Solar Heating System.* Paper presented at the summer meeting of ASAE and CSAE, June 24–27, University of Manitoba, Winnipeg, 1979.

61. Boulard, T.; Razafinjohany, E. B.; Jaffrin, A.; Fabre, A. Performance of a Greenhouse Heating System with a Phase Change Material. *Agricul. Forest Meteo.* **1990,** *52,* 303–318.

62. Hüseyin, B.; Aydın, D. Performance Analysis of a Latent Heat Storage System with Phase Change Material for New Designed Solar Collectors in Greenhouse Heating. *Solar Energy* **2009,** *83,* 2109–2119.

63. Xinxina, S.; Zhirong, Z.; Hongli, W.; Qinghai, S. Field Measurement and Analysis of Performance of Solar Greenhouse with Compound Phase Change Material Wall. *J. Agricult. Mechanizat. Res.* **2010,** *03,* 1–10.

64. Gutherz, J. M.; Schiler, M. E. A Passive Solar Heating System for the Perimeter Zone of Office Buildings. *Energy Sources* **1991,** *13,* 39–54.

65. Nagano, K.; Mochida, T.; Iwata, K.; Hiroyoshi, H.; Domanski, R.; Rebow, M. *Development of New PCM for TES of the Cooling System.* Terrastock, 8th International Conference on Thermal Energy Storage, Syuttgart, B. M. and Hahne E. W. P. ed.; Vol. 2, 345–350, 2000.

66. Takeshi, K. Research on Using the PCM for Ceiling Board. IEA ECESIA, Annex 17, 4th workshop, Indore, India, 2003.

67. Stritih, U. *An Experimental Model of Thermal Storage System for Active Heating or Cooling of Buildings.* IEA ECESIA, Annex 17, 6th workshop, Arvika, Sweden, 2004.

68. Xu, X.; Zhang, Y.; Di, H.; Lin, K.; Yang, R. *Experimental Study on the Thermal Performance of Shape-stabilized Phase Change Material Floor Combined with Solar Energy.* IEA ECESIA, Annex 17.7[th] workshop, Beijing, China, 2004.

69. Huang, M. J.; Eames, P. C.; Norton, B. Phase Change Materials for Limiting Temperature Rise in Building Integrated Photo Voltaic. *Solar Energy* **2006**, *80*, 1121–1130.

70. Hasan, A.; McCormack, S. J.; Huang, M. J.; Norton, B. Evaluation of Phase Change Materials for Thermal Regulation Enhancement of Building Integrated Photovoltaics. *Solar Energy* **2010**, *84*, 1601–1612.

71. Pasupathy, A.; Athanasius, L.; Velraj, R.; Seeniraj, R. V. Experimental Investigation and Numerical Simulation Analysis on the Thermal Performance of a Building Roof Incorporating Phase Change Material (PCM) for Thermal Management. *Appl. Therm. Engineer.* **2008**, *28*, 556–565.

72. Haghshenaskashani, S.; Pasdarshahri, H. Simulation of Thermal Storage Phase Change Material in Buildings. *World Acad. Sci., Engineer. Technol.* **2009**, *58*, 1–5.

73. Zhou, G.; Yang, Y.; Wang, X.; Zhou, S. Numerical Analysis of Effect of Shape-Stabilized Phase Change Material Plates in a Building Combined with Night Ventilation. *Appl. Energy* **2009**, *86*, 52–59.

74. Raj, A. A. V.; Velraj, R. Review on Free Cooling of Buildings Using Phase Change Materials. *Renew. Sustain. Energy Rev.* **2010**, *14*, 2819–2829.

75. Castell, A.; Martorell, I.; Medrano, M.; Pérez, G.; Cabeza, L. F. Experimental Study of using PCM in Brick Constructive Solutions for Passive Cooling. *Energy Build.* **2010**, *42*, 534–540.

76. Bogdan; Diaconu, M. Thermal Energy Savings in Buildings with PCM-Enhanced Envelope: Influence of Occupancy Pattern and Ventilation. *Energy Build.* **2011**, *43*, 101–107.

77. Karlessi, T.; Santamouris, M.; Synnefa, A.; Assimakopoulos, D.; Didaskalopoulos, P.; Apostolakis, K. Development and Testing of PCM Doped Cool Colored Coatings to Mitigate Urban Heat Island and Cool Buildings. *Build. Environ.* **2011**, *46*, 570–576, 2011.

# CHAPTER 2

# Thermal Energy Storage in Phase Change Materials and Its Applications

MANOJ K. SINGH[1*], L. SYAM SUNDAR[2], M. B. PEREIRA[2], and ANTONIO C. M. SOUSA[2]

[1]Department of Physics under School of Engineering and Technology, Central University of Haryana, Haryana 123031, India

[2]Centre for Mechanical Technology and Automation (TEMA–UA), Department of Mechanical Engineering, University of Aveiro, Aveiro, Portugal

*Corresponding author. E-mail: manojksingh@cuh.ac.in

## ABSTRACT

The effective way to store the thermal energy is by using the phase change materials (PCMs). The latent heat thermal storage systems are widely used in solar engineering, heat pump, heating and cooling applications in buildings, and so forth. Large number of PCMs are available that melt and solidify at a wide range of temperatures, which attracts huge applications. This chapter summarizes the investigation and analysis of the available thermal energy storage systems incorporating PCMs for use in different applications.

## 2.1 INTRODUCTION

There is a large demand for energy in the industries all over the world. It is somehow difficult to meet the energy demand with conventional methods. Nonconventional source of energy, such as solar energy is one of the promising sources of energy, which is available in all the places in the world. The ongoing research aims to develop devices for better usage of nonconventional source of energy to achieve the energy demand. A better idea is to develop energy storage device, which comes under new source of energy[1,2]. Phase change materials (PCMs) are the best examples for storing the energy and which can be converted into the required form whenever necessary. The energy storage in the device and conversion can save the wastage of energy and the capital cost of the industry. For example, in the thermal power plant, the energy storage can improve the efficiency of the plant and reduce the power generation cost. The PCM technology is in development stage, and its application is limited.

This chapter focuses on the energy storage methods, types of PCMs, properties, and applications to various sectors.

### 2.1.1  *ENERGY STORAGE METHODS*

There are several methods to store the energy such as, mechanical, electrical, and thermal energy as explained later.[3] Various types of energy storage are shown in Figure 2.1.

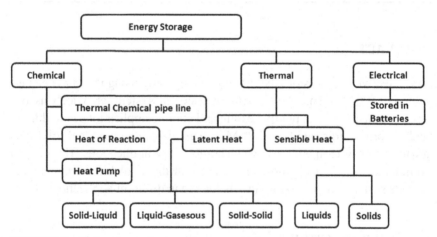

**FIGURE 2.1**    Various type of energy storage systems.

### 2.1.1.1  *MECHANICAL ENERGY STORAGE*

The mechanical energy storage systems are flywheel, compressed air storage, and pumped hydropower storage. The flywheels are used for intermediate energy storage. The compressed air storage and pumped hydropower storage are used for large-scale energy storage. The stored energy is discharged whenever the power is required.

## 2.1.1.2 ELECTRICAL ENERGY STORAGE

The electrical energy storage systems are batteries. The storage of electrical energy in the battery is possible by supplying the electricity, and it is discharged when the chemical energy is converted into mechanical energy. The commonly used batteries are lead acid, Nickel–Cadmium, and Lithium–ion.

## 2.1.1.3 THERMAL ENERGY STORAGE

The thermal energy as stored in terms of internal energy of a material in the form of sensible heat, latent heat, and thermochemical combination of sensible and latent heat. The temperature raise of a material in solid or liquid is called as sensible heat storage. Some commonly used materials for sensible heat storage is shown in Table 2.1. Water is the best example for the sensible heat storage, which has high specific heat. The latent heat storage is based on the heat absorption or release when a storage material undergoes a phase change from solid to liquid or liquid to gas, or vice versa.

a) The sensible heat storage is expressed as

$$Q = m \, C_p \, (T_f - T_i) \qquad (2.1)$$

where, $Q$ is amount of heat storage, $m$ is the mass of the latent heat storage material, $T_f$ and $T_i$ are the final and initial temperatures.

b) The latent heat storage is expressed as

$$Q = m \, [C_{ap} \, (T_m - T_i) + a_m \, \Delta h_m + C_{lp} \, (T_f - T_m)] \qquad (2.2)$$

where, $C_{ap}$ is the average specific heat between $T_i$ and $T_f$, $C_{lp}$ is the average specific heat between $T_m$ and $T_f$.

## 2.1.1.4 THERMOCHEMICAL ENERGY STORAGE

In the thermochemical storage systems, the heat stored depends on the amount of storage material, the endothermic heat of reaction, and the extent of conversion. In addition to the thermal heat storage techniques mentioned earlier, latent heat thermal energy storage is particularly attractive due to its ability to provide high energy storage density and its characteristics to store heat at constant temperature corresponding to the phase transition temperature of PCM. Generally in this case, the phase change occurs between solid–solid, solid–liquid, solid–gas, liquid–gas, and vice versa. For solid–solid, the heat stored in the material is transformed from one crystalline to another. Commonly used materials are pentaerythritol (latent heat of fusion 323 kJ/kg), pentaglycerine (latent heat of fusion 216 kJ/kg), $Li_2SO_4$ (latent heat of fusion 214 kJ/kg), and $KHF_2$ (latent heat of fusion 135 kJ/kg).

For solid–gas and liquid–gas, the higher latent heat of phase transition is possible but their large volume changes on phase transition are associated with the containment problems and rule out their potential utility in thermal storage systems. The solid–liquid can have the smaller latent heat than liquid–gas; however, these transformations involve only a small change of order of 10% in volume. The solid–liquid have proved to be economically attractive for use in thermal energy storage systems. The latent heat energy storage systems can fulfill the following three components: (1) PCMs, (2) containers materials, and (3) heat exchangers.

**TABLE 2.1**   List of Selected Solid–Liquid Materials for Sensible Heat Storage.

| Solid/fluid | | Temperature range (°C) | Specific heat (J/kg K) | Density (kg/m³) |
|---|---|---|---|---|
| Solids | Rock | 20 | 879 | 1600 |
| | Concrete | 20 | 880 | 1900–2300 |
| Fluids | Water | 0–100 | 4190 | 1000 |
| | Engine oil | 160 | 1880 | 867 |
| | Propanol | 120 | 2400 | 809 |
| | Ethanol | 78 | 2400 | 790 |
| | Octane | 125 | 2400 | 704 |
| | Isotunaol | 100 | 3000 | 808 |
| | Isopentanol | 150 | 2200 | 831 |

## 2.2   LATENT HEAT STORAGE MATERIALS

Latent heat storage materials are classified as PCMs, where the thermal energy transfer occurs when a material changes from solid state to liquid state, or liquid state to solid state, which is known as change of its state (i.e., phase change). The conventional storage PCMs are solid–liquid, where the temperature rise is caused by absorbing heat. In the nonconventional (sensible) storage materials, the PCM absorbs and releases heat at a nearly constant temperature; they can store 5–14 times more heat per unit volume than the sensible storage materials such as water, masonry, or rock. A large number of PCMs are known to melt with a heat of fusion in any required range. The thermo-physical, kinetics, and chemical properties are important for the design of thermal storage systems.

### 2.2.1  THERMAL PROPERTIES

The thermal properties such as suitable phase-transition temperature, high latent heat of transition, and good heat transfer of PCMs are important. The high thermal conductivity would assist the charging and discharging of the energy storage.

### 2.2.2  PHYSICAL PROPERTIES

The physical properties such as favorable phase equilibrium, high density, small volume change, and low vapor pressure are important. The small volume changes on phase transformation and small vapor pressure at operating temperatures are used to reduce the containment problem.

### 2.2.3  KINETIC PROPERTIES

The kinetic properties such as no supercooling and sufficient crystallization rate are important. Supercooling has been a troublesome aspect of PCM development, particularly for salt hydrates. Supercooling of more than a few degrees will interfere with proper heat extraction from the store and 5°C–10°C supercooling can prevent it entirely.

### 2.2.4  CHEMICAL PROPERTIES

The chemical properties such as long-term chemical stability, compatibility with materials of construction, no toxicity, and no fire hazard are important.

The PCM can suffer from degradation by loss of water of hydration, chemical decomposition, or incompatibility with materials of construction. PCMs should be nontoxic, nonflammable, and nonexplosive for safety.

### 2.2.5 ECONOMICS

The low cost and large-scale availability of the phase change materials is also very important, that means abundantly available.

## 2.3 CLASSIFICATION OF PHASE CHANGE MATERIALS

The organic, inorganic, and eutectic PCMs in large numbers are available in any required temperature range. The organic and inorganic chemical materials can be identified as PCM from the point of view melting temperature and latent heat of fusion. However, except for the melting point in the operating range, majority of PCMs do not satisfy the criteria required for an adequate storage media as discussed earlier. For example, metallic fins can be used to increase the thermal conductivity of PCMs, supercooling may be suppressed by introducing a nucleating agent or a cold finger in the storage material, and incongruent melting can be inhibited by use of suitable thickness. The description of the latent heat thermal energy storage systems using PCMs are described later. Figure 2.2 illustrates the classification of PCMs.

### 2.3.1 ORGANIC PHASE CHANGE MATERIALS

Paraffin and non-paraffins are classified as organic materials. These materials include congruent melting, which means melt and freeze repeatedly without phase segregation and consequent degradation of their latent heat of fusion, self-nucleation means they crystallize with little or no supercooling and are usually noncorrosive.

#### 2.3.1.1 PARAFFINS

Paraffin wax consists of a mixture of mostly straight chain n-alkanes $CH_3-$ $(CH_2)-CH_3$. The crystallization of the $(CH_3)$-chain release a large amount of latent heat both at melting point and latent heat of fusion increase with chain length. The paraffin qualifies as heat of fusion storage materials due to their availability in a large temperature range. Due to cost consideration, however, only technical grade paraffins may be used as PCMs in latent heat storage systems. Paraffin is safe, reliable, predictable, less expensive, and noncorrosive. They are chemically inert and stable below 500°C, show little volume changes on melting, and have low vapor pressure in the melt form. For these properties of the paraffins, system-using paraffins usually have very long freeze–melt cycle.[4]

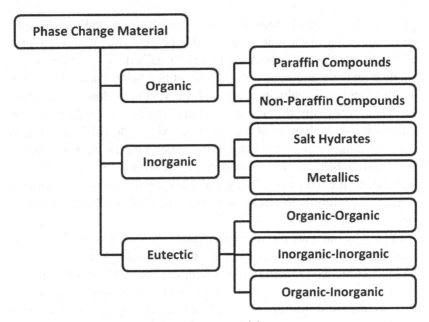

**FIGURE 2.2** Classification of phase change materials.

## 2.3.1.2 NON-PARAFFINS

The non-paraffin organic materials are the phase change materials with highly varied properties. Each of these materials will have its own properties unlike the paraffins, which have very similar properties. This is the largest category of candidate's materials for phase change storage. Abhat et al.[5] and Buddhi and Sawhney[6] have conducted an extensive survey of organic materials and identified a number of esters, fatty acids, alcohols and glycols suitable for energy storage. These organic materials are further subgroups as fatty acids and other non-paraffin organic.

### 2.3.2  INORGANIC PHASE CHANGE MATERIALS

The inorganic materials are further classified as salt hydrate and metallics.

#### 2.3.2.1  SALT HYDRATES

The solid–liquid transformation of salt hydrates is actually a dehydration of hydration of the salt, although this process resembles melting or freezing thermodynamically. A salt hydrate usually melts to either to a salt hydrate with fewer moles of water, that is, at the melting point the hydrate crystals breakup into anhydrous salt and water, or into a lower hydrate and water. Due to density difference, the lower hydrate (or anhydrous salt) settles down at the bottom of the container. Most salt hydrates also have poor nucleating properties resulting in supercooling of the liquid before crystallization beings. One solution to this problem is to add a nucleating agent, which provides the nuclei on which crystal formation is initiated. Another possibility is to retain some crystals, in a small cold region, to serve as nuclei.[7,8]

The salt hydrates are the most important PCMs, which have been extensively studied for their use in latent heat thermal energy storage systems. The most attractive properties of salt hydrates are: (1) high latent heat of fusion per unit volume, (2) relatively high thermal conductivity (almost double of the paraffins), and (3) small volume changes on melting. They are not very corrosive, compatible with plastics, and only slightly toxic. Many salt hydrates are sufficiently inexpensive for the use in storage.[9]

Three types of the behavior of the melted salts can be identified: congruent, incongruent, and semi-congruent melting.

   i)   Congruent melting occurs when the anhydrous salt is completely soluble in its water of hydration at the melting temperature.
   ii)  Incongruent melting occurs when the salt is not entirely soluble in its water of hydration at the melting point.
   iii) Semi-congruent melting the liquid and solid phases in equilibrium during a phase transition is of different melting composition because

of conversion of the hydrate to a lower-hydrated material through loss of water.

The major problem in using salt hydrates, as PCMs are the most of them, which are judged suitable for use in thermal storage, melts incongruently.

The problem of incongruent melting can be tackled by one of the following means: (1) by mechanical stirring, (2) by encapsulating the PCM to reduce separation, (3) by adding of the thickening agents that prevent setting of the solid salts by holding it in suspension, (4) by use of excess water so that melted crystals do not produce supersaturated solution, and (5) by modifying the chemical composition of the system and making incongruent material congruent. To overcome the problem of salt segregation and supercooling of salt hydrates, scientists of General Electric Co., NY[10] suggested a rolling cylinder heat storage system. The system consists of a cylindrical vessel mounted horizontally with two sets of rollers. A rotation rate of 3 rpm produced sufficient motion of the solid content (1) to create effective chemical equilibrium, (2) to prevent nucleation of solid crystals on the walls, and (3) to assume rapid attainment of axial equilibrium in long cylinders. Some of the advantages of the rolling cylinder method as listed by Herrick et al.[11] are: (1) complete phase change, (2) latent heat released was in the range of 90%–100% of the theoretical latent heat, (3) repeatable performance over 200 cycles, (4) high internal heat transfer rates, and (5) freezing occurred uniformly.

## 2.3.2.2 METALLICS

This category includes the low melting metals and metal eutectics. These metallics have not yet been seriously considered for PCM technology because of weight penalties. However, when volume is a consideration, they are likely candidates because of the high heat of fusion per unit volume. They have high thermal conductivities, so fillers with added weight penalties are not required. The use of metallics poses a number of unusual engineering problems. A major difference between the metallics and other PCMs is their high thermal conductivity. Some of the features of these materials are as follows: (1) low heat of fusion per unit weight, (2) high heat of fusion per unit volume, (3) high thermal conductivity, (4) low specific heat, and (5) relatively low vapor pressure.

### 2.3.3   EUTECTICS

Minimum-melting composition of two or more components is called as eutectics, in which it melts and freezes congruently forming a mixture of the component crystals during crystallization. Eutectic nearly always melts and freezes without segregation since they freeze to an intimate mixture of crystals, leaving little opportunity for the components to separate. On melting both components liquefy simultaneously, again with separation unlikely. Some segregation PCM compositions have sometimes been incorrectly called eutectics, since they are minimum melting.

## 2.4   AVAILABLE THERMAL ENERGY STORAGE SYSTEMS

### 2.4.1   SOLAR WATER-HEATING SYSTEMS

The solar water heating systems are used to heat the water using solar insulation, which is simple in construction and inexpensive.[12] Prakesh et al.[13] built and analyzed a water heater containing a layer of PCM filled at the bottom. In the sunshine hours, the water is heated due to solar radiation, and the water transfers heat to the PCM below it. The PCM collects energy from the water in the form of latent heat and melts. In the sunset hours, the hot water is withdrawn and is substituted by cold water, which gains energy from the PCM. Bansal and Buddhi[14] studied theoretically, by considering a cylindrical storage unit in the closed loop with a flat plate collector under charging and discharging mode and they selected as a paraffin wax (p-116) and stearic acid as PCMs. Chaurasia et al.[15] made comparative study of solar energy storage systems based on the latent heat and sensible heat technique to preserve the solar heated hot water for night time. Ghoneim[16] made comparison between different-sized latent heat storage vessels and sensible heat storage in a water tank with different degree of stratification and its experimental diagram is schematically depicted in Figure 2.3.

Bajnoczy et al.[17] studied the two-grade heat storage system (60°C–30°C and 30°C–20°C) based on calcium chloride hexahydrate and calcium chloride tetrahydrate and also they studied the storage capacity changes during the cycles and possible use of a solar energy storage system for domestic water heating system. Kamiz Kayguz et al.[18] had conducted an experimental and theoretical study to determine the performance of phase change energy storage materials for solar water heating systems. $CaCl_2.6H_2O$ was used as PCM. Rabin et al.[19] also studied a solar collector with storage for water heating having salt hydrate as a PCM. The results of parametric studies on the effect of the transition temperature and of the thickness layer of the salt-hydrate PCM on the thermal performance of the charging process are also presented. Sharma et al.[20] designed, developed and performance evaluate of a latent heat storage unit for evening and morning hot water requirements, using a box type solar collector. Mettawee and Assassa[21] investigated the thermal performance of a compact PCM solar collector based on latent heat storage. In this collector, the absorber plate–container unit performs the function of both absorbing the solar energy and storing PCM and the schematic diagram is shown in Figure 2.4. Cabeza et al.[22] selected as granular PCM–graphite compound of about 90 vol. % of sodium acetate trihydrate and 10 vol. % graphite as a PCM and constructed solar pilot plant at the University of Lleida to test the PCM behavior in real conditions.

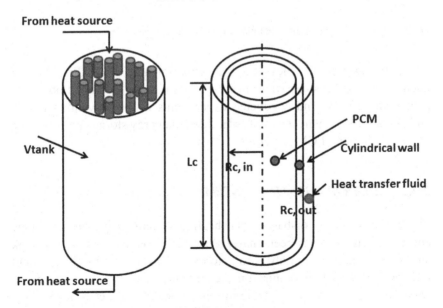

**FIGURE 2.3** A cylindrical shell with PCM storage.

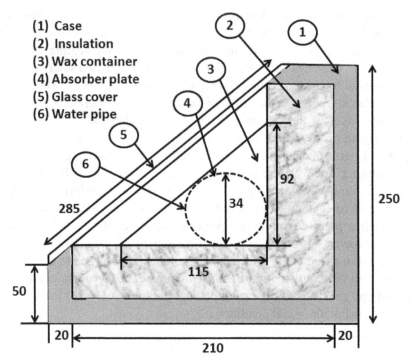

(1) Case
(2) Insulation
(3) Wax container
(4) Absorber plate
(5) Glass cover
(6) Water pipe

**FIGURE 2.4**    Schematic representation of latent heat storage system.

Canbazoglu et al.[23] developed an experimental setup (Fig. 2.5) of open-loop passive solar water-heating system using sodium thiosulfate pentahydrate-PCM and conducted experiments in the November month and obtained an enhancement of solar thermal energy storage performance without using any PCM materials.

## 2.4.2   SOLAR AIR HEATING SYSTEMS

The performance of air-based solar heating systems using phase change energy storage unit has been analyzed by Morrison and Khalik[24], Jurinak and Khalik[25] and their study is to determine the effect of the PCM latent heat and melting temperature on the thermal performance of air-based solar heating systems by considering the collector's effective area was 1 m² and its total volume was divided into five sectors. Ghoneim and Klein[26] compared theoretically the performance of phase change and sensible heat storage for air- and water-based solar heating systems using sodium sulfate decahydrate and paraffins as PCMs and observed the similar results of Jurnik and Khalik[25]. Enibe[27] developed natural convection solar air heater with PCM and evaluated its performance under no-load conditions that was tested under natural environmental conditions involving ambient temperature variations in the range of 19°C–41°C and daily global irradiation in the range 4.9–19.9 MJ/m² and the schematic experimental diagram is shown in Figure 2.6. Zhou et al.[28] investigated numerically the performance of a hybrid heating system combined with thermal storage by shape-stabilized phase change material (SSPCM) plates. A direct gain passive-solar house in Beijing is considered: It includes SSPCM plates as inner linings of walls and the ceiling.

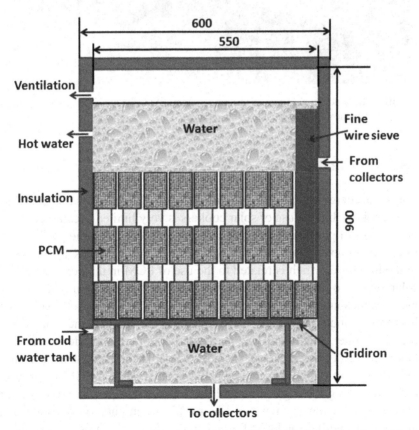

**FIGURE 2.5** Detailed cross-sectional view of the cylindrical heat storage tank combined with PCM.

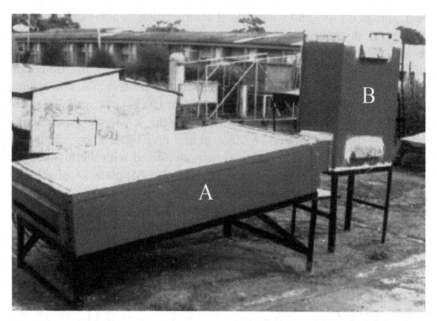

**FIGURE 2.6** Photograph of the air heating system. (A) Collector assembly with energy storage and air-heating subsystems; (B) heated space.

### 2.4.3 SOLAR COOKERS

The solar energy can also be used in the cooking process that is called as solar cookers, but the use of solar cookers is very limited because cooking process is not possible during the night time. If the solar energy is stored in the solar cooker, then its utilization is more. Some studies are available in the literature, which is related to the use of PCM in the box type of the solar cooker to prepare the food during the night times. Domanski et al.[29] used magnesium nitrate hexahydrate ($Mg(NO_3)_2.6H_2O$) as a PCM for the heat storage material for a box type solar cooker to prepare the food during the night time. Buddhi and Sahoo[30] coated commercial grade stearic acid below the absorbing plate of the box type solar cooker. Sharma et al.[31] developed box type solar cooker using acetamide as a PCM to prepare the food during the night time and the schematic representation of their experimental setup is shown in Figure 2.7. In another study, Buddhi et al.[32] used acetanilide as PCM for box type solar cooker to store the larger quantity of heat. In another study, Sharma et al.[33] used erythritol as a PCM for the solar cooker based on an evacuated tube solar collector and the experimental diagram is shown in Figure 2.8.

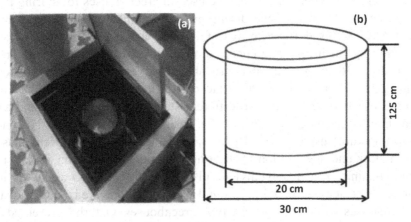

**FIGURE 2.7** (a) Solar cooker photo and (b) storage vessels.

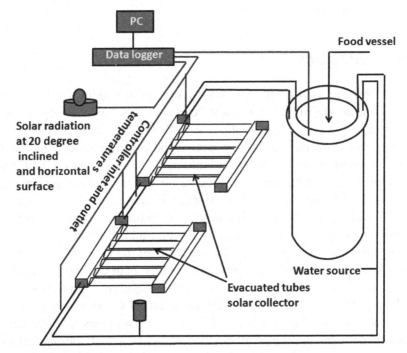

**FIGURE 2.8** The diagram of solar cooker based on evacuated tube solar collector with PCM storage unit.

### 2.4.4   SOLAR GREEN HOUSE

Phase change materials can also be used in green houses for storing the solar energy for curing and drying process and plant production. Kern and Aldrich[34] used 1650 kg of $CaCl_2.6H_2O$ in aerosol cans (each can of 0.74 kg) to investigate energy storage possibilities both inside and outside a 36 m²-ground area greenhouse covered with tedlar-coated fiberglass. Hung and Toksoy[35] had designed and constructed a latent heat storage system with two different stacking configurations and air baffling as an integrated part of the greenhouse solar system. Nishina and Takakura[36] used $Na_2SO_4.10H_2O$ with some additives to prevent phase separation and degradation for heating a greenhouse in Japan. Takakura and Nishina[37] tested polyethylene glycol and $CaCl_2.6H_2O$ as PCMs in greenhouse heating for 7.2 m² ground area. They compared conventional greenhouses with PCM storage type greenhouses. Ozturk[38] presented a seasonal thermal energy storage using paraffin wax as a PCM with the latent heat storage technique was attempted to heat the greenhouse of 180 m² floor area.

### 2.4.5   BUILDINGS

The PCMs are used for thermal energy storage in buildings since before 1980. Considering the advantages of PCM materials, they are implemented in Trombe wall, wallboards, shutters, under-floor heating systems, and ceiling boards of the building for heating and cooling applications. There are two applications of PCMs in buildings, (1) using natural heat, that is, solar energy for heating or night cold for cooling, (2) using manmade heat or cold sources. In any case, storage of heat or cold is necessary to match availability and demand with respect to time and also with respect to power. Ghoneim et al.[39] studied a Trombe wall and was recently added to the south facade to investigate if the effect of the PCM can be used all year long in Mediterranean weathers to reduce both cooling and heating demands. Bourdeau[40] tested two passive storage collector walls using calcium chloride hexahydrate (melting point 29°C) as a phase change material. He concluded that an 8.1 cm PCM wall has slightly better thermal performance than a 40-cm thick masonry wall. Experimental and theoretical tests were conducted to investigate the reliability of PCMs as a Trombe wall[41–43] used sodium sulfate decahydrate (melting point 32°C) as a PCM in south-facing Trombe wall. They also reported that Trombe wall with PCM of smaller thickness was more desirable in comparison to an ordinary masonry wall for providing efficient thermal energy storage.

The wallboards are cheap and widely used in a variety of applications, making them very suitable for PCM encapsulation. However, the principles of latent heat storage can be applied to any appropriate building materials. The idea of improving the thermal comfort of lightweight buildings by integrating PCMs into the building structure has been investigated in various research projects over several decades given by Stritih and Novak[44], Kedl and Stovall[44], Salyer and Sircar[45], Shapiro et al.[46] and Shapiro[47]. In most of these studies they applied macro-capsules or direct immersion processes, which both turned out to present several drawbacks. Due to these problems, none of these PCM products was successful in the wider market. The new option to microencapsulate PCMs, a key technology which overcomes many of these problems, may make PCM products accessible for the building industry.

The PCM shutters containing PCM is placed outside of window areas. During daytime they are opened to the outside and exposed to solar radiation, heat is absorbed and PCM melts. Heat is absorbed and PCM melts, during the night time the PCM shutters are closed so that the heat from the PCM can be radiate into the rooms..

The floor of the building is also an important part for heating and cooling of buildings. Athienies and Chen[48] investigated the transient heat transfer in floor heating systems by considering the influence of the cover layer and incident solar radiation on floor temperature distribution and on energy consumption. Complete and partial (area) carpets were considered as well as hardwood cover layers over concrete or gypcrete (gypsum–concrete mixture) thermal storage.

Ceiling boards are the important part of the roof, which are utilized for the heating and cooling in buildings. Bruno[49] developed a system, which stored coolness in phase change material in off-peak time and released this energy in peak time. The effects of the peak-cut control of air-conditioning systems using PCM for ceiling board in the building were also tried. The melting point of the PCM used was of the range 20°C–30°C, which was almost equal to the room temperature suitable for the purpose. Sharma et al.[50] explained in detail about the storage and use of latent heat materials in the review paper.

## 2.5   CONCLUSION

This chapter is focused on the available thermal energy storage technology with PCMs with different applications. These technologies are very beneficial for the humans as well as for the energy conservation. The current research in this particular field, with the main focus being on the assessment of the thermal properties of various PCMs, is explained in this chapter. The heat storage applications used as a part of solar water heating systems, solar air heating systems, solar cooking, solar green house, space heating and cooling application for buildings, off-peak electricity storage systems, and waste heat recovery systems also present the melt fraction studies of the few identified PCMs used in various applications for storage systems with different heat exchanger container materials.

## ACKNOWLEDGMENT

The authors, LSS and MKS, acknowledge Foundation for Science and Technology (FCT, Portugal) for the financial support received through the grant: SFRH/BPD/100003/2014. The author ACMS acknowledges the 2017– Visiting Scientist Fellowship awarded to him under the Chinese Academy of Sciences President's International Fellowship Initiative. TEMA/DEM researchers also acknowledge FCT grant UID/EMS/00481/2013-FCT and the infrastructures support CENTRO-01-0145-FEDER-022083.

## KEYWORDS

- **thermal energy storage**
- **latent heat**
- **phase change material**
- **solar energy**

## REFERENCES

1.  Garg, H. P.; Mullick, S. C.; Bhargava, A. K. *Solar Thermal Energy Storage*. D. Reidel Publishing Co, 1985.

2. Project Report. *Energy Conservation through Thermal Energy Storage*. An AICTE project.

3. Khartchenko, N. V. *Advanced Energy Systems*. Institute of Energy Engineering and Technology University: Berlin, 1997.

4. Abhat, A. Low Temperature Latent Heat Thermal Energy Storage: Heat Storage Materials. *Sol. Energy* **1981**, *30* (4), 313–332.

5. Buddhi, D.; Sawhney, R. L. In *Proceedings on Thermal Energy Storage and Energy Conversion*, 1994.

6. Lane, G. A.; Glew, D. N. In *Heat of Fusion System for Solar Energy Storage*. Proceedings of the Workshop on Solar Energy Storage Subsystems for the Heating and Cooling of Buildings. Virginia: Charlothensville, 1975; pp. 43–55.

7. Herrick, S.; Golibersuch, D. C. In *Quantitative Behavior of a New Latent Heat Storage Device for Solar Heating/Cooling Systems*. General International Solar Energy Society Conference, 1978.

8. Lane, G. A.; Rossow, H. E. In *Encapsulation of Heat of Fusion Storage Materials*. Proceedings of the Second South Eastern Conference on Application of Solar Energy, 1976; pp. 442–455.

9. Biswas, R. Thermal Storage Using Sodium Sulfate Decahydrate and Water. *Sol. Energy* **1977**, *99*, 99–100.

10. Furbo, S. *Heat Storage Units Using a Salt Hydrate as Storage Medium Based on the Extra Water Principle*. Report no. 116. Technical University of Denmark, 1982.

11. Herrick, S. A Rolling Cylinder Latent Heat Storage Device for Solar Heating/Cooling. *ASHRAE Tans.* **1979**, *85*, 512–515.

12. Tanishita, Int Solar Energy Eng, Meibourne, Paper 2/73, 1970.

13. Prakash, J.; Garg, H. P.; Datta, G. A. Solar Water Heater with a Built-in Latent Heat Storage. *Energy Convers. Manag.* **1985**, *25*, 51–56.

14. Bansal, N. K.; Buddhi, D. An Analytical Study of a Latent Heat Storage System in a Cylinder. *Sol. Energy* **1992**, *33*, 235–242.

15. Chaurasia, P. B. L. In *Phase Change Material in Solar Water Heater Storage System*. Proceedings of the 8th International Conference on thermal energy storage. 2000.

16. Ghoneim, A. A. Comparison of Theoretical Models of Phase Change and Sensible Heat Storage for Air and Water Solar Heating Systems. *Sol. Energy* **1989**, *42*, 209–230.

17. Bajnoczy, G., et al. Heat Storage by Two Grade Phase Change Material. *Periodica Polytecinica Ser Chem Eng.* **1999**, *43*, 137–147.

18. Kayugz, K., et al. Experimental and Theoretical Investigation of Latent Heat Storage for Water Based Solar Heating Systems. *Energy Conver. Manag.* **1995**, *36* (5), 315–323.

19. Rabin. Y.; Bar-Niv, I.; Korin, E.; Mikic, B. Integrated Solar Collector Storage System Based on a Salt Hydrate Phase Change Material. *Sol. Energy* **1995**, *55* (6), 435–444.

20. Sharma, A.; Sharma, A.; Pradhan, N.; Kumar, B. Performance Evaluation of a Solar Water Heater Having Built in Latent Heat Storage Unit, IEA, ECESIA Annex 17. Advanced Thermal Energy Storage through Phase Change Materials and Chemical Reactions—Feasibility Studies and Demonstration Projects. 4th Workshop, Indore, India, March 21–24, 2003; pp. 109–115.

21. Mettawee, E-B. S.; Assassa, G .M. R. Experimental Study of a Compact PCM Solar Collector. *Energy* **2006**, *31*, 2958–2968.

22. Cabeza, L. F.; Ibanez, M.; Sole, C.; Roca, J.; Nogues, M. Experimentation with a Water Tank Including a PCM Module. *Solar Energy Mater. Solar Cells* **2006**, *90*, 1273–1282.

23. Canbazoglu, S.; Sahinaslan, A.; Ekmekyapar, A.; Gokhan, Aksoy, Y.; Akarsu, F. Enhancement of Solar Thermal Energy Storage Performance Using Sodium Thiosulfate Pentahydrate of a Conventional Solar Water Heating System. *Energy Build* **2005**, *37*, 235–242.

24. Morrison, D. J.; Abdel, Khalik, S. I. Effects of Phase Change Energy Storage on the Performance of Air-Based and Liquid-Based Solar Heating Systems. *Sol. Energy* **1978**, *20*, 57–67.

25. Jurinak, J. J.; Adbel, Khalik, S. I. On the Performance of Air-Based Solar Heating Systems Utilizing Phase Change Energy Storage. *Sol. Energy* **1979**, *24*, 503–522.

26. Ghoneim, A. A.; Klein, S. A. The Effect of Phase Change Material Properties on the Performance of Solar Air-Based Heating Systems. *Sol. Energy* **1989**, *42* (6), 441–447.

27. Enibe, S. O. Performance of a Natural Circulation Solar Air Heating System with Phase Change Material Energy Storage. *Ren. Energy* **2002**, *27*, 69–86.

28. Zhou, G.; Zhang, Y.; Zhang, Q.; Lin, K.; Di, H. Performance of a Hybrid Heating System with Thermal Storage Using Shape-Stabilized Phase Change Material Plates. *Appl. Energy* **2007**, *84* (10), 1068–1077.

29. Domanski, et al. Cooking During Off Sunshine Hours Using PCMs as Storage Media. *Fuel Energ. Abstr.* **1995**, *36* (5), 348.

30. Buddhi, D.; Sahoo, L. K. Solar Cooker with Latent Heat Storage Design and Experimental Testing. *Energy Convers. Manage* **1997**, *38* (5), 493–498.

31. Sharma, S. D.; Buddhi, D.; Sawhney, R. L.; Sharma, A. Design, Development and Performance Evaluation of a Latent Heat Unit for Evening Cooking in a Solar Cooker. *Energy Convers. Manage* **1997**, *38* (5), 493–498.

32. Buddhi, D.; Sharma, S. D.; Sharma, A. Thermal Performance Evaluation of a Latent Heat Storage Unit for Late Evening Cooking in a Solar Cooker Having Three Reflectors. *Energy Convers. Manage* **2003**, *44* (6), 809–817.

33. Sharma, S. D.; Iwata, T.; Kitano, H.; Sagara, K. Thermal Performance of a Solar Cooker Based on An Evacuated Tube Solar Collector with a PCM Storage Unit. *Sol. Energy* **2005**, *78*, 416–426.

34. Kern, M.; Aldrich, R. A. Phase Change Energy Storage in a Greenhouse Solar Heating system. ASME paper no. 79-4028. Am Soc. Agric. Eng., St. Joseph, MI, 1979.

35. Hung, K.; Toksoy, M. Design and Analysis of Green House Solar System in Agricultural Production. *Energy Agric.* **1983**, *2* (2), 115–136.

36. Nishina, H.; Takakura, T. Greenhouse Heating by Means of Latent Heat Storage Units. Acta Hort (Energy in Protected Cultivation Ill). 148, 1984; pp. 751–754.

37. Takakura, T.; Nishina, H. A Solar Greenhouse with Phase Change Energy Storage and a Microcomputer Control System. *Acta Hort* (Energy in Protected Cultivation) **1981**, *115*, 583–590.

38. Huseyin, Ozturk, H. Experimental Evaluation of Energy and Energy Efficiency of a Seasonal Latent Heat Storage System for Greenhouse Heating. *Energy Convers. Manag.* **2005**, *46*, 1523–1542.

39. Ghoneim, A. A.; Klein, S. A.; Duffie, J. A. Analysis of Collector–Storage Building Walls Using Phase Change Materials. *Sol. Energy* **1991**, *47* (1), 237–242.

40. Bourdeau, L. E. In *Study of Two Passive Solar Systems Containing Phase Change Materials for Thermal Storage*. Hayes J, Snyder R, Eds. Proceedings of Fifth National Passive Solar Conference, October 19–26, Amherst. Newark, Delaware: American Solar Energy Society, 1980; pp. 297–301.

41. Swet, C. J. In *Phase Change Storage in Passive Solar Architecture*. Proceedings of the 5th National Passive Solar Conference, 1980; pp. 282–286.

42. Ghoneim, A. A.; Kllein, S. A.; Duffie, J. A. Analysis of Collector–Storage Building Walls Using Phase Change Materials. *Sol. Energy* **1991**, *47* (1), 237–242.

43. Chandra, S.; Kumar, R.; Kaushik, S.; Kaul, S. Thermal Performance of a Non A/C Building with PCCM Thermal Storage Wall. *Energy Convers. Manage* **1985**, *25* (1), 15–20.

44. Stritih, U.; Novak, P. Solar Heat Storage Wall for Building Ventilation. World Renewable Energy Congress (WREC), 1996; pp. 268–271.

45. Kedl, R. J.; Stovall, T. K. Activities in Support of the Wax-Impregnated Wallboard Concept. Thermal Energy Storage Researches Activity Review. New Orleans, Louisiana, USA: US Department of Energy, 1989.

46. Salyer, I. O.; Sircar, A. K. In *Phase Change Material for Heating and Cooling of Residential Buildings and Other Applications*. Proceedings of 25th Intersociety Energy Conservation Engineering Conference, 1990; pp. 236– 243.

47. Shapiro, M. M.; Feldman, D.; Hawes, D.; Banu, D. In *PCM Thermal Storage in Wallboard*. Proceedings 12th Passive Solar Conference, Portlan, 1987; pp. 48–58.

48. Shapiro, M. Development of the Enthalpy Storage Materials, Mixture of Methyl Stearate and Methyl Palmitate. Subcontract report to Florida Solar Energy Center, 1989.

49. Athienitis, A.; Chen, Y. The Effect of Solar Radiation on Dynamic Thermal Performance of Floor Heating Systems. *Sol. Energy* **2000**, *69* (3), 229–237.

50. Sharma, A.; Tyagi, V. V.; Chen, C. R.; Buddhi, D. Review on Thermal Energy Storage with Phase Change Materials and Applications. *Renew. Sustain. Energy Rev.* **2009**, *13*, 318–345.

# CHAPTER 3

# High Temperature Energy Storage and Phase Change Materials: A Review

MADHU SHARMA* and DEBAJYOTI BOSE

*School of Engineering, University of Petroleum and Energy Studies, Energy Acres, Bidholi, Dehradun 248007, Uttarakhand, India*

*Corresponding author. E-mail: madhusharma@ddn.upes.ac.in*

## ABSTRACT

Energy storage is key to successful resource utilization, coupled with storage material properties, fuel energy density, and the type of heat energy stored. In this review, different energy storage materials and media are evaluated for their performance efficiency, with perspectives on active and passive storage. Further, when considering latent heat storage, phase change materials are utilized, which accounts for low storage media economics. Materials which can account for such storage are reviewed, with emphasis on various properties, as well as applications during peak load shifting. The effect of heat transfer and performance enhancement is also discussed, with various encapsulation methods, along with implementations and scope of thermal energy storage for the future.

## 3.1 INTRODUCTION TO ENERGY STORAGE

Energy is in many ways exchangeable currency of technology. In today's generation, countries and their economies are based on ample and reliable energy supplies. As population increases, the need for more and more energy resources aggravates, and this results in increased demand for energy. Simultaneously, concerns regarding power generation will increase, for example, environmental concern like acid precipitation and stratospheric ozone depletion. Resources that cause minimum environmental impact are of significance. With increasing importance for environmental protection and rising energy costs, the world is turning to renewable energy systems to contribute significantly in meeting society's needs for more efficient, environmentally benign energy. So, some form of thermal energy storage (TES) is needed for an additional utilization of energy sources.[1] The main objective is to analyze and develop various technologies used to store energy in thermos-solar plants and to figure out which storing materials are used.

Thermal energy storage has the rising future for the productive use of thermal energy equipment and for facilitating large-scale switching. Energy storage is the trapping of energy that has been produced for a particular time for use at a later time to perform any productive operation. Basically, the device used for storing energy is called accumulator or battery. In applications like building and industrial processes, most of the energy consumed is in the form of thermal energy, and accordingly demand varies during day or from one day to next day. Thus, TES systems can help balance energy demand and supply on weekly or seasonal basis.[2] They can also help in reducing peak demand, energy consumption, carbon dioxide emission and costs, while increasing overall efficiency of energy system. Currently, TES is becoming particularly important for electricity storage in concentrating solar power plant (CSP), where solar heat can be stored for electricity production when sunlight is not available.[3]

Now considering the case for renewable sources, such as solar thermal power plants, they neither contribute to any polluting emissions nor have environmental concerns that are of concern with the conventional power plants: fossil-fuel or nuclear-based power generation plants.[4-5] The solar thermal power plants produce electricity in the same manner as conventional power plants but uses solar energy as the input. There are four key elements that are required in solar thermal power plants, namely, solar concentrator, receiver, transport/storage media system, and power conversion device, where the TES is the key element. But, only a few solar thermal power plants have tested high temperature TES systems. Thermal energy storage systems have the potential of increasing the effective use of thermal energy equipment. These systems are useful for correcting the supply and demand side mismatch.

There are two types of TES systems: sensible storage systems and latent storage systems. The energy released or absorbed by a body with the variation in temperature is called the sensible heat. Conversely, the heat required for the change of phase of a substance, be it from solid to liquid or from liquid to gas, is known as the latent heat. Another way of storing the heat is through the use of reversible endothermic chemical reactions. A complete storage system can be defined by three processes: charging, storing, and discharging. Several factors have to be kept in view while designing a thermal storage system, the key issue being the thermal capacity. The cost of the thermal storage system also has to be considered in the design process.[6–7] The cost of the TES system mainly depends on the following factors: the storage material, the heat exchanger for charging or discharging the system along with the cost of the enclosure of the TES system.

The important technical requirements that are to be kept in mind are: high energy density of the storage material, heat transfer rate between the fluid and storage medium, chemical stability of the storage medium, compatibility of the fluid with the heat exchanger and the storage medium, thermal losses (which should be low), and ease of control. An overview of the storage is shown in Figure 3.1.

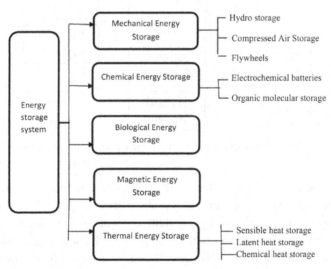

**FIGURE 3.1** Classification of energy storage systems.

### 3.1.1 THERMAL ENERGY STORAGE

In humid region the requirement for energy becomes much higher when the sun's luminescence is at its peak, so for these regions the concentrated solar power plant (CSP) would act better to meet the peak energy demand.

Codification of TES can be done according to the storage media or by notion of storage system. There are various technologies available for utilizing the solar energy for energy production, out of which CSP is gaining more popularity since the thermal energy produced can be stored. The TES is advantageous since it stores energy in a cheaper and more efficient way.

As shown in Figure 3.2, energy can be stored in various ways, such as, sensible heat, latent, or by combining these two ways, out of these, energy stored in latent heat is more economical and efficient since storing thermal energy in these requires smaller-sized tank, also the phase change in the latent heat is more productive because the operating temperature difference is much lower as compared to sensible heat TES, which eventually lead the way to better plant productivity.[8]

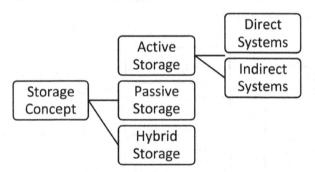

**FIGURE 3.2**   Classification of energy storage system according to concept (or notion).

Active storage systems stores are frequently connected as an aide to the system while passive storage systems stores are put together to build environment. An active storage system is signalized by forced convection heat transfer into the storage material, and the storage medium itself circulates through a heat exchanger; it can be a steam generator or solar receiver. Active storage system is categorized into direct system (the heat transfer fluid (HTF) collects the solar heat) and indirect system (for storing the heat and it is a second medium).

Passive storage systems are dual-medium storage systems.[9] The heat transfer fluid passes through the storage for charging and discharging from solid material, and in this storage system heat transfer medium itself does not circulate. So, the dual storage medium is for sensible heat storage materials. Hybrid storage systems are two storage systems, using a combination of both sensible and latent heat materials characteristics, to improve the storage characteristics of the systems. The TES is effectively utilized for storage of thermal energy derived from solar, industrial waste, process heat, and high temperature gas reactor for space heating, heating of process, and so forth.

Further, thermal energy can also be classified according to the storage media material interaction as shown in Figure 3.3. These are: (1) sensible heat media, in which the energy associated is released by a material as the temperature decreases or absorbed by a material as temperature increases, (2) latent heat media, which is associated with the phase change, and (3) chemical heat media, which is basically done through the use of reversible endothermic chemical reactions. The sensible heat media is subdivided into solid (requires high temperature concrete and ceramics) and liquid (requires steam, molten salts, mineral oils, synthetic oils).[10-11] The latent heat media is subdivided into gas–liquid, solid–gas, solid–liquid (organic substance requires paraffin and fatty acids and also an inorganic substance requires hydrated salts), solid–solid. The chemical heat media requires ammonia, metal oxide/metal, iron carbonate, magnesium oxide, methane/water, and calcium carbonate.

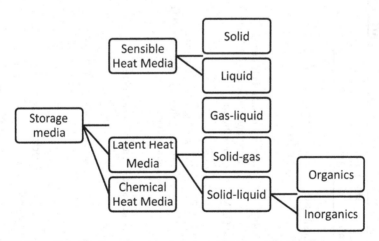

**FIGURE 3.3** Classification of energy storage systems according media.

Therefore, the thermal energy materials will be categorized depending on the TES media: Materials to store by sensible heat, materials to store by latent heat, materials to store by chemical heat. Further, the benefits and demerits of various storage technologies are discussed in Tables 3.1 and 3.2.

**TABLE 3.1**  Overview of Active Storage Systems with Their Relative Advantage and Disadvantage.

| | Active Storage | | |
| --- | --- | --- | --- |
| | Direct system | Indirect system | |
| **Direct steam generation** | **Two tanks** | **Two tanks** | **Cascaded tanks** |
| **Advantages** | | | |
| Intermediate heat transfer fluid and steam generation is not required. | Both cold and hot (HTF) stored separately. | Both cold and hot HTF stored separately. | Utilization of PCM storage with better storage capacity. |
| Plant configuration is quite simple. | Increase in solar field output temperature increases Rankine cycle efficiency. | Low risk approach. | Temperature is more uniform with time. |
| Lower investment and O&M costs. | Reducing the size of thermal storage system by raising the HTF temperature. | The material flows between hot and cold tanks, not through parabolic troughs. | |
| Operates at high temperatures, to increase the power cycle efficiency. | | | |
| **Disadvantages** | | | |
| Pipe installation cost is increased. | Material cost is high. | Cost of TES material is very high. | Huge increase in cost due to number of storage tanks and HTF loops. |
| Auxiliary equipment maintenance. | Small temperature difference between hot and cold fluid. | Heat exchanger is needed between HTF and TES. | |
| Difficult to control solar fields under solar radiation transients. | Electricity cost is high. As freeze point is high, there is high risk of storing fluid. | Efficiency is less as compared to direct systems. | Simulation-based systems, practically not yet integrated. |

**TABLE 3.2**   Perspectives of Passive Storage Systems.

| Concrete/ceramics | |
| --- | --- |
| **Passive storage** | **One tank (thermocline with filler materials)** |

**ADVANTAGES:**

- Cost of thermal energy storage media is low.
- Cost of storage tanks reduce, due to this system uses only one tank.
- Heat transfer rates into and out of the solid medium is high.
- Between the heat exchanger and the storage material, low degradation of HT.
- Cost of the filler material is low.

**DISADVANTAGES:**

- Increase of cost of heat exchanger.
- Long-term instability.
- Freeze point of most molten salts formulations is relatively high
- More difficult to separate the hot and cold HTF.
- It was an inefficient power plant and system is riskier with respect to performance.

**Active direct systems.** In this system, the heat transfer fluid and storage medium are same, therefore the cost of the system reduces as well as the working temperature can be increased to a better level. This improves the efficiency and performance of the plant and reduces the electrical cost.[12] The system is basically a two-tank direct system, that is, it has two tanks, one of the tanks is used to store the hot heat transfer fluid, which is supplied, and the other tank is used to store the discharged heat transfer fluid.

The advantages of this system are as follows:

a.   The hot and cold heat storage materials are separated inside the system.
b.   It can perform on high working temperatures.

The system has certain disadvantages as well, which are as follows:

a.   High cost of heat exchanger and heat transfer fluid.
b.   Relative small temperature difference between hot and cold fluid in storage system.

**Active indirect systems.** This system works on two concepts of working with heat transfer fluid and the storage medium.[12] First one is the two-tank indirect system, which works like the direct system with a heat exchanger between the heat transfer fluid and the storage material.

The second concept is using a single tank system where the hot and the cold fluids are in the same tank and are separated by stratification. This zone is called as thermocline zone. In this system, the hot fluid is at the top and the cold fluid is at the bottom so a heat exchanger is also required to transfer the heat between the heat transfer fluid and storage material. The advantage of single-tank system is reduction in the cost due to elimination of one tank; however, the disadvantages are as follows:

a.  Separation of hot and cold fluid becomes difficult.
b.  While using molten salts, a minimum system temperature is maintained in order to avoid freezing and salt dissociation.
c.  Outlet temperature being high enough causes losses in the solar field.
d.  The design of the plant becomes complex thus inefficient.

**Passive systems.** These systems are also dual-medium storage systems known as Regenerators. The heat transfer medium passes through the storage for charging and discharging the system, the storage medium itself does not circulate. Passive storage systems are solid storage systems, that is, made up of concrete, castable materials, and PCM. While using concrete, the rate of heat transfer increases due to better contact between concrete and piping. It also gives the facility to handle the material and low degradation of heat transfer between heat exchanger and the storage material. While if PCM is used, the storage ratio is increased due to higher heat of fusion than specific heat of materials.

There are three kinds (or modes) of TES. These are: (1) Sensible heat storage (SHS)—it is based on storing thermal energy by heating or cooling a liquid or solid storage medium (e.g., water, sand, molten salts, rocks) with water being the cheapest option, (2) Latent heat storage (LHS) using phase change materials (e.g., from a solid state into liquid state),[11–12] and (3) Thermo-chemical storage (TCS) using chemical reactions to store and release thermal energy. SHS is inexpensive as compared to PCM and TCS systems, and it is mostly applicable to domestic system, district heating, and industrial needs. TES can be either centralized or distributed systems as shown in Fig. 3.4.

For SHS, the thermal energy can be stored in the change of temperatures of the substances as shown in Fig. 3.4. During this process, these substances experience a change in internal energy. Sensible TES consists of a storage medium, a container, and inlet/outlet devices, and can be made by solid or liquid media. The solid media are usually used in packed beds with a fluid to exchange the heat. It is known as a dual storage system.

The advantage of a dual system is the use of inexpensive solids for storage materials. The different materials used for the solid medium are: rock, sand, or concrete.

Due to the difference in density of the hot and cold fluids, the liquid media maintain the natural stratification. This temperature gradient is a desired phenomenon. The different materials used as liquid medium are silicon and synthetic oils and nitrites in salts.

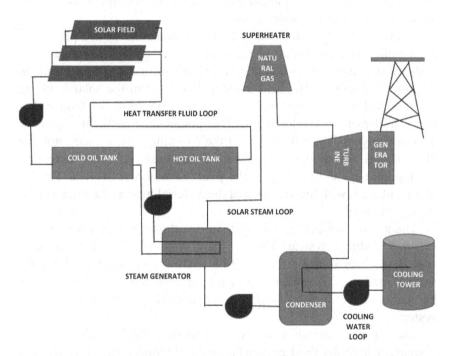

**FIGURE 3.4** System uses one tank in which cold HTF approaches from the steam generator and another tank in which hot HTF approaches directly from solar receiver before it is supplied to the steam generator. In this system, the main benefit is that the hot and cold HTF are stored separately. The drawback is that there is a need of a second tank. The storage tank is directly synchronized with HTF pressure levels. By eliminating the second tank the storage volume reduces for single tank system, but it is difficult to separate hot and cold HTF in a single tank system. The HTF naturally stratifies in the tank from coolest layer at the bottom to the warmest layer at the top due to the difference in density between hot and cold fluid. Maintaining the thermal stratification requires controlled charging and discharging procedures to avoid mixing. Filling the storage tank with second solid storage material (rock, iron, sand, etc.) can help achieve the stratification.[12]

For latent heat storage systems, the thermal energy can be stored isothermally in some substances called the phase change materials (PCMs). The storage systems using the PCMs can be reduced in size, compared to single-phase SHS. The PCMs allow large amounts of energy to be stored in relatively small sizes, which results in some of the lowest storage media costs.

For chemical heat storage, it is desired that the chemical reactions involved are completely reversible. This mechanism has high energy storage densities and heat-pumping capability.[10] The chemical heat storage systems have relatively lower costs.

In the case of high temperature storage, the solar power plants can be classified into two types: active or passive systems. The active direct systems are those where forced convection heat transfer takes place to the storage material. The heat transfer fluid used in the solar field can also be used as a storage material. This dual usage of the fluid makes the overall plant cost-effective, as it eliminates the need for expensive heat exchangers. The advantages of the two-tank solar systems are that the cold and hot storage materials are stored separately, thus increasing the Rankine cycle efficiency of the plant to the order of 40%. But the disadvantage is with the high cost of the material used as the fluid and the storage material.

The active indirect storage systems are those in which we find the two-tank indirect system. The two-tank system is called the indirect system, as heat transfer fluid is different from the fluid used for the storage medium. The advantage of the indirect system is that both the hot and cold heat transfer fluid is stored separately, similar to the direct system.

The passive storage systems are generally dual medium storage systems, where the fluid passes through the storage only for charging and discharging. These are generally dual medium storage systems and are solid in nature. The advantages of these systems are better usage of PCM storage capacities, reduction of costs, and improvement of storage ratio.

Latent heat storage systems are the storage systems that are based on PCMs with solid to liquid transition. These systems are considered to be efficient than the sensible thermal storage systems. It is because, the PCMs operate with small temperature differences between charging and discharging. The most important phase change that is applicable in these types of storage systems are the solid to liquid phase change.

The energy storage systems using PCMs have the ability to improve the energy efficiency, storing and releasing of the thermal energy. But, the PCMs have low thermal conductivity that proves to be a drawback, which leads to slow charging and discharging rates.

To improve the thermal conductivity of the PCMs, the use of composite latent heat storage materials (CLHSM) can be made, where the good thermal conductivity of an additive material and the high latent heat capacity of the PCMs are combined. One of the most common materials added to PCMs to increase their low thermal conductivity is graphite.

### 3.1.2   LATENT HEAT STORAGE MATERIALS

Thermal energy can be stored isothermally, that is, at constant temperature as latent heat of phase change, that is, as heat of fusion (solid–liquid transition), heat of vaporization (liquid–vapor) or heat of solid–solid crystalline phase transformation. Substances having these features are categorized as PCMs, as the latent heat of fusion is high compared to sensible heat of liquid and solid state of materials.[7-9] The size of the storage system utilizing PCMs can be reduced compared to single phase SHS; however, media selection and heat transfer design are more difficult, performance of materials can degrade with low temperature, salts after a moderate number of freeze melt cycles.

A proposal by LUZ International, evaluation of an innovative phase change salt concept to solar community that used a series of salts in a cascade design is shown in Table 3.3.

**TABLE 3.3**   Different Types of Medium and Its Phase Change Occurrence.[13]

| Storage medium | Heat capacity (kWh$_t$/m$^3$) | Media cost ($/kWh$_t$) |
| --- | --- | --- |
| NaNO$_3$ | 125 | 4 |
| KNO$_3$ | 156 | 4 |
| KOH | 85 | 24 |

The latent heat storage materials are basically of two types: having a cylinder-tube geometry and having a packed bed. The cylinder-tube geometry is the most widely studied. There have been various studies going on across the world, where researchers have modeled the cylinder-tube geometry using various modeling techniques to achieve better thermal efficiency of the TES system. Some of the modeling techniques that have been carried out with the cylinder-tube geometry are discussed later. He and Zhang numerically solved a mathematical model describing the unsteady freezing problem coupled with forced convection. They also showed the importance of the PCMs' thickness while modeling. Buschle et al. described the model using the "Modelica" software, where the storage tube was discretized in the axial direction, and the PCM around the storage tube elements was discretized in the radial direction.[14] Another study used a numerical model to simulate cascaded latent heat storage (CLHS) configurations using "Dymola/Modelica."[15]

One study developed a finite element model to simulate the cyclic thermal process involved as a result of alternate freezing and melting processes.[16] Cui et al. analyzed the energy transfer of the heat receiver cavity by developing a heat balance model of the solar heat receiver, a cavity radiation mathematical model, and a working fluid tube heat transfer model. They developed a numerical model of the heat exchanger tube. They also presented a model composed of three different phase change temperature materials, together with the corresponding physical model. Yimer and Adami developed a two-dimensional transient analytical model based on the enthalpy method.[17] Hoshi et al. developed a double-tube latent heat TES model system.[18] They presented a classification of high-temperature PCM according to heat capacity, phase change temperature, thermal conductivity and cost. Morisson et al. reported a detailed model of heat transfer and fluid flow for numerical simulation of latent heat storage unit.[19]

Lafdi et al. developed a model to study carbon foams saturated with PCM to be used as TES.[20] In the case of packed bed type of storage material, Yagi and Akiyama studied a single sphere and packed bed. They developed a model for a single spherical capsule and also conducted heat transfer simulation for a packed bed process of spherical capsules. They studied the flow of gas in the packed bed with the SIMPLE method.[21] Jalalzadeh-Azar et al. developed a packed bed computational model, where they made analysis with the help of the second law of thermodynamics.[22]

The study also carried out material stability tests for the assessment of the PCM.

### 3.1.3   CHEMICAL ENERGY STORAGE

Chemical reaction is the third storage mechanism. In this type, chemical reaction involved should be completely reversible. Solar receiver produces heat that is used to excite an endothermic chemical reaction. Heat can be recovered completely by the reversed reaction only if this reaction is completely a reversible reaction controlled by a catalyst.[10–12] Advantages of TES in a thermochemical reaction are high storage energy density. At close to ambient temperature for indefinitely long duration having heat pumping capability. Disadvantages include, thermodynamic properties of the reaction components are unpredictable, expensive, toxicity, and flammability. Although RTRs have several advantages concerning their thermodynamic characteristics, development is at a very early stage. To date, no viable prototype plant has been built.

## 3.2   PHASE CHANGE ENERGY STORAGE

The principle is that when heat is applied to the material it changes its phase from solid to liquid by storing the heat known as latent heat of fusion or from liquid to vapor known as latent heat of vaporization. When the stored heat is extracted by the load, the material will again transform its phase from one solid phase to another.[23] Solid–vapor and liquid–vapor transitions have large amount of heat of transformation but large changes in volume make the system very complex.

The advantage of heat storage through phase change is compactness. Any latent heat TES system should have at least three main components: a suitable phase change material in the desired temperature range, a containment for the storage substance, and a suitable heat carrying fluid. Due to its high cost, latent heat storage is more likely to find application when high energy density capacity is required. Energy is required at a constant temperature or within a small range of temperature for loads. Compactness is very important in order to limit the containment cost and at the same time the losses are more or less proportional to the surface area.

### 3.2.1   PHASE CHANGE MATERIALS

A detailed study of PCMs have been carried out, among them salt hydrates are studied extensively because these materials have a high amount of latent heat, are nonflammable materials, and are economical as well.[24] These salt hydrates are categorized under inorganic PCMs that have certain limitations such as, they lead to corrosion with the container materials, also the resolidification is improper. These limitations of inorganic PCMs lead to the study of organic PCMs.

Table 3.4 explains the various properties to be considered while selecting PCMs for a particular application.

**TABLE 3.4**   Different Properties of PCM.

| Thermophysical properties | Kinetic properties | Chemical properties |
| --- | --- | --- |
| The melting temperature should be in the temperature range in which it is to be operated. | In order to increase the rate of charging and discharging of energy storage the thermal energy should be high in both the phases, that is, solid and liquid. | It should be chemically stable. |
| Higher latent heat per volume. | To avoid the supercooling of the liquid phase the nucleation rate is made high. | Completely reversible. |
| The specific heat should be high to add extra heat through sensible heat. | | It should not degrade after melt cycles. |

### 3.3   PHASE CHANGE MATERIAL DEVELOPMENT

The current interest in PCM technology on solar heat storage for space heating can probably be traced back to its earlier respective work just after World War II. More volume efficient alternate approaches than SHS, invariably led to the advent of latent heat systems. The early studies of latent heat storage focused primarily on the fusion-solidification of low cost, easily available salt hydrates. After the phase change, they have a tendency to supercool, and the components do not melt congruently. So, these phenomena of supercooling and phase separation often determined the thermal behavior of these materials and caused random variation or progressive drifting of the transition zone over repeated phase change cycles.[22–23] Although certain significant advances were made till then, but major hurdles still remained toward the development of reliable and practical-based storage systems using salt hydrates and similar inorganic substances.

The inherent problems in inorganic PCMs have led to an interest in a new class of materials: low volatility, anhydrous organic substances such as polyethylene glycol, fatty acids, and their derivatives and paraffins. These materials were rejected earlier because they happen to be more expensive than common salt hydrates, and they have comparatively lower heat storage capacity per unit volume. It has been realized by now that these materials have certain advantages like physical and chemical stability, good thermal behavior, and adjustable transition zone.

Not all PCMs can be used for latent heat storage. An ideal PCM should have a high heat of fusion and thermal conductivity, high specific heat capacity, small volume change, noncorrosive, nontoxic, and possess little or no decomposition or supercooling.[25] In building applications, only PCMs that have a phase transition close to human comfort temperature close to 20°C can be used. Table 3.5 shows the possible candidates from the many hydrated salts and organic PCMs for application in buildings.

### 3.3.1 PHASE CHANGE THERMAL STORAGE FOR PEAK LOAD SHIFTING

When electricity supply and demand are out of phase, then one of the possible solutions to this problem can be the development of an energy storage system. Distributed thermal storage materials integrated with a building could shift most of the load coming from the residential air conditioners from peak to off-peak time periods.[23] As a result, capital investment in peak power generation equipment could be effectively reduced for power utilities and then could be reflected in cheap service to customers.

**TABLE 3.5**   Hydrated Salts and Organic PCMs.[26]

| PCM | Melting point (°C) | Heat of fusion (kJ/kg) |
|---|---|---|
| CaCl2.6H2O | 29.7 | 171 |
| Calcium chloride hexahydrate | | |
| 45% CH3(CH2)8COOH | 17–21 | 143 |
| 55% CH3(CH2)10COOH | | |
| 45/55 Capric–lauric acid | | |
| CH3(CH2)11OH | 17.5–23.3 | 188.8 |
| Dodecanol | | |
| KF 4H2O | 18.5–19 | 231 |
| Potassium fluoride tetrahydrate | | |
| CH3(CH2)16COO(CH2)3CH3 | 18–23 | 140 |
| Butyl stearate | | |
| CH3(CH2)16CH3 | 22.5–26.2 | 205.1 |
| Tech. grade octadecane | | |
| CH3(CH2)12COOC3H7 | 16–19 | 186 |
| Propyl palmitate | | |

The building of integrated thermal storage would enable customers and consumers to take advantages of lower utility rates during off-peak hours where power utilities are offering time of day rates. PCM wallboard can thus be used as a load management device for passive solar applications. A 120 m² house could save up to 4 GJ a year (or 15% of the annual energy cost). The optimal diurnal heat storage occurs with a melt temperature 1°C–3°C above average room temperature.[25] Even certain claims determine that PCM wallboards can save up to 20% of house space conditioning cost.

### 3.3.2   *FIRE RETARDATION OF PCM-TREATED CONSTRUCTION MATERIALS*

The flammability requirements and stringent safety codes in the recent years have been imposed on building materials to protect these from fire hazards.[27]

Following are some approaches that have been applied and measured successfully in laboratory tests to fire retard PCM imbibed plasterboard:

a.  Addition of nonflammable surface to the plasterboard (e.g., aluminum foil and rigid polyvinyl chloride film)

b.  Sequential treatment of plasterboard, first in PCM and then in an insoluble liquid fire retardant (e.g., Fyrol CEF). The insoluble fire retardant displaces some part of the PCM while some remains on the surface thereby imparting self-extinguishing characterization to the plasterboard.

c.  It is anticipated that brominated hexadecane and octadecane when combined with antimony oxide in plasterboard, the product would self-extinguish itself.

d.  Certain fire retardant surface coatings can be used to prevent effectively the wicking action of the plasterboard paper covers.

Flammability test evaluation of the burning characteristics of ordinary gypsum wallboard impregnated with approximately 24% organic PCM shows that flames spread and smoke development classifications were determined in a Steiner tunnel, and heat and smoke release rates were determined by cone calorimeter. When the comparison of the test results with similar data for other building materials are examined, it indicated the possibility of reduction in the flammability of energy storing wallboard by incorporation of a flame retardant.

### 3.3.3   ORGANIC PCMs

Organic compounds have very low thermal conductivity (from 0.1 to 0.7 W/mK), hence requiring certain processes and mechanisms to enhance heat transfer in order to achieve reasonable rates of heat output (W). Table 3.6 presents the thermo-physical properties of the compounds identified for further analysis based on their relatively low price, their quoted stability from the review, and enthalpy of phase change. Some of the saturated fatty acids, sugar alcohols, carboxylic acids, amides ad alkanes appeared to be promising in this temperature range. Urea is not a promising compound in its pure state because of its instability in its molten state, but some of its eutectic mixtures have suitable properties for latent heat storage.

**TABLE 3.6** Properties for Materials which can be Suitable for Improving PCM Performance.[24-26]

| Compound | Tm (°C) | ΔHm (kJ/kg) | Cps (kJ/kg.K) | Cpl (kJ/kg. K) | λs (W/m.K) | λl (W/m. K) | ρs (kg/m³) | Vexp (m³/m³) | Edensity (kW h/m³) |
|---|---|---|---|---|---|---|---|---|---|
| Paraffin wax | 0–90 | 150–250 | 3.00 | 2.00 | 0.2 | 0.2 | 880–950 | 12–14 | 50–70 |
| Acetamide | 82 | 260 | 2.00 | 3.00 | 0.40 | 0.25 | 1160 | 13.9 | 93 |
| Lauric acid | 44 | 212 | 2.02 | 2.15 | 0.22 | 0.15 | 1007 | 13.6 | 66 |
| Formic Acid | 8 | 277 | 1.00 | 1.17 | 0.30 | 0.27 | 1227 | 12.0 | 96 |
| Palmitic acid | 61 | 222 | 1.69 | 2.20 | 0.21 | 0.17 | 989 | 14.1 | 67 |
| Stearic acid | 54 | 157 | 1.76 | 2.27 | 0.29 | 0.17 | 940 | 9.9 | 49 |
| Acetic Acid | 17 | 192 | 1.33 | 2.04 | 0.26 | 0.19 | 1214 | 13.5 | 71 |

## 3.4 HEAT TRANSFER ENHANCEMENT METHODS

Certain low thermal conductive PCMs can seriously affect the storage system charge and discharge rates. To deal with this limitation, extended metal surfaces, conductive powders or conductive matrices are proven to be effective in increasing the PCMs heat transport properties, leading to a more uniform temperature within the PCM and better charge and discharge effectiveness for the latent het storage container/system.

### *3.4.1 EXTENDED METAL SURFACES*

One of the most widely used heat transfer enhancement techniques is to increase the heat transfer area by adding extended metal surfaces known as fins. Different studies have been made on modeling the phase change process with various fin geometries. A compact horizontal tube in tube container using erythritol as the PCM, using axial fins to enhance heat transfer. For this system, melting/solidifying properties that would provide a suitable heat source for driving an absorption cooling system illustrates two common fin geometries widely used in the literature. An example of such system is shown in Figure 3.5.

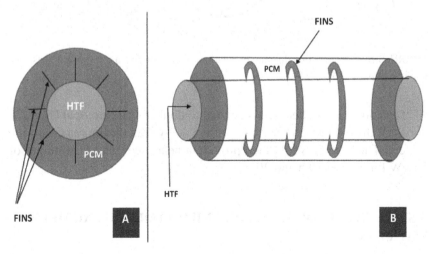

**FIGURE 3.5** (a) Cross section of a tube in tube heat storage unit with longitudinal fins, and (b) schematic representation of a tube in tube heat storage unit with annular fins.[28]

### 3.4.2    HEAT TRANSFER ENHANCEMENT USING CARBON

Exfoliated graphite, also known as expanded graphite (EG), with a thermal conductivity ranging from 24 to 470 W/m-K, has the potential to increase the global PCM thermal conductivity with low volume ratios (usually around 10%–15%). EG is generally obtained from the oxidation of natural graphite with a mixture of nitric and sulfuric acid, followed by drying in an oven and rapid heating in a furnace at 800°C to 900°C to obtain rapid expansion. The PCM is impregnated into the EG under vacuum, this prevents the formation of air gaps within the EG/PCM composite. This technique is the most effective procedure currently used to enhance the PCM thermal conductivity. It also provides a shape-stabilized (SS) form to the PCM since the pore cavities (as shown in Fig. 3.6) can withstand the thermal expansion typical during phase change and prevent leakage of molten organic material.

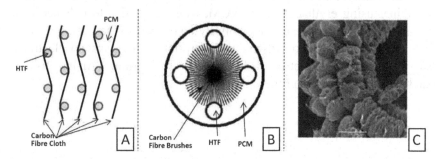

**FIGURE 3.6**    Cross-section diagram of a carbon fiber cloth (A), a carbon fiber brush (B) and EG viewed with SEM (C).[29]

The container studied was predicted to double its heat output (from 25 to 50 kWth) if a carbon fiber cloth of 0.8% v/v was incorporated into the system. The system would then provide a nearly constant heat output of 50 kW for around 10 h and 20 min.

## 3.5    THERMAL CONDUCTIVITY ENHANCEMENT USING METAL MATRICES

Using sparse metal matrices is another significant way to increase thermal conductivity within a PCM container, structure that would also provide multiple nucleation points. Steel wool is a more feasible method to improve thermal conductivity of a PCM, compared to expandable graphite, but does not provide a shape-stabilized solution; since it is not as compactable as graphite. Figure 3.7 presents two approaches used to effectively enhance heat transfer within a PCM.

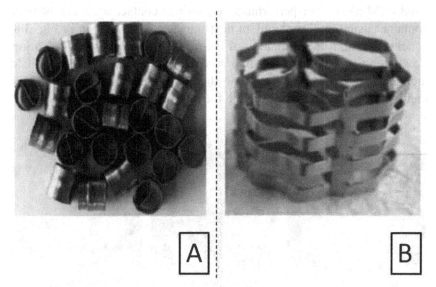

**FIGURE 3.7** Heat transfer effectiveness can be increased with materials such as (A) stainless steel and (B) aluminum rings.[19-21]

### 3.5.1 USING CONDUCTIVE POWDERS

Including small percentages by volume of metallic particles (aluminum, copper, silver, nickel), or graphite can also increase thermal diffusion within low thermal conductivity PCMs. It would also have the added benefit of increasing the number of potential nucleation points, potentially enhancing crystallization within the PCM. However, the conductive material could lose its miscibility when the PCM is in the molten state (due to differences in density), separating from the storage material and sinking to the base of the container. This could be prevented by including gelling agents in the PCM with a consequent reduction in the PCM volume ratio.

### 3.5.2   *DIRECT HEAT TRANSFER TECHNIQUES*

Another technique to increase the heat transfer would be to provide direct contact between the PCM and the heat transfer fluid. This would provide an effective increase in heat transfer during the melting process since the convective nature of the heat transfer fluid would act directly on the solid PCM phase. The performance of a direct contact latent heat storage container using erythritol and an heat transfer oil (as shown in Fig. 3.8) concluded that at the beginning of the melting process the oil has a low flow rate due to the block of solid erythritol, the top surface of the PCM melts faster than the bottom due to the higher heat transfer rate and the melting time varies effectively with the oil flow rate.

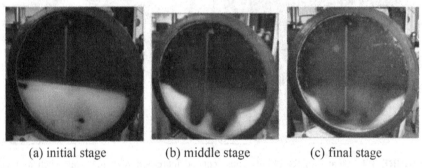

(a) initial stage             (b) middle stage             (c) final stage

**FIGURE 3.8**   Temporal variation of the melting process in a direct contact heat transfer container using erythritol as the PCM and oil as the heat transfer fluid.[21]

To overcome the initial blocking of the fluid flow path when the PCM is in the solid state, the insertion of electric heaters is an effective way is shown in Figure 3.9, and concluded that the overall energy spent on melting the initial flow pathways was 5% of the total thermal energy stored.

Limited experimental studies of PCM wallboard determine few general rules relating to the dynamics of PCM wallboard. The significant conclusion is that a PCM wall is capable of capturing a large proportion of solar radiation incident on the walls or roof of a building. They are also capable of minimizing the effect of large fluctuations in the ambient temperature on the inside temperature of the building because of the high thermal mass of the PCM walls. For shifting the heating and cooling loads to off-peak electricity periods, they can be very effective. During new construction and rehabilitation of a building gypsum wallboard impregnated with PCM could be installed in place or ordinary wallboard. Thermal storage will be provided by it, that is distributed throughout the building enabling passive solar design and off peak cooling with frame construction.[18] For installation of PCM wallboard in place of ordinary wallboard, little or no additional cost will be incurred.

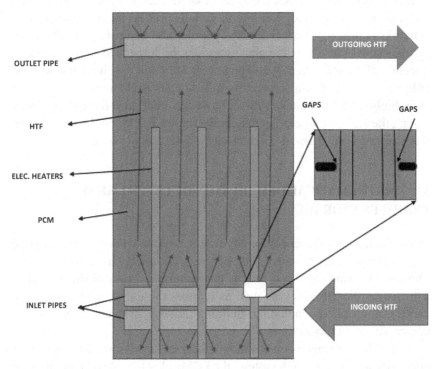

**FIGURE 3.9** Schematic cross section showing the locations of the electric heaters in the inlet pipes.[22]

Phase change materials have the potential to store large amounts of energy within a smaller temperature range when compared to common SHS materials. Due to the low thermal conductivities of many PCMs, poor rates of thermal diffusion within the PCM can seriously affect the storage system charge and discharge rates that can be achieved.

A comprehensive review of PCMs melting between 0 and 250°C has been made and the thermo-physical properties of the materials having the most appropriate properties presented. Below 100°C, organic compounds and salt hydrates are the most interesting materials. Eutectic mixtures with Urea seem promising around 100°C, and in the range from 130°C up to 1 250°C eutectic mixtures of inorganic salts appear the most promising PCMs. A mixture of sodium and potassium formates melting around 170°C appears attractive due to is relatively low price and moderate latent heat of fusion. A review of potential indirect latent heat storage containers and systems suitable for integration with various process heating and cooling networks is also reported. Due to its geometrical versatility, encapsulated systems seem more feasible since they can be integrated to any existing system without major technical constrains, although they have lower PCM volume ratios.[11–15] Compact systems offer larger isothermal stages due to their higher PCM volume ratios; however, heat transfer enhancement among the PCM is imperative to achieve reasonable thermal power output rates, since PCMs thermal conductivity can be a major issue.

## 3.6  MATHEMATICAL DESCRIPTION OF LATENT HEAT TRANSFER STORAGE

Due to increase in awareness for environmental safety and increase in price of conventional energy sources, use of renewable energy has also increased. The solar thermal storage is one of the important aspect of this scenario. It increases the efficiency as well as reduces the use of other sources in various applications including building heating systems, thermo-solar power plants, and so forth.

Latent heat storage is one of the methods to store thermal energy other than SHS and TCS. In LHS system, the PCMs play the role of a medium for storing energy and releasing it when required. These materials store heat energy when they change from solid form to liquid form, from liquid form to gaseous form or from solid form to solid form (change of one crystalline form into another without a physical phase change). Then release that energy when they have the reverse phase changes.

The storage capacity of a LHS system in the case of solid–liquid transformation is given by eq 3.1

$$Q = \int_{Ti}^{Tm} m.\,Cp.\,dT + m.\,a_m.\,\Delta hm + \int_{Tm}^{Tf} m.\,Cp.dT \qquad (3.1)$$

Where, *Ti* is the initial temperature *Tm* is the melting temperature
m is the mass of heat storage medium *Cp* is the specific heat
a$_m$ is the fraction melted
$\Delta hm$ is the heat of fusion per unit mass (J/kg)

The latent storage materials have larger volumetric energy storage capacity because the heat of fusion or heat of evaporation is much greater than the specific heat. Also the absorption and release of the energy stored takes place at constant temperature, which makes the choices easier to use in different applications. To get a feasible and effective PCM, the following properties are to be considered namely, physical, thermal, kinetic, chemical, and economical.
The physical properties include:

a.　High density.
b.　Small volume change to facilitate construction of heat exchangers.
c.　Low vapor pressure to avoid problems with heat exchangers.
d.　Favorable phase equilibrium to facilitate heat storage.

The thermal properties include:

a.　High thermal conductivity to provide minimum temperature gradients and maintain charge and discharge of heat.
b.　High latent heat of transition to occupy minimum possible volume.
c.　Suitable phase-transition temperature.

The kinetic properties include:

a.　Proper crystallization rate.
b.　No supercooling as it makes difficult to control the heat transfer.

The chemical properties include:

a.　Nonexplosive.
b.　Nontoxic.
c.　No fire hazard.
d.　Long-term chemical stability.

The economical properties include:

a.　Cost-effective.
b.　Abundant and available easily.

Although it is difficult to get all the properties in a material, methods such as use of fins or composite materials in the form of matrixes are used to get a good system design.

Organic compounds consist of two types: paraffin compounds and non-paraffin compounds.

The paraffin compounds are made by saturated hydrocarbons having a general formula—$C_nH_{2n+2}$. These hydrocarbons are obtained from the petroleum distillation process. The non-paraffin compounds are made by esters, alcohols, and glycols. Inorganic compounds are divided into two types; salt hydrates and metallics. Salt hydrates are the alloys of inorganic salts and water, which forms a crystalline solid of general formula of $AB.nH_2O$. Metallics are the low melting metals and metal eutectics. They have got high thermal conductivity, relatively low vapor pressure, and high heat of fusion per unit volume.[16]

Eutectics are a composition of two or more minimum melting components, which melt and freeze identically and form a mixture of crystals while crystallization occurs. Eutectic mixture contain three types: organic–organic, inorganic–inorganic, and inorganic–organic.

**Eutectic mixtures of inorganic salts.** As discussed earlier, the inorganic salts have a melting temperature between 250°C and 1680°C and heat of fusion from 68 to 1041 J/g. When a substance solidifies, it reaches a temperature that is lower than the melting temperature and causes no crystallization rate and nucleating properties that lead to lack of crystal nucleus to solidify. This is called supercooling phenomena.[25]

A challenging feature of inorganic salts is change of volume at phase transition. It is due to the lack of information of temperature dependences of density. The lifetime of the material is another feature to consider. In space applications the lifetime can be about months and years, but in power plants the expected lifetime is around 20–30 years. The compatibility between salt and constructional materials should also be tested, as no chemical reaction or corrosion test should be done.

The container is made up of alloy, which is easily processed, accessible, and cheap. Also the heat transfer area is extended between the heat carrier and PCM by using fins or by encapsulation of PCM. Another overview of these PCMs is presented in Figure 3.10.

**Metallics.** These metals and alloys can be used as high-temperature storage materials because they have high heat conductivity, less corrosion activity, and lesser cost. An alloy made by the composition of aluminum and silicon (AlSi$_{12}$) has been used for the development of isothermal electric heater because it has a melting temperature of 576°C and heat of fusion of 560 J/g.

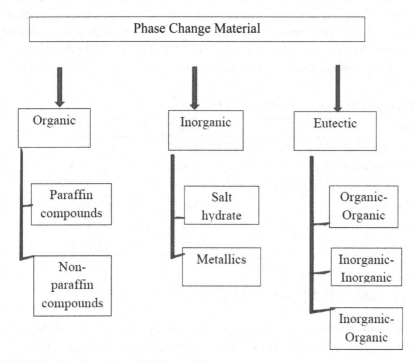

**FIGURE 3.10**   Overview of PCM materials.

There are various properties of PCMs that affect the charging and discharging time (properties are shown in Table 3.7), out of them low thermal conductivity is of prime consideration. Therefore, various techniques have been developed in order to improve it. Some of them are finned tubes, metal matrix, high-conductive particles dispersed in PCMs, micro encapsulation of PCM or shell and tube.

## 3.7   ENCAPSULATION OF PCM

Since the PCM has a lower thermal conductivity, therefore, in order to improve the heat transfer rate, encapsulation of PCM is done in conducting materials. For encapsulating the PCM, the containment should meet certain requirements such that it should be corrosion resistant and thermally stable, should be strong enough to avoid interaction with the surrounding, provide enough surfaces for heat transfer to take place, and should check that the change in volume is controlled when the phase change is occurring.[27–28] There are various types of containments present with different shapes that affect the heat transfer rate of PCMs such as cylindrical shape, heat pipes, and rectangular containers.

**TABLE 3.7**   Thermophysical Properties of AlSi12 Alloy.

| Property | Value | Unit |
|---|---|---|
| Heat capacity of solid | 1038 | J/g.K |
| Heat capacity of liquid | 1741 | J/g.K |
| Temperature of phase transition | 576 | °C |
| Heat of fusion | 560 | J/g |
| Density | 270 | g/cm³ |
| Thermal conductivity | 160 | W/m.K |

### 3.7.1  ENCAPSULATION METHODS

Various methods for encapsulating PCM were developed with an objective of making a storage system that should have following characteristics:

a.  Can withstand high temperatures.
b.  To provide better heat transfer rate.
c.  Should not react with high temperature fluid and molten PCM.
d.  Should be economical.
e.  Should be strong enough to hold the PCM inside during melting and solidification.

In order to make encapsulation of PCM with these desirable properties, two methods were developed:

**Shell development on a preformed PCM pellet:** In this PCM encapsulation method, coating of PCM pellets is done of cylindrical and spherical shape by various coating materials. The coating done can be of polymethylmethacrylate (PMMA) or sodium silicate. The coating of sodium silicate is preferred since it can withstand high temperatures. It is economical, nonflammable, as well as forms good bonding with glass, metals, and ceramics. It is present in two states; dry and liquid, in dry form as white powder and in liquid as solution in water. Sodium silicate has excellent adhesive property, which is used for pelletizing, refractory cements, and coating. There are various ways by which the sodium silicate can be dried and cured depending upon their application.

By varying the silica to sodium oxide ratios, three different types of commercially available sodium silicate are obtained. These are shown in Table 3.8.

**TABLE 3.8** Commercially Available Sodium Silicates.[30]

| Type | $SiO_2$% | $H_2O$% | Weight ratio |
|------|----------|---------|--------------|
| N    | 8.90     | 62.4    | 3.22         |
| K    | 11.00    | 57.3    | 2.88         |
| RU   | 13.85    | 52.95   | 2.40         |

In order to fabricate pellets of spherical and cylindrical shapes of 10mm diameter, pressure ranging from 1–5 ton was applied by using a stainless steel pellet press and hydraulic press for about 3 min to 5 min.[30] The coating on the PCM pellet can be done either by PMMA or sodium silicate application, in PMMA coating method, the coating is applied with a brush and the pellet is completely covered, then it is cured in UV light for about 15 min and then second coating of sodium silicate is applied. There are three techniques by which sodium silicate is applied: Rotating drum coating, brush coating, and dip coating.

In rotating drum coating, the coating is applied through a spray atomizer nozzle, in this technique, the process time required is less because the coating of pellets is done in one go. While in brush coating, the procedure and thickness of coating is controllable, whereas in order to get uniformity in coating, dip coating is preferred.

**Coating of sodium silicate.** The principle technique of curing sodium silicate is by moisture loss. Various methods of curing were investigated such as, microwave heating, drying at varying relative humidity, addition of pigments like cuprous oxide, zinc oxide, and spray drying.

**Rotating–drum drying.** In this mechanism hot air is passed through front opening and the drum rotates at a certain speed and pellets are coated in one go. The steps involved in drum coating are shown in Figure 3.11. There are, however, certain drawbacks with this coating mechanism, such that during the process sodium silicate deposits on the drum surface, which is undesirable since it leads to nonuniform coating on pellet, so this mechanism is not preferred widely.

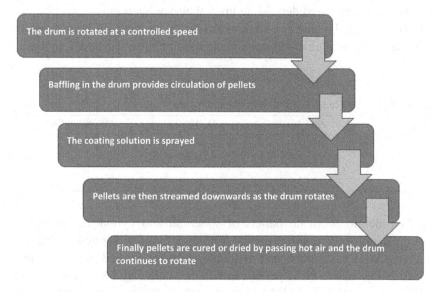

**FIGURE 3.11**   Overview of the drum drying method.

**Curing at room temperature in ambient air.** In this curing mechanism, a thin layer of sodium silicate was applied on PMMA coating and was left to dry overnight. It was observed that a hard coating of glass was formed on the PMMA, further to know the temperature-withstanding ability of sodium silicate coating, further heat was given up to 300°C and it was found that the layer was not good enough and lost its adhesive property at higher temperatures, hence it was concluded that PMMA layer was porous and does not act as a good barrier between sodium silicate and PCM.

**Preformed PCM shell.** Earlier in shell development mechanism a shell was developed; in this encapsulation method a new approach was made, that is, preformed shell. The advantage of this mechanism is that the time required for fabricating was lesser compared to previous method, also one material was required for coating unlike in the earlier one; due to these advantages this mechanism is economical.

For this preformed shell, carbon steel was selected due to its high conductivity property, strength, and low cost. In order to select and judge the thermal cycling ability of the system and harmony of steel, a capsule of cylindrical shape was forged. In the carbon steel sodium chloride and potassium chloride were filled and the ends were blocked with end caps and sodium silicate, then for 10 cycles the system was thermally cycled and the result was obtained for different temperatures and thermal cycles.

Further it was observed that since the capsule is in contact with air at higher temperature, it has led to scaling of capsule on its outer surface, also corrosion occurs inside the container due to the contact with molten salt. Hence a need to protect the steel shell against this corrosion emerged, which led to the development of protection of coating.

## 3.8  THERMAL STORAGE FOR PARABOLIC TROUGH POWER PLANTS

Thermal storage can use sensible or latent heat mechanism or heat emitted from a chemical reaction. Sensible heat stores energy by increasing the temperature of liquid or solid. Whereas latent heat is the energy storage from heat transition from a solid to liquid state.

The high and low temperature limits are given for each material, combined with average mass density and heat capacity, lead to a volume-specific heat capacity in kWh per cubic meter. Table 3.1 explains the approximate costs of storage media in dollars per kilogram, finally arriving at unit costs in $/kWh.

From Table 3.9 the average thermal conductivity has strong influence on heat transfer design and heat transfer surface requirements of storage system, especially for solid media. High volumetric heat capacity is on priority as it leads to lower storage system size, reducing external piping and structural costs.

**Solid media.** Solid will be used in packed beds for thermal storage that will require a fluid to exchange heat. When the heat capacity of fluid is very low then the solid material is used for storage, but the capacity cannot be negligible if the fluid is liquid and the system is called as dual storage system. We can easily extract storage energy from warmer strata, and cold fluid can be extract from colder strata, and supply into the collector field. The advantage of dual system is that the least expensive

**TABLE 3.9** Characteristics of Solid and Liquid Sensible Heat Storage Materials and Latent Heat Storage Media for SEGS Plants.[31]

| Storage medium | Temperature | | Average density (kg/m³) | Average heat conductivity (W/m.K) | Average heat capacity (kJ/kgK) | Volume specific heat capacity (kWh$_t$/m³) | Media costs per kg ($/kg) | Media costs per kWh$_t$ ($/kWh$_t$) |
| --- | --- | --- | --- | --- | --- | --- | --- | --- |
| | Cold (°C) | Hot (°C) | | | | | | |
| **Solid media** | | | | | | | | |
| Cast iron | 200 | 400 | 7200 | 37.0 | 0.56 | 160 | 1.00 | 32.0 |
| Cast steel | 200 | 700 | 7800 | 40.0 | 0.60 | 450 | 5.00 | 60.0 |
| Silica fire bricks | 200 | 700 | 1820 | 1.5 | 1.00 | 150 | 1.00 | 7.0 |
| **Liquid media** | | | | | | | | |
| Mineral oil | 200 | 300 | | 0.12 | 2.6 | 55 | 0.30 | 4.2 |
| Synthetic oil | 250 | 350 | | 0.11 | 2.3 | 57 | 3.00 | 43.0 |
| Silicone oil | 300 | 400 | | 0.10 | 2.1 | 52 | 5.00 | 80.0 |
| **Phase change media** | | | | | | | | |
| NaNO₃ | 308 | | | 0.5 | 200 | 125 | 0.20 | 3.6 |
| KNO₃ | 333 | | | 0.5 | 267 | 156 | 0.30 | 4.1 |
| KOH | 380 | | | 0.5 | 150 | 85 | 1.00 | 24.0 |

solids such as rock, sand, or concrete for storage materials are used in conjunction with more expensive heat transfer fluids like thermal oil. Therefore, pressure drop as well as parasitic energy consumption may be high in dual system.

The cold to hot temperature limits of few solid media are greater than that could be used in a solar electricity generation systems (SEGS) plant, because the maximum outlet temperature limit for parabolic trough solar field is about 400°C.

From Table 3.10, it is concluded that sand rock oil combination is eliminated because it is limited to 300°C. Reinforced concrete and salt have low cost and acceptable heat capacity. Silica and magnesia fire bricks offer no advantages over concrete and salt and are usually identified by high temperature. Cast iron offers a very high heat capacity and thermal conductivity but the cost is high.

**TABLE 3.10** Effect on Solid Media by Imposing this Temperature Limit on the Storage Medium Temperature Range, Media Costs and the Unit Heat Capacities.[31]

| Storage medium | Heat capacity (kWh$_t$/m³) | Media cost ($/kWh$_t$) |
|---|---|---|
| Reinforced concrete | 100 | 1 |
| NaCl (solid) | 100 | 2 |
| Cast iron | 160 | 32 |
| Cast steel | 180 | 150 |
| Silica fire bricks | 60 | 18 |
| Magnesia fire bricks | 120 | 30 |

**Liquid media.** It maintains natural thermal stratification as there is a difference in density between hot and cold fluid. For the requirement of hot fluid be supplied to the upper part of storage system during charging and cold fluid be extracted from bottom part during discharging. This can be done by stratification device such as mantle heat exchange, and so forth. The temperature range of heat transfer fluid in a SEGS plant ranges between 300°C and 400°C approximately. Applying these limitations on temperature gives the result as depicted in Table 3.11.

Both the oils and salts are feasible in heating thermal storage. Generally, salts having higher melting point and parasitic heating is required to keep the salts liquid at night, during plant shutdowns. Silicone oil have environmental benefits as they are nonhazardous and are expensive, whereas synthetic oils are hazardous. Nitrites in salts present potential corrosion problems, for the required temperature they are acceptable.

**TABLE 3.11** The Effect on Liquid Media by Imposing this Temperature Limit on the Storage Medium Temperature Range, Media Costs and the Unit Heat Capacities.

| Storage medium | Heat capacity (kWh$_t$/m³) | Media cost ($/kWh$_t$) |
|---|---|---|
| Synthetic oil | 57 | 43 |
| Silicone oil | 52 | 80 |
| Nitrite salts | 76 | 24 |
| Nitrate salts | 83 | 16 |
| Carbonate salts | 108 | 44 |
| Liquid sodium | 31 | 55 |

### 3.8.1 STATE-OF-THE-ART TES SYSTEMS IN SOLAR THERMAL PLANTS

Seven out of eight TES systems in solar thermal electric plants have been prototype nature and only one being a commercial unit. Table 3.12 explains the characteristics of existing units.

All are SHS systems: two of them are single tank oil thermocline systems, four are single medium two-tank systems and rest are dual medium single tank systems.

A 30 MWe SEGS system having efficiency of 35% would require about 260 MWht up to 3 h storage capacity in order to put the size of these systems in perspective. Till now this is the largest among any other solar thermal electric storage system.

Round trip efficiency and the cost per unit of thermal energy delivery are the two main characteristics of storage systems. Round trip efficiency is defined as the ratio of useful energy recovered from the storage system to the amount of energy extracted from heat from heat source.

More than 90% of round trip efficiency were measured in many of the systems listed in the above-mentioned table. But still some systems were low as 70%. Both oil and molten salt systems were technically feasible, but encountered various problems due to the errors in design, construction, and operation. The SEGS 1 storage system cost was around dollars $25/kWt in 1984, in which oil represented 42% of the cost of investment.18–20 For the operation of SEGS plants up to 400°C the oil cost around eight

**TABLE 3.12** Features of Existing Units for Solar Thermal Energy Storage.[15-17]

| Project | Type | Storage medium | Cooling loop | Nominal temperature Cold (°C) | Hot (°C) | Storage concept | Tank volume (m³) | Thermal capacity (MWh_t) |
|---|---|---|---|---|---|---|---|---|
| Irrigation pump, Coolidge, AZ, USA | Parabolic trough | Oil | Oil | 200 | 228 | 1 tank thermocline | 114 | 3 |
| IEA-SSPS Almeria, Spain | Parabolic trough | Oil | Oil | 225 | 295 | 1 tank thermocline | 200 | 5 |
| SEGS 1 Daggett, CA, USA | Parabolic trough | Oil | Oil | 240 | 307 | Cold tank / Hot tank | 4160 / 4540 | 120 |
| IEA-SSPA Almeria, Spain | Parabolic trough | Oil cast iron | Oil | 225 | 295 | 1 dual medium tank | 100 | 4 |
| Solar One Barstow, CA, USA | Central receiver | Oil/sand/rock | Steam | 224 | 304 | 1 dual medium tank | 3460 | 182 |
| CESA-1 Almeria, Spain | Central receiver | Liquid Salt | Steam | 220 | 340 | Cold tank / Hot tank | 200 / 200 | 12 |
| Themis Targasonne, France | Central receiver | Liquid Salt | Liquid salt | 250 | 450 | Cold tank / Hot tank | 310 / 310 | 40 |
| Solar Two Barstow, CA, USA | Central receiver | Liquid salt | Liquid salt | 275 | 565 | Cold tank / Hot tank | 875 / 875 | 110 |

times more than SEGS, while there are also other parameters such as total system investment, inflexibility compared to backup system, and bigger tank size.

### 3.8.2 SUMMARY OF WORK PERFORMED BEFORE 1990

In this the most relevant investigations reviewed carried out prior to 1990. Geyer (1991) provides an overview of the applicability of thermal storage to solar power plants.[7-9] Table 3.13 shows the storage systems but from these only few were investigated in detail.

**TABLE 3.13**    Report on Thermal Energy Storage as Adapted from Geyer.

| TES concepts | Storage type | Status* | Assessment |
|---|---|---|---|
| Sensible active | Two-tank oil | T | Basic concept, state of the art. |
| | HITEC | T | 2 variants analyzed based on existing PSA/THEMIS designs. |
| | Thermocline | T | Proved on pilot scale, no advantages over basic two-tank system. |
| Sensible DMS | Oil/cast iron | T | Proved on pilot scale, no advantages over basic two-tank system. |
| | Oil/steel | LR | Used in chipboard presses. |
| | Oil/concrete | MR | Several variants analyzed. |
| | Oil/solid salt | MR | Several variants analyzed. |
| PCM | Oil/PC Salts | HR | Several cascade arrangements analyzed. |
| Chemical | Oil/metal hybrids | HR | Early stage of development, no lead concepts, no cost data. |

*Nomenclature: T, tested; LR, low risk; MR, medium risk; HR, high risk

### 3.9   LIFE CYCLE ASSESSMENT OF STORAGE MATERIALS

PCM materials, because of their potential in heats storage application, are evaluated across thermal research areas. There are three different phases that were evaluated.[14-19] The first was LUZ design using five PCMs in a series or cascade design (SERI 1989), the second one was designed by a Spanish company, INITEC, which also used five PCMs but in different configuration of heat exchanger. The third one was designed by German companies, Siempelkamp and Gertec (SGR) and uses concrete for high temperature with three PCMs.

**TABLE 3.14** Comparison of Embodied Life Cycle Greenhouse Gas Emissions from the Materials used in Thermocline TES Designed to Supply Six Hours of TES for 50 MW CSP Plants.[18]

| Material | Material mass in kg | Emissions in MTCO[2] |
|---|---|---|
| Calcium silicate | 25,700 | 1.6 |
| Carbon steel | 45,6000 | 654 |
| Mineral wool | 15,8000 | 212 |
| Stainless steel | 3080 | 16 |

**TABLE 3.15** Characteristics of Storage Tanks and Liquid Media.[19]

| Characteristics | Values |
|---|---|
| Storage capacity | 600 MWh |
| Mass of molten salts | 5500 Tons |
| Thermal gradient between tanks | 260 degrees Celsius |
| Molten salts used | $NaNO_3$ and $KNO_3$ |
| Percentage of $NaNO_3$ and $KNO_3$ | 60/40 respectively |
| Temperature of hot fluid tank | 550 degrees Celsius |
| Temperature of cold fluid tank | 290 degrees Celsius |
| Density of mixture of salts | 1740 $Kg/m^3$ |
| Specific heat capacity | 1437.6 KJ/kg k |
| Total volume of storage unit | 4335 cubic meters |
| **Characteristics** | **Values** |
| Weight of steel | 279 tons |
| Thickness of lateral material insulation | 125 mm |
| Roof insulation | 125 mm |

**TABLE 3.16** The Following Table Gives the Comparison of Embodied Life Cycle Greenhouse Gas Emissions from the Materials Used in Two Tanks TES Designed to Supply Six Hours of TES for 50 MW CSP Plants.[19]

| Material | Material mass in kg | Emissions in $MTCO_2$ |
|---|---|---|
| Calcium silicate | 51,300 | 3.2 |
| Carbon steel | 885,000 | 1270 |
| Mineral wool | 283,000 | 382 |
| Stainless steel | 6110 | 31.7 |

### 3.9.1   OVERVIEW OF PCM PERFORMANCE

For the SEGS conditions, the storage system designs based on five concepts were developed by Dinter et al. in 1990. The results are summarized and presented here giving thermal storage capacity, overall system volume, utilization, and specific costs in $/kWht of capacity.

Aspects of temperature differences within the HTF fluid and between HTF fluid and solid storage medium were described in earlier discussions. The temperature difference within the medium itself is one of the important aspects of the storage design.[17–22] In a two-tank liquid system the whole fluid is heated up to a charged temperature so as to make the entire storage medium utilized. Temperature gradient required in solid system for thermal conduction through the media itself prevents full use of the material. In this case, if the entire solid medium were heated to the full charging temperature, then 100% utilization would be achieved. That leads to the potential storage capacity, may be 2–3 times, higher than the practical storage capacity.

With respect to volume, the salt and the concrete media fill approximately between 5200 m³ and 6900 m³ of space respectively, on the other hand, the PCM system and the molten salt need 2600 m³. The length of the concrete system would be 41 m as compared to 15 m length for the PCM system if the cross sectional area perpendicular to the flow measured 13 m by 13 m. Poor potential capacity is the main reason for the larger size of concrete and solid salt system. Only 36% of its full potential capacity is utilized in concrete system.[25–28] Instead, the utilization factor is up to 100% for molten salt and PCM system. The advantage of concrete system is very low cost that leads to low system cost even there is more structure needed for larger volume system.

Approximate storage cost in this assessment is between $25 to $50/kWht (in the order of $65 to $130/kWht). 270 and 450 MWHt capacity of TES units having a capital cost of 6.8 MUSD and 11.3 MUSD respectively. A symposium workshop (SERI 1989) on TES systems for SEGS plants, held in 1989 and sponsored by solar energy research institute (SERI—now the national renewable energy laboratory—NREL), discussed several of the options presented earlier. In this, the workshop mainly focused on phase change material concepts. In the agenda both SHS and chemical storage were included.

In view of sensible heat storage, approaches taken so far, are materialized in lieu of their cost-effectiveness. Some of the issues required more detailed design for design concept such as thermal expansion, heat transfer configuration, potential leakage and heat exchange optimization.

Latent heat storage was in the priority for primitive state of development, after analysis it is obtained that shell and tube heat exchanger and a system of enclosed particles of phase change salts were worthy of exploration, having more potential in terms of cost effectiveness and lower success probability in later approach.

### 3.9.2 OBSERVATIONS AND RECOMMENDATIONS

According to the literature the following observations are made:

1. In 1990s there was no significant development in the field of TES systems. But important contributions have been done work on candidate systems previously identified.
2. Nitrate salt eutectic contained in prototype two-tank molten salt system was tested successfully over 1½-year testing period.
3. For trough applications lower melting point molten salt systems should be explored. A one tank thermocline system is riskier in terms of performance while two-tank system has low risk approach but one tank system offers reductions in cost.
4. Center for Solar Energy and Hydrogen Research (ZSW), Germany has carried out various laboratory scale testing on various PCM modules.
5. In 1990, prototype construction proposal and testing of 1–2 MWh prototype concrete steel storage system was submitted to EU.
6. The design and development of parabolic trough application for the storage of chemical was not found significantly advanced in the past decade, still a few evaluations have been carried out further.

With the above-mentioned observations the following conclusions can be derived:

1. Molten salts and concrete system were of high priority as candidates for near-term deployment in terms of cost estimates and current

progress. The other system that can be used additionally for longer-term development is PCM.

2.  Prototype system development should be the near-term research to refine the designs and for the bases for valid cost estimates and performance field implementation is employed.

### 3.9.3  COMPARISON OF STORAGE SYSTEM TYPES INCLUDING ECONOMIC ASPECTS

The main problem with water storage system is the corrosion for long operation periods. Another problem with this is the volume of water storage may be very large, which makes the whole system very heavy. With packed bed storage there is no corrosion or storage or scale forming problem.

The comparison of these three systems has been given for $10^6$ kJ with 40 degree temperature difference. The water storage system occupies a volume 80 times more than the volume occupied by the phase change system, which is four times more than the amortization period of phase change systems.

Phase change systems are the most expensive but also the most compact types having least using periods because of the material deformation and degradation problem. Because of their compactness, the total initial costs are small.

### 3.9.4  MATERIALS FOR HEAT TRANSFER RATE IMPROVEMENT

In solar thermal power generation working fluid is not used for energy storage. Hence, to transfer the heat from the PCM to working fluid an indirect heat exchanger is needed. But the major problem in PCMs is their insufficient thermal conductivity. Hence, the main task in heat exchanger is to increase the heat conductivity of storage material with effective cost.

The suitable technique to overcome this problem is sand, which uses the concept of fins to enhance the heat transfer in the storage material. By mounting the fins vertical to the axis tubes heat transfer area is increased. The materials that are considered for the fins are mentioned with their properties:

*   The graphite foil has thermal conductivity of 150 W/(m.K) with density of 1000 kg/m$^3$.

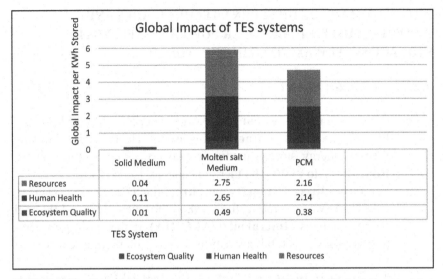

**FIGURE 3.12** Despite higher energy storage capacity, from the analysis, it is evident that the molten salt storage system has the highest impact on environment. Therefore, it needs to be substituted by any of the other two systems that are studied. Even though the storage capacity is less, the solid medium TES system has the least impact on the environment and has reduced construction complexity.

- The aluminum has thermal conductivity of 200 W/(m.K) with density of 2700 kg/m³.
- The stainless steel has thermal conductivity of 20 W/(m.K) with density of 7800 kg/m³.
- The carbon steel has thermal conductivity of 30 W/(m.K) with density of 7800 kg/m³. From the data provided, graphite foil or aluminum is used as fins. Aluminum fins are applicable up to 330°C.
- The fins made of graphite have superior thermal conductivity.
- They have good stability in nitrites and nitrate with temperature up to 250°C.
- The galvanic corrosion does not occur in contact with steel tubes.

When comparing the performance of the materials mentioned for charging state versus time the graphite maintains uniform charging rate for long period of time.[21-25] Hence, it improves the thermal conductivity of the PCM.

## 3.10  SUMMARY OF DIFFERENT TECHNOLOGIES AND MATERIALS USED IN THE SOLAR POWER SYSTEM WITH STORAGE SYSTEM EXISTING IN THE WORLD

### 3.10.1  TROUGH PLANT

a.  For **Active direct system (two tank)** experienced by SSPS DCS, PSA, Spain in 1981, **Mineral oil** is used as Heat transfer fluid and thermal energy storage media. It's operating temperature ranges from 180°C to 290°C. Its thermal capacity is 0.5 $Mwh_{th}$.

b.  For **active direct system (two tank)** conducted at SEGS1, Dagget, CA, USA, was operating from 1984–2001. It's HTF and thermal energy media is **Mineral oil (CALORIA)**. It's operating temperature is 307°C. Its total capacity is 14$Mw_e$. Its thermal capacity is 115 $Mw_{th}$.

c.  The **active indirect (two tank)** in Extresol 1-SENER was scheduled in 2010. The same project was scheduled by SOLANA, USA in 2011. **Molten salts** were used as TES media and **synthetic oil** as HTF. The total capacity of the system is 50 $Mw_e$.

d.  For **active indirect (two tank)** system done by ANDASOL 1, Spain in 2008. The TES media they used are **molten salts** of **(60% of $NaNO_3$ + 40% of $KNO_3$)**. Its HTF used is **steam**. The operating temperature ranges from 384°C for hot tank and 291°C for cold tank. Its thermal capacity is 1010 $Mw_{th}$.

e.  For passive system done by LS3-SSPS-PSA, Spain in 2004, the TEs media used is high temperature concrete versus catastable ceramics. Its HTF is mineral oil with thermal capacity of 0.48 $MW_{th}$.

### 3.10.2  PCM

• For **cascaded PCM storage passive system** by LUZ in 1990 used **synthetic oil** as HTF. The TES media used are $MgCl_2/KCl/NaCl$; **KOH; $KNO_3$; $NaNO_3$.** The operating temperature ranges from 345°C to 295°C. The thermal capacity is 875 $Mw_{th}$.

The same experiment for sensible PCM was conducted by DLR-ZSW in 1993.

### 3.10.3   SOLAR DISH

• The **ammonia synthesis** at theoretical and laboratory level was done in 1998 by using **ammonia** as HTF and TES media. Its operating temperature is 750°C.

### 3.10.4   CENTRAL RECEIVER PLANT

• For **active indirect single tank** system by solar one, USA in 1982–1988 used **steam** as HTF and **mineral oil+sand+rock** as TES media. Its operating temperature is 304–224°C.
• For **active direct (two tank)** by SOLAR TRES, Spain from 2002–2007 used **molten salts (NaNO$_3$+KNO$_3$)** as HTF and TES media.

## KEYWORDS

• **latent heat**
• **TES**
• **storage**
• **heat transfer**
• **PCM**

## REFERENCES

1. Park, K. S.; Ni, Z.; Côté, A. P.; Choi, J. Y.; Huang, R.; Uribe-Romo, F. J.; Chae, H. K.; O'Keeffe, M.; Yaghi, O. M. Exceptional Chemical and Thermal Stability of Zeolitic Imidazolate Frameworks. *Proc. Natl. Acad. Sci.* **2006**, *103* (27), 10186–101891.
2. Gilman, J. W. Flammability and Thermal Stability Studies of Polymer Layered-Silicate (Clay) Nanocomposites1. *Appl. Clay Sci.* **1999**, *15* (1–2), 31–49.
3. Kuravi, S.; Trahan, J.; Goswami, D. Y.; Rahman, M. M.; Stefanakos, E. K. Thermal Energy Storage Technologies and Systems for Concentrating Solar Power Plants. *Prog. Energy Combust. Sci.* **2013**, *39* (4), 285–319.
4. Bose, D.; Kandpal, V.; Dhawan, H.; Vijay, P.; Gopinath, M. Energy Recovery with Microbial Fuel Cells: Bioremediation and Bioelectricity. In *Waste Bioremediation*, Springe: Singapore, 2018; pp. 7–33.

5. Bose, D.; Dhawan, H.; Kandpal, V.; Vijay, P.; Gopinath, M. Bioelectricity Generation from Sewage and Wastewater Treatment using Two-Chambered Microbial Fuel Cell. *Int. J. Energy Res.* **2018,** *42* (14), 4335–4344.

6. Mawire, A.; Taole, S. H. A Comparison of Experimental Thermal Stratification Parameters for an Oil/Pebble-Bed Thermal Energy Storage (TES) System during Charging. *Appl. Energy* **2011,** *88* (12), 4766–4778.

7. Liao, Z.; Xu, C.; Ren, Y.; Gao, F.; Ju, X.; Du, X.; A Novel Effective Thermal Conductivity Correlation of the PCM Melting in Spherical PCM Encapsulation for the Packed Bed TES System. *Appl. Therm. Eng.* **2018,** *135,* 116–122.

8. Lim, K.; Che, J.; Lee, J. Experimental Study on Adsorption Characteristics of a Water and Silica-Gel Based Thermal Energy Storage (TES) System. *Appl. Therm. Eng.* **2017,** *110,* 80–88.

9. Safari, A.; Saidur, R.; Sulaiman, F. A.; Xu, Y.; Dong, J. A Review on Supercooling of Phase Change Materials in Thermal Energy Storage Systems. *Renew. Sustain. Energy Rev.* **2017,** *70,* 905–919.

10. Rea, J. E.; Oshman, C. J.; Olsen, M. L.; Hardin, C. L.; Glatzmaier, G. C.; Siegel, N. P.; Parilla, P. A.; Ginley, D. S.; Toberer, E. S. Performance Modeling and Techno-Economic Analysis of a Modular Concentrated Solar Power Tower with Latent Heat Storage. *Appl. Energy* **2018,** *217,* 143–152.

11. Haillot, D.; Pincemin, S.; Goetz, V.; Rousse, D. R.; Py, X. Synthesis and Characterization of Multifunctional Energy Composite: Solar Absorber and Latent Heat Storage Material of High Thermal Conductivity. *Sol Energ Mat. Sol. Cells* **2017,** *161,* 270–277.

12. Pelay, U.; Luo, L.; Fan, Y.; Stitou, D.; Rood, M. Thermal Energy Storage Systems for Concentrated Solar Power Plants. *Renew. Sustain. Energy Rev.* **2017,** *79,* 82–100.

13. Gajbhiye, P.; Salunkhe, N.; Kedare, S.; Bose, M. Experimental Investigation of Single Media Thermocline Storage with Eccentrically Mounted Vertical Porous Flow Distributor. *Sol. Energy* **2018,** *162,* 28–35.

14. Pakalka, S.; Valančius, K.; Čiuprinskas, K.; Pum, D.; Hinteregger, M. Analysis of Possibilities to Use Phase Change Materials in Heat Exchangers-Accumulators. In *Environmental Engineering. Proceedings of the International Conference on Environmental Engineering. ICEE* (Vol. 10, pp. 1-8). Vilnius Gediminas Technical University, Department of Construction Economics & Property, 2017.

15. Michels, H.; Pitz-Paal, R. Cascaded Latent Heat Storage for Parabolic Trough Solar Power Plants. *Sol. Energy* **2007,** *81* (6), 829–837.

16. Gong, Z. X.; Mujumdar, A. S. Finite-Element Analysis of Cyclic Heat Transfer in a Shell-and-Tube Latent Heat Energy Storage Exchanger. *Appl. Therm. Eng.* **1997,** *17* (6), 583–591.

17. Cui, Y.; Liu, C.; Hu, S.; Yu, X. The Experimental Exploration of Carbon Nanofiber and Carbon Nanotube Additives on Thermal Behavior of Phase Change Materials. *Sol. Energy Mater. Sol. C.* **2011,** *95* (4), 1208–1212.

18. Hoshi, A.; Mills, D. R.; Bittar, A.; Saitoh, T. S. Screening of High Melting Point Phase Change Materials (PCM) in Solar Thermal Concentrating Technology Based on CLFR. *Sol. Energy* **2005,** *79* (3), 332–339.

19. Morrison, D. J.; Abdel-Khalik, S. I. Effects of Phase-Change Energy Storage on the Performance of Air-Based and Liquid-Based Solar Heating Systems. *Sol. Energy* **1978,** *20* (1), 57–67.

20. Lafdi, K.; Mesalhy, O.; Elgafy, A. Graphite Foams Infiltrated with Phase Change Materials as Alternative Materials for Space and Terrestrial Thermal Energy Storage Applications. *Carbon* **2008,** *46* (1), 159–168.

21. Yagi, J.; Akiyama, T. Storage of Thermal Energy for Effective Use of Waste Heat from Industries. *J. Mater. Process. Technol.* **1995,** *48* (1–4), 793–804.

22. Jalalzadeh-Azar, A. A.; Steele, W.G.; Adebiyi, G. A. Heat Transfer in a High-Temperature Packed Bed Thermal Energy Storage System—Roles of Radiation and Intraparticle Conduction. *J. Energy Resour. Techno.* **1996,** *118* (1), 50–57.

23. Farid, M. M.; Khudhair, A. M.; Razack, S. A.; Al-Hallaj, S. A Review on Phase Change Energy Storage: Materials and Applications. *Energy Convers. Manag.* **2004,** *45* (9–10), 1597–1615.

24. Zalba, B.; Marın, J. M.; Cabeza, L. F.; Mehling, H. Review on Thermal Energy Storage with Phase Change: Materials, Heat Transfer Analysis and Applications. *Appl. Therm. Eng.* **2003,** *23* (3), 251–283.

25. Sharma, A.; Tyagi, V. V.; Chen, C. R.; Buddhi, D. Review on Thermal Energy Storage with Phase Change Materials and Applications. *Renew. Sustain. Energy Rev.* **2009,** *13* (2), 318–345.

26. Baetens, R.; Jelle, B. P.; Gustavsen, A. Phase Change Materials for Building Applications: A State-of-the-Art Review. *Energy Build.* **2010,** *42* (9), 1361–1368.

27. Cabeza, L. F.; Castell, A.; Barreneche, C. D.; De Gracia, A.; Fernández, A. I. Materials used as PCM in Thermal Energy Storage in Buildings: A Review. *Renew. Sustain. Energy Rev.* **2011,** *15* (3), 1675–1695.

28. Sarı, A.; Karaipekli, A. Thermal Conductivity and Latent Heat Thermal Energy Storage Characteristics of Paraffin/Expanded Graphite Composite as Phase Change Material. *Appl. Therm. Eng.* **2007,** *27* (8–9), 1271–1277.

29. Mondal, S. Phase Change Materials for Smart Textiles–An Overview. *Appl. Therm. Eng.* **2008,** *28* (11–12), 1536–1550.

30. Chang, C. C.; Tsai, Y. L.; Chiu, J. J.; Chen, H. Preparation of Phase Change Materials Microcapsules by Using PMMA Network-Silica Hybrid Shell via Sol-Gel Process. *J. Appl. Polym. Sci.* **2009,** *112* (3), 1850–1857.

31. Herrmann, U.; Kearney, D. W. Survey of Thermal Energy Storage for Parabolic Trough Power Plants. *J. Sol. Energy Eng.* **2002,** *124* (2), 145–152.

# CHAPTER 4

# Characterization Techniques of Phase Change Materials: Methods and Equipment

KARUNESH KANT[1,2*], AMRITANSHU SHUKLA[1], and ATUL SHARMA[1]

*¹Non-Conventional Energy Laboratory, Rajiv Gandhi Institute of Petroleum Technology, Jais, Amethi 229304, India*

*²Department of Mechanical Engineering, Eindhoven University of Technology, 5600 MB-Eindhoven, Netherlands*

*\*Corresponding author. E-mail: k1091kant@gmail.com*

## ABSTRACT

This chapter summarizes the techniques to characterize phase change materials for thermal energy storage applications. The most relevant properties include giving the answer to physical, thermal, and technical requirements. The chapter provides characterization methods and tools that are used to characterize latent heat, melting temperature, specific heat, thermal conductivity, durability, and cyclability of the PCMs. The chapter also describes various characterization standards for different thermophysical properties of PCMs.

## 4.1 INTRODUCTION

The continuous growth of population needs higher energy supply to fulfill the demand of energy for daily use. The fossil fuels are the main source of energy supply, which leads to increase in fuel prices and emission of greenhouse gasses. The incessant upsurge in the level of greenhouse gas emissions and the hike in fuel costs are the driving forces behind efforts to efficiently utilize various sources of renewable energy. The direct solar radiation is considered to be one of the most promising sources of renewable energy. Scientists and researchers all over the world are in the search of new and renewable energy sources that can help to supply energy for day-to-day uses. The thermal energy storage using phase change materials (PCM) could play a vital role for an efficient and economic utilization of thermal energy in the industries, as well as in power stations to generate power based on new conversion methods and renewable energy resources. Furthermore, integration of PCM in lightweight buildings is predicted to be a useful way to smoothen indoor temperature variations and reduce overall heating or cooling demand.

Characterization of PCM usually includes quantifying the thermophysical properties such as thermal conductivities and heat capacities of the solid and liquid phases, as well as transition temperatures and latent heat. Furthermore, for PCM experiencing melting over a range of temperature, enthalpy-temperature function is required too. Characterization is, thus, carried out using different samples and experimental devices. For instance, thermal conductivity and thermal diffusivity can be measured by the hot plate method and the flash method, respectively. Dynamic hot probes methods allow simultaneous determination of thermal conductivity and capacity. As for specific heat and latent heat, separate and specific differential scanning calorimetric (DSC) tests are usually used. Transition temperatures are better determined using differential thermal analysis (DTA) methods and enthalpy-temperature function estimation requires DSC tests in isothermal step mode (Richardson, 1997; Rudtsch, 2002) of the number of apparatus/tests required for complete characterization, one notices that the tests based on DTA/DSC devices generally require very small samples (some few millilitres), so that they become inappropriate for testing heterogeneous materials with large-size representative volumes. Such a problem could be partially overcome using the T-History method (Zhang and Jiang, 1999; Hong et al., 2004), a cheap and easy way for the determination of latent heats and specific heats. Unfortunately, a T-History method is unable to reliable estimation of transition temperatures and enthalpy-temperature functions (Gunther et al., 2009).

## 4.2 CHARACTERIZATION OF PCM

Characterization of PCM usually involves measurement of thermosphysical properties such as thermal conductivities and heat capacities of the solid and liquid phases, as well as transition temperatures and latent heat. Furthermore, for PCM undergoing melting over a range of temperature, enthalpy-temperature function is required too. Characterization is, thus, carried out using different samples and experimental devices. For instance, thermal conductivity and thermal diffusivity can be measured, respectively, by the hot plate method and the flash method. Dynamic hot probes methods allow simultaneous determination of thermal conductivity and capacity. As for specific heat and latent heat, separate and specific DSC tests are usually used. Transition temperatures are better determined using DTA methods and enthalpy-temperature function estimation requires DSC tests in isothermal step mode. Number of apparatus/tests are required for complete characterization; one notices that the tests based on DTA/ DSC devices generally require very small samples (some few millilitres), so that they become inappropriate for testing heterogeneous materials with large-size representative volumes. Such a problem could be partially overcome using the T-History method, a cheap and easy way for the determination of latent heats and specific heats. Unfortunately, T-History method is unable for reliable estimation of transition temperatures and enthalpy-temperature functions.

## 4.3 MATERIAL CHARACTERIZATION STANDARD

As the knowledge of the behavior of heat storage materials is very essential to design an application, it is important to characterize these materials. The important characteristics of heat storage materials are their heat capacity, the thermal conductivity, the density, and viscosity—all in dependency on temperature. All these parameters are necessary to size a thermal storage or to develop heat exchanger to charge and discharge such storages. Simulations are also very often used to analyze applications or components of it and their interaction with the storage material. Such simulations will not be valid if the used material data is not describing its behavior in a correct way, so also for this purpose good and reliable results from the characterization are needed.

The listed points in the subsequent sections summarize the standards defined by measured parameters of materials that are also important when characterizing PCMs (Gschwander et al., 2011).

### 4.3.1   MELTING POINT

i.   *ISO 2207:1980:* The International Standard organization gives a characterization standard for the determination of congealing point. The specified procedure in this method allows us to find the temperature at which the testing sample being cooled. At that temperature, the sample product may be at or close to the solid phase. In the case of petrolatum products, the congealing is associated with the formation of a gel structure as the sample cools off.

ii.  *ASTM D87-07a:* American Society for Testing and Material had developed the standard test method for determination of melting point. It is unsuitable for waxes of the petrolatum group, microcrystalline waxes, or blends of such waxes with paraffin wax or scale wax.

iii. *ASTM D4419-90 (2005):* American Society for Testing and Material developed a standard test method for measurement of transition temperatures using DSC. In this test technique, the transition temperatures of petroleum waxes, including microcrystalline waxes, by DSC can be determined.

### 4.3.2   HEAT CAPACITY

i.   *ISO 11357-4:2005:* This characterization standard specifies the methods for determining the specific heat capacity of plastics by DSC.

### 4.3.3   ENTHALPY

i.   *ISO 11357-3:199:* This characterization standard is given by International Standard Organization for the determination of melting temperature and enthalpy of melting and crystallization by DSC.

ii.  *ASTM E793-06:* American Society of Testing and Materials has given a standard test method for determination of test enthalpies of fusion and crystallization by use of DSC. This characterization standard is applicable to solid samples in granular form or in any fabricated shape from which an appropriate specimen can be cut, or to liquid samples that crystallize within the range of the instrument.

### 4.3.4   THERMAL CONDUCTIVITY

i.   *ISO 22007-2:2008:* The International Standard Organization developed a test procedure for the determination of thermal conductivity and thermal diffusivity. The measurements of thermal conductivity and thermal diffusivity can be made in gaseous and vacuum environments at a range of temperatures and pressures.

ii.   *ISO 22007-4:2008:* In this test standard, the thermal conductivity and thermal diffusivity is determined by laser flash method. This method is based upon the measurement of the rise in temperature at the rear face of the thin-disc specimen produced by a short energy pulse on the front face. The method can be used for homogeneous solid plastics as well as composites having an isotropic or orthotropic structure.

iii.   *ISO 13787:2003:* ISO 13787:2003 establishes a procedure for the determination and verification of declared thermal conductivity, as a function of temperature, of thermally insulating materials and products used for the insulation of building equipment and industrial installations.

### 4.3.5   DENSITY

i.   *ASTM D 1481-02 (2007):* This standard test method is used for the measurement of density and specific gravity of viscous materials by Lipkin Bicapillary Pycnometer.

ii.   *ISO 3675:1998:* This test standard is used for density measurement of crude petroleum and liquid petroleum products by hydrometer method.

iii.   *ISO 12185:1996:* This standard gives a method for the measurement of density of the product using an oscillation U-tube densitometer.

### 4.3.6   VISCOSITY

i.   *ASTM D 2534-88(2007):* This test method covers the determination of the coefficient of kinetic friction for a petroleum wax coating or wax-based hot melt coating when sliding over it.

ii.   *ASTM D 2669-06*

*iii.  ASTM D1986-91(2007)*
*iv.  ASTM F766-82(2005)*

### 4.3.7  THERMAL STABILITY

*i.   ASTM E537-07:* A standard test method for the thermal stability of chemicals is DSC. This test method describes the ascertainment of the presence of enthalpy changes in a test specimen, using minimum quantities of material, approximates the temperature at which these enthalpy changes occur, and determines their enthalpies (heats) using DSC or pressure DSC.

### 4.3.8  COMPATIBILITY

*i.   ISO 175:1999:* Methods of test determination of the effects of immersion in liquid chemicals.
*ii.*  ISO 62:2008
*iii.* ISO 4611:2008

## 4.4  ADVANCEMENT IN CHARACTERIZATION OF PCMs

The performance of a thermal energy storage system is directly associated with the phase transition properties of PCM. It has been pointed out (Tyagi and Buddhi, 2007) that the data supplied by the producers could be incorrect, doubtful, and over optimized. Therefore, it is essential to make measurements so as to get the correct phase change properties of PCM. Numerous thermos-physical properties measurement techniques such as DSC, T-history method, and DTA exists; however, DSC is the most common practice in one of them and therefore would be extensively discussed.

### 4.4.1  DIFFERENTIAL SCANNING CALORIMETER (DSC)

It is an analytical technique developed by Watson in 1962. The equipment that is used for this purpose is named as DSC. This equipment has the ability to directly measure the energy storage capacity during melting and allow accurate measurement of heat capacity. The temperatures and heat flow associated with material changes as a function of time are measured in controlled environment (Brown, 1998). The qualitative and quantitative data about physical and chemical changes is measured that involve endothermic or exothermic processes (DSC, 2010). According to David et al. (2011), the name DSC is very clear:

• Calorimetry: The measurement of the quantity of heat absorbed or released of a sample with the change in temperature.
• Differential: The measurement has been carried out on sample with respect to reference sample with known properties.
• Scanning: The thermal excitation with a linear temperature ramp.

The analysis procedure in pictorial form is shown in Figure 4.1 and is explained herein.

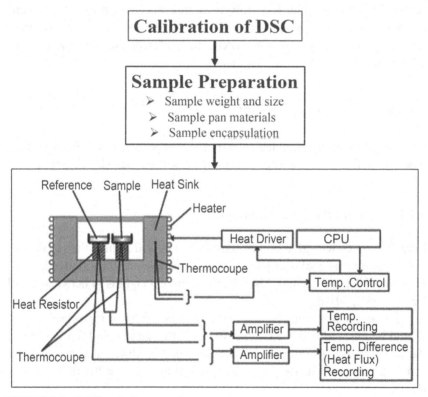

**FIGURE 4.1** DSC analysis protocol.

### 4.4.1.1   CALIBRATION

The purpose of calibration is to reduce any measurement uncertainty by ensuring the accurateness of the apparatus (DSC, 2010). For DSC, temperature and heat flow standardizations must be done from time to time (DSC, 2010). Naturally, indium is used as a standardizing material. Baseline slope and offset calibrations require warming an empty pan through estimated temperature range in the experiment to obtain precise results (DSC, 2010). Standardization is used to flatten the baseline and zero the heat flow signal. It is essential to be pointed out here that the type of purge gas and flow rate affects calibration. Consequently, nitrogen, which is inert, cost-effective, least affected by changes in flow rate due to its low thermal conductivity, and provides good sensitivity is preferred (Instruments, 2012). Furthermore, it removes humidity (if any) and oxygen from the cell that may have accrued over time. The too slow flow rate may cause moisture accumulation and early aging of the cell; however, too fast flow rate may cause excessive noise. For nitrogen, the preferred flow rate is 50 mL/min (DSC, 2010; Instruments, 2012).

### 4.4.1.2   SAMPLE PREPARATION

Sample preparation includes selecting the appropriate weight and size, type and material, and encapsulation of the sample pan (DSC, 2010).

### 4.4.1.3   SAMPLE WEIGHT AND SIZE

Larger samples will give higher sensitivity but will reduce the resolution. Consequently, the goal is to achieve heat flow rate in the transition of interest in between 0.1 and 10 mW. Usually, the sample weight is in between 5 and 20 mg (DSC, 2010).

The sample size must be made as thin as possible for reproducibility, it should be ensured that the contact between the sample and the bottom of the pan is good. The powdered samples must be distributed consistently across the bottom of the pan, which will minimize the thermal gradient (DSC, 2010).

### 4.4.1.4  SAMPLE PAN MATERIALS AND CONFIGURATION

The selection of pan material and its configuration depends on numerous factors such as temperature range experienced during the thermal properties measurement of the sample and reactivity of the sample with the pan material. In general, Aluminium pans can be used in the temperature range 180 to 600°C. However, if the sample has the ability to react with aluminium than platinum, copper, or gold pans can be opted for (DSC, 2010). As far as the sample pan configuration is concerned, the pan can be hermatic, non-hermatic, or open (DSC, 2010). However, hermatic pan (Fig. 4.2) is preferred due to obvious advantages such as better thermal contact, reduced thermal gradient in the sample, higher internal pressure resistance due to air tight seal, and preservation of sample for further study (DSC, 2010). In addition, when contact with cell atmosphere or the reaction of the sample gas is required, hermatic pan can be used by making a pin hole in the lid before sealing. After deciding the sample pan material and configuration and keeping the sample in the sample pan, both the hermatic and non-hermatic pan should be sealed with the sample encapsulating press (Fig. 4.3). The sample and reference pan are then loaded into the DSC cell.

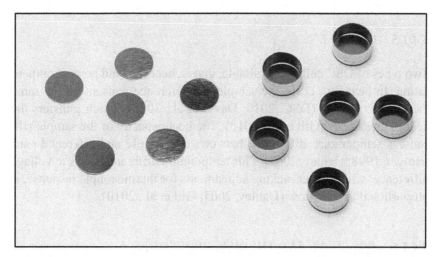

**FIGURE 4.2**  Sample pan and lid set (DSC, 2014).

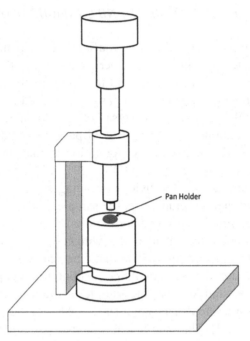

**FIGURE 4.3**   Schematic of sample encapsulating press used to hermetically seal sample pan (DSC, 2014).

### 4.4.1.5   DSC CELL

Two types of DSC cells are available, that is, heat flux and power compensating. In heat flux DSC, the sample and reference pans sit on the same thermoelectric disc (DSC, 2010; David et al., 2011), which transfers the heat to these pans (Gill et al., 2010). The heat capacity of the sample will cause a temperature difference between the sample and reference pans (Brown, 1998; Danley, 2003). This temperature difference leads to voltage difference, which, after making adjustments for thermocouple response, is proportional to heat flow (Danley, 2003; Gill et al., 2010).

### 4.4.1.6   DSC CURVE AND THERMAL PROPERTIES

The thermal properties such as phase change temperature and thermal energy stored in unit weight can be determined by using commercially available software such as TA Universal Analysis 2000 (Instruments, 2001). The typical DSC plot is shown in Figure 4.4. The phase change temperature is divided into starting, peak, and ending temperatures. The starting and ending temperatures are the temperatures at the intersection of extrapolated baseline and the tangents to the DSC curve drawn at the inflection points to the left and right side of the peak while the peak temperature is the temperature at the peak point of DSC curve. The thermal heat stored in the unit weight of PCM is obtained by dividing the integrated area between the baseline and the DSC curve with a temperature rising rate in the DSC test. This value is calculated automatically by the software (Zhang et al., 2005). It is worthy to mention here that the user can choose different types of baselines such as linear, sigmoidal horizontal, sigmoidal tangent, and extrapolated to calculate the thermal energy stored in the unit weight of PCM (Instruments, 2001; Marangoni and Wesdorp, 2013).

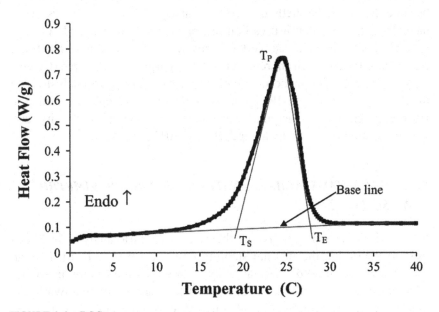

**FIGURE 4.4** DSC curve.

There are several studies that are based on the DSC characterization of PCM. The thermos-physical characterization of PCM is mainly done using DSC. The DSC can measure melting temperature, latent heat of fusion, and specific heat. Recently, Sharma and colleagues (Sharma et al., 2013) developed binary mixture of fatty acids as PCM for the application of buildings. The melting temperature and latent heat of fusion are characterized using DSC. The study had been based on the characterization of PCM. The characterized PCM are in the temperature range of 20–30°C and good amount of latent heat of fusion.

### 4.4.2   CHEMICAL COMPATIBILITY ANALYSIS—FOURIER TRANSFORM INFRARED SPECTROSCOPY (FT-IR)

For PCM-based building applications, FT-IR has been successfully used to determine the chemical compatibility between the components of composite PCM, that is, the PCM and the container/supporting material/ encapsulation. With the invention of world's first commercial FT-IR spectrometer (Model FTS-14), in 1969 (Barth and Haris, P 2009), FT-IR became the preferred method of spectral analysis. The term FT-IR originates from the fact that it uses Fourier transform, a diverse and versatile analytical technique, to convert the raw data into an actual spectrum. Since the infrared spectrum represents the fingerprint of a sample (Coates, 2000), therefore, it can be positively used for qualitative analysis. The major advantage of FT-IR includes high speed (Flegett advantage), improved sensitivity (Jacquinot advantage), mechanical simplicity, and self-calibration (Connes advantage) (FTIR, 2017).

### 4.4.3   THERMAL STABILITY ANALYSIS THERMO-GRAVIMETRIC ANALYSIS (TGA)

For PCM-based applications, the thermal gravimetric analyser has been used to determine the thermal stability of the composite PCM. It is an experimental method of thermal analysis in which the change in weight of a material is observed as a function of temperature or time while the material is subjected to a controlled temperature program in a controlled environment (Coats and Redfern, 1963; Elmer, 2004). Several researchers have used this technique to ensure that the composite PCM is stable in the working temperature range (Karaipekli and Sarı, 2011; Sarı and Karaipekli, 2012). Also, it is used to ensure that the components of composite PCM are decomposing separately.

### 4.4.4 MEASUREMENT OF THERMAL CONDUCTIVITY OF PCM/ COMPOSITE PCM

In order to improve the performance of PCM in terms of heat transfer, it is necessary to determine the thermal conductivity of the PCM/composite PCM. The thermal conductivity of a sample can be measured by steady-state and nonsteady steady methods (Memon, 2014). In the steady-state method, the measurements are made when the temperature of the material does not vary with time, that is, at the constant signal. Therefore, the signal analysis is simple and straight forward. In non-steady state method, the measurements are made during the process of heating up, that is, the signal is studied as a function of time. It can be performed more quickly; however, in general, the mathematical analysis of the data is more difficult.

The temperature history methods are utilized to determine the melting point, degree of super-cooling, heat of fusion, specific heat, and thermal conductivity of several PCM samples simultaneously. It is especially useful for the selection of lots of candidate PCMs or for the preparation of new PCMs for use in practical systems. Figure 4.5 represents experimental setup for the measurement of thermos-physical properties of PCM. The advantages of this kind of experimental system are as follows: (i) Using conventional tubes as PCM sample containers makes measurement convenient and means that the phase-change process of each sample can be clearly observed. (ii) One is able to measure several samples during a test (the number of samples measured in a test run depends upon the number of channels of the data acquisition system).

### 4.4.5 SCANNING ELECTRON MICROSCOPE

A scanning electron microscope (SEM) is a type of electron microscope that produces images of a sample by scanning it with a focused beam of electrons. The electrons interact with atoms in the sample, producing various signals that contain information about the sample's surface topography and composition. The electron beam is generally scanned in a raster scan pattern, and the beam's position is combined with the detected signal to produce an image. SEM can achieve resolution better than 1 nanometer. Samples can be observed in a high vacuum, in low vacuum, in wet conditions (in environmental SEM), and at a wide range of cryogenic or elevated temperatures.

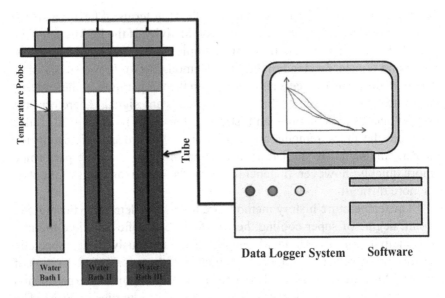

**FIGURE 4.5**　A schematic diagram of the experimental setup.

### 4.4.6　*DIFFERENTIAL THERMAL ANALYSER*

DTA is a technique that dates back to Le Chatelier and has in the past yielded valuable information on physical, chemical, and mechanical changes in substances or systems as they are heated. It has been characteristic of the advances in instrumentation that many classical procedures have been revived or completely overhauled to yield new results of unprecedented sensitivity and versatility. Such is the case with the Du Pont 900 Differential Thermal Analyser (Müller, 1963).

A DTA consists of a sample holder comprising thermocouples, sample containers, a ceramic or metallic furnace, a temperature programmer, and a recording system. The basic configuration is the two thermocouples that are connected in a differential arrangement and connected to a high gain low noise differential amplifier. One thermocouple is placed in an inert material such as $Al_2O_3$, while the other is placed in a sample of the material under study. As the temperature is increased, there will be a brief deflection of the voltage if the sample is undergoing a phase transition. This occurs because the input of heat will raise the temperature of the inert substance, but be incorporated as latent heat in the material changing phase.

### 4.4.7   RHEOMETER

The rheometer is a laboratory device that is used to measure the technique in which a liquid, suspension, or slurry flows in response to applied forces. It is used for those fluids that cannot be defined by a single value of viscosity and therefore require more parameters to be set and measured than is the case for a viscometer. Viscosity is a measure of the resistance of a fluid, which is being deformed by the shear stress. Stress is the measure of internal force applied to an object. Shear stress is the stress that is applied parallel to the face of an object or material. In every day terms, viscosity is "thickness or internal friction." Viscometer is an instrument used to measure the viscosity of a fluid. It measures the rheology of the fluid. Rheology is the study of the flow of matter, primarily in liquid state. The term *rheometer* comes from the Greek word *rheo*, meaning flow, and rheometer is a device for "measuring flow."

### 4.5   SUMMARY AND CONCLUSIONS

This chapter presents a comprehensive overview of various techniques for the characterization of thermal properties of PCMs. The discussion is made on the characterization of thermal properties such as thermal conductivity, specific heat and latent heat of fusion, and chemical stability. The melting temperature, specific heat, and latent heat of fusion can be characterized by the using DSC. The DSC technique is the best technique to measure the melting temperature, specific heat, and energy stored in the materials as it has higher accuracy as compared to other methods like temperature history methods. The chemical compatibility of the mixture used as thermal energy storage can be checked by FTIR and the thermal stability can be checked by TGA. The thermal conductivity of the PCM can be measured by temperature history method. Mainly in this chapter, the various techniques of PCM characterization and characterization standard are described, which is very useful in development of new and thermal energy storage as PCMs.

## KEYWORDS

- **characterization**
- **thermo-gravimetric analysis (TGA)**
- **scanning electron microscope (SEM)**
- **differential thermal analysis (DTA)**

## REFERENCES

1. Barth, A.; Haris, P. I. Advances in Biomedical Spectroscopy. In *Bilogical and Biomedical Infrared Spectroscopy*; Barth, A., Haris, P. I., Eds.; IOS Press BV: Netherlands, 2009.

2. Coates, J. Intrepretation of Infrared Spectra, A Practical Approach. In *Encyclopedia of Analytical Chemistry*; Meyers, R. A., Ed.; John Wiley & Sons Ltd: Hoboken, New Jersey, 2000.

3. Coats, A. W.; Redfern, J. P. Thermogravimetric Analysis A Review. *J. Analyst* **1963**, *88*, 906–924, doi: 10.1038/1731011b0.

4. Danley, R. L. New Heat Flux DSC Measurement Technique. *Thermochimica Acta* **2003**, *395*, 201–208, doi: 10.1016/S0040-6031(02)00212-5.

5. David, D.; Johannes, K.; Roux, J-J., et al. A Review on Phase Change Materials Integrated in Building Walls. *Renew. Sustain. Energy Rev.* **2011**, *15*, 379–391, doi: 10.1016/j.rser.2010.08.019.

6. DSC. *Guide to Selection of Differential Scanning Calorimetry (DSC)*, 2014, Sample Pans.

7. DSC. *Differential Scanning Calorimeter Operator's Manual*, 2010.

8. Elmer, P. Thermogravimetric Analysis (TGA): A Beginner's Guide. Perkin Elmer, 2004.

9. FTIR. Introduction to Fourier Transform Infrared Spectrometry. A Thermo Electron Bussines, 2017, pp. 1–8.

10. Gill, P.; Moghadam, T. T.; Ranjbar, B. Differential Scanning Calorimetry Techniques: Applications in Biology and Nanoscience. *J. Biomol. Techniq.* **2010**, *21*, 167–193.

11. Gschwander, S.; Cabeza, L. F.; Chui, J. Development of a Test-Standard for PCM and TCM Characterization Part 1 : Characterization of Phase Change Materials Compact Thermal Energy Storage: Material Development for System Integration; 2011.

12. Gunther, E; Hiebler, S.; Mehling, H.; Redlich, R. Enthalpy of Phase Change Materials as a Function of Temperature: Required Accuracy and Suitable Measurement Methods. *Internat. J. Thermophys.* **2009**, *30*, 1257–1269, doi: 10.1007/s10765-009-0641-z.

13. Hong, H.; Kim, S. K.; Kim, Y. S. Accuracy Improvement of T-History Method for Measuring Heat of Fusion of Various Materials. *Internat. J. Refrigerat.* **2004**, *27*, 360–366, doi: 10.1016/j.ijrefrig.2003.12.006.

14. Instruments, T. *Thermal Advantage Universal Analysis Operator's Mannual*, 2001.

15. Instruments, T. A. *Thermal Analysis*. Brochure TA Instruments, *29*, 2012.

16. Karaipekli, A.; Sarı, A. Preparation and Characterization of Fatty Acid Ester/Building Material Composites for Thermal Energy Storage in Buildings. *Energy Build.* **2011,** *43,* 1952–1959, doi: 10.1016/j.enbuild.2011.04.002.
17. Marangoni, A.; Wesdorp, L. *Structure and Properties of Fat Crystal,* Second; CRC Press, Boca Raton, Florida, 2013.
18. Memon, S. A. Phase Change Materials Integrated in Building Walls: A State of the Art Review. *Renew. Sustain. Energy Rev.* **2014,** *31,* 870–906, doi: 10.1016/j.rser. 2013.12.042.
19. Brown, Michael E. *Handbook of Thermal Analysis and Calorimetry,* 1st ed.; Elsevier B.V., Amsterdam, The Netherlands, 1998.
20. Müller, R. H. Differential Thermal Analysis. *Anal. Chem.* **1963,** *35,* 103A–105A, doi: 10.1021/ac60197a799.
21. Richardson, M. J. Quantitative Aspects of Differential Scanning Calorimetry. *Thermochimica Acta* **1997,** *300,* 15–28, doi: 10.1016/S0040-6031(97)00188-3.
22. Rudtsch, S. Uncertainty of Heat Capacity Measurements with Differential Scanning Calorimeters. *Thermochimica Acta* **2002,** *382,* 17–25, doi: 10.1016/S0040-6031(01) 00730-4.
23. Sarı, A.; Karaipekli, A. Fatty Acid Esters-Based Composite Phase Change Materials for Thermal Energy Storage in Buildings. *Appl. Therm. Engineer.* **2012,** *37,* 208–216, doi: 10.1016/j.applthermaleng.2011.11.017.
24. Sharma, A.; Shukla, A.; Chen, C. R. R.; Dwivedi, S. Development of Phase Change Materials for Building Applications. *Energy Build.* **2013,** *64,* 403–407 doi: 10.1016/j. enbuild.2013.05.029.
25. Tyagi, V. V.; Buddhi, D. PCM Thermal Storage in Buildings: A State of Art. *Renew. Sustain. Energy Rev.* **2007,** *11,* 1146–1166, doi: 10.1016/j.rser.2005.10.002.
26. Zhang, D.; Zhou, J.; Wu, K.; Li, Z. Granular Phase Changing Composites for Thermal Energy Storage. *Solar Energy* **2005,** *78,* 471–480, doi: 10.1016/j.solener.2004.04.022.
27. Zhang, L. Z.; Jiang, Y. Heat and Mass Transfer in a Membrane-Based Energy Recovery Ventilator. *J. Memb. Sci.* **1999,** *163,* 29–38, doi: 10.1016/S0376-7388(99)00150-7.

## CHAPTER 5

# Heat Transfer Studies of PCMs to Optimize the Cost Efficiency for Different Applications

AMRITANSHU SHUKLA[1*], KARUNESH KANT[1*,2] and ATUL SHARMA[1]

[1]Non-Conventional energy laboratory, Rajiv Gandhi Institute of Petroleum Technology, Jais, Amethi 229304, India

[2]Department of Mechanical Engineering, Eindhoven University of Technology, 5600 MB-Eindhoven, Netherlands

*Corresponding author. E-mail: ashukla@rgipt.ac.in, k1091kant@gmail.com

### ABSTRACT

Heat transfer analysis of phase change materials (PCMs), is a challenging task due to their unique thermochemical properties and the complexity of their operation. The aim of this chapter is to provide information about mathematical relations that are used to investigate the thermal performance of PCM. The chapter provides an up-to-date information regarding the modeling of thermophysical properties of the PCM, which is varying with their phase change. Various relations are suggested to improve the thermophysical properties of PCM with additives. Further, the studies have been discussed, which are used these relations to investigate the performance of the systems integrated with PCM to optimize the cost efficiency, material and design of the system.

## 5.1 INTRODUCTION

With the current penetration of renewable energy, such as wind and solar, into the electrical grid, there is an inevitable need to increase the dispatching ability of these power sources. Solar energy is one of the most attractive energy sources in the world. As a kind of clean renewable energy option, it is receiving a lot of attention due to the finite quantities of fossil fuels and the pollution drawback associated with them. The thermal energy storage systems are an essential feature to make an efficient use of solar energy due to the inherent intermittency of this energy source. These systems allow making use of thermal energy accumulated in hours of high solar radiation in moments of lower solar radiation, reducing the mismatch between the supply and demand of the energy. Phase change materials (PCMs) provide an effective way of accumulating thermal energy, due to their high capacity to store heat at a constant or near to constant temperature (Galione et al., 2011). Thermal energy storage (TES) could be classified into three forms, based on the energy storage process mechanism: sensible heat storage, latent heat storage, and thermochemical heat storage. In sensible heat storage systems, energy is stored and released by raising or lowering the temperature of a material. The amount of energy stored depends on the heat capacity of the material and the temperature difference applied to the material. While sensible heat storage systems rely on large temperature differences, latent heat-based storage systems can store energy under nearly isothermal conditions, by utilizing the large quantity of energy required to induce a change of state within a material. In a similar way, thermochemical energy storage systems use reversible endothermic/exothermic reactions to store and release energy. To reduce the cost of a TES system, a material with a high energy storage density should be used. While thermochemical systems have the highest energy storage density of the three forms of TES systems, research in this area is still in the early stages and uncertainties in thermodynamic properties, phase change process, as well as reaction kinetics, limit its usage.

Heat storage systems using PCMs are an effective way of storing thermal energy due to the high energy storage density and the isothermal nature of the storage process. Latent heat storage systems have been widely used in building envelopes, residential heating and cooling, solar engineering, and spacecraft thermal control applications (Sharma et al., 2009). In recent years, the utilization of PCMs has been also considered in the thermal control of compact electronic devices.

This chapter describes the mathematical modeling of PCMs and its application. The mathematical formulation of the PCM is the reduction of the physical problem to a set of either algebraic or differential equations subject to certain assumptions. The process of modeling of physical systems coupled with PCMs in the real world should generally follow the path illustrated schematically in the Figure 5.1.

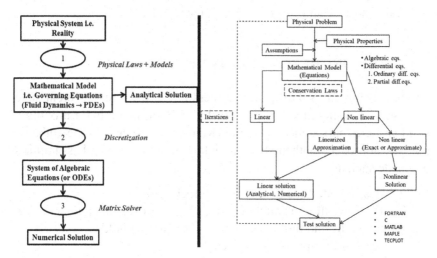

**FIGURE 5.1** Numerical solution procedure.

## 5.2 THE STEPHAN PROBLEM

The change of one phase to another of any a material can be labeled by a certain kind of boundary value problems for partial differential equations, where phase transition boundary can change with time. The problem has been first studied by Clapeyron and Lame in 1831 at the time of analyzing the foundation of the Earth's crust though it cooled. In that case, the spherical geometry problem was simplified to a one-dimensional semi-infinite slab (Lamé and Clapeyron, 1831). Franz Neumann independently found this solution, who introduced it in his lectures notes of 1835–1840 (Brillouin, 1930). The Slovene physicist Jozef Stefan introduced the general class of such problems in 1889 (Stefan, 1898) in relation to problems of ice formation. The presence of an explanation was showed by Evans in 1951 (Evans, 1951), though the individuality was showed by Douglas in 1957 (Douglas, 1957). Very rare analytical descriptions are obtainable in the closed method. The problems are mainly applicable to the one-dimensional cases of an infinite or semi-infinite region having simple initial and boundary conditions and having constant thermal properties. Under these situations, these exact solutions usually take the form of functions of the single variable $x/t^{1/2}$ and are known as similarity solutions (Crank, 1987; Hill, 1987). A collection of similarity solutions and references is to be found in (Carslaw and Jaeger, 1959; Lunardini, 1981).

## 5.3   MODELING HEAT AND MASS TRANSFER

At the melting and solidification process of PCMs, three different states coexist, that is, liquid-solid and "mushy" region, involving a mixture of solid particles surrounded by liquid (Kant et al., 2016a). In the enthalpy-porosity approach, mathematical modeling of there is several assumptions are made such as:

   i.    Heat transfer due to radiation is neglected.
   ii.   The flow after the melting of PCMs is Newtonian, laminar, incompressible, and the Boussinesq approximation holds.
   iii.  Viscous dissipation is neglected.
   iv.   Thermophysical properties are constant.
   v.    The PCM material is pure, homogeneous, and isotropic.
   vi.   The volumetric expansion is neglected upon melting.
   vii.  Natural convection occurs in the liquid phase.

On the basis of above-mentioned assumptions, the governing equations for the 2-D heat transfer melting process are as follows:
In solid region:

$$\frac{\partial T}{\partial t} = \alpha_s \left[ \frac{\partial^2 T}{\partial x^2} + \frac{\partial^2 T}{\partial y^2} \right]. \tag{5.1}$$

In the liquid region:

$$\frac{\partial u}{\partial x} + \frac{\partial v}{\partial y} = 0. \tag{5.2}$$

$$\rho_l \left( \frac{\partial u}{\partial t} + u\frac{\partial u}{\partial x} + v\frac{\partial u}{\partial y} \right) = -\frac{\partial p}{\partial x} + \mu\left( \frac{\partial^2 u}{\partial^2 x} + \frac{\partial^2 u}{\partial^2 y} \right) + Su \tag{5.3}$$

$$\rho_l \left( \frac{\partial v}{\partial t} + u\frac{\partial v}{\partial x} + v\frac{\partial v}{\partial y} \right) = -\frac{\partial p}{\partial x} + \mu\left( \frac{\partial^2 u}{\partial^2 x} + \frac{\partial^2 u}{\partial^2 y} \right) + \rho g \beta\left(T - T_{ref}\right) + Sv \tag{5.4}$$

$$\left(\frac{\partial T}{\partial t}+u\frac{\partial T}{\partial x}+v\frac{\partial T}{\partial y}\right)=\alpha_1\left(\frac{\partial^2 u}{\partial^2 x}+\frac{\partial^2 u}{\partial^2 y}\right)-\frac{L\partial f}{c_1\partial t}. \tag{5.4}$$

A parameter $S$ in the momentum equations (Brent et al., 1988) is appropriately defined that sets the velocities in the solid regions. The basic principle is to reduce gradually the velocities from a finite value in the liquid to zero in the solid, over the computational cell that undergoes a phase change. This can be achieved by assuming that such cells behave like a porous media with porosity equal to the liquid fraction. In order to achieve this behavior, an appropriate definition of $S$ is:

$$S=-\frac{C(1-f)^2}{(f^3+b)}, \tag{5.5}$$

which is the Carman–Kozeny relation. In this model, $f=1$ in the liquid region while $f=0$ in the solid region while it takes a value between 0 and 1 in the mushy zone. The constant $C$ has a large value to suppress the velocity as the cell becomes solid and $b$ is a small constant used to avoid a division by zero when a cell is fully located in the solid region, namely $f=0$. The choice of the constants is arbitrary. However, the constants should ensure sufficient suppression of the velocity in the solid region and should not influence the numerical results significantly. In this work, $C=1\times10^9$ kg/m$^3$s and $b=0.0003$ are used (Viswanath and Jaluria, 1999).

As the gravity terms in the momentum equations may have an effect on the fluid flow, the density in the buoyancy forces is dependent on temperature based on the Boussinesq approximation. The relevant dimensionless parameters, the Rayleigh, Stefan, and Prandtl numbers are as follows, respectively.

$$Ra=\frac{\rho^2 c_l g\beta L_y^3(T_w-T_F)}{\mu k_l} \tag{5.6}$$

$$Ste=\frac{c_l(T_w-T_F)}{L} \tag{5.7}$$

$$\Pr=\frac{c_l\mu}{k_l}. \tag{5.8}$$

## 5.4   MODELING PCMs PROPERTIES

The variation of thermos-physical properties of PCMs with temperature can be given as follows:
   Density:

$$\rho_{PCM}(T) = \rho_{solid} + (\rho_{liquid} - \rho_{solid}).B(T), \tag{5.9}$$

where

$$B(T) = \begin{cases} 0 & , & T < (T_m - \Delta T) \\ (T - T_m + \Delta T)/(2\Delta T), & (T_m - \Delta T) \leq T \langle (T_m + \Delta T). \\ 1 & , & T \rangle (T_m + \Delta T) \end{cases} \tag{5.10}$$

   Equation 5.10 shows that $B$ is zero when the PCM is in solid phase and 1 when it is in the liquid phase. $B$ linearly grows from zero to $1$ between the two states (Biwole et al., 2013). The specific heat and latent heat of fusion for the PCM can be written as:

$$C_{pPCM}(T) = C_{psolid} + (C_{pliquid} - C_{psolid}).B(T) + L_f D(T), \tag{5.11}$$

where

$$D(T) = e^{\left(\dfrac{-(T-T_m)^2}{\Delta T^2} \Big/ \sqrt{\pi.\Delta T^2}\right)}. \tag{5.12}$$

   Function $D$ is a smoothed Delta Dirac function which is zero everywhere except in interval $[T_m - DT, T_m + DT]$. It is centered on $T_m$ and its integral is 1. Its main role is to distribute the latent heat equally around the mean melting point. The thermal conductivity of the PCM depending on its phase is:

$$k_{PCM}(T) = k_{solid} + (k_{liquid} - k_{solid}).B(T). \tag{5.13}$$

   The thermal energy storage capacity of PCMs embedded with foam is given by:

$$(\rho C_p)_{eff} = \varepsilon(\rho C_p)_{PCM} + (1-\varepsilon)(\rho C_p)_{foam} \tag{5.14}$$

$$(\rho C_p)_{eff} = \varepsilon(\rho C_p)_{PCM} + (1-\varepsilon)(\rho C_p)_{foam} + \frac{(\rho L_f)_{PCM}}{\Delta T} \tag{5.15}$$

$$k_{eff} = \varepsilon k_{PCM} + (1-\varepsilon) k_{foam}, \tag{5.16}$$

where $\varepsilon$ is the foam porosity.

A number of complex expression with greater accuracy has been developed for the effective thermal conductivity of foam embedded PCM (Lemlich, 1978; Weaver and Viskanta, 1986; Calmidi and Mahajan, 1999) known as Maxwell-Garnett effective medium theory in which the matrix is considered to be the foam and the inclusions are considered to the PCM.

$$k_{eff} = k_{foam} \frac{k_{PCM}(1+2\varepsilon) + 2k_{foam}(1-\varepsilon)}{k_{PCM}(1-\varepsilon) + k_{foam}(2+\varepsilon)} \tag{5.17}$$

$$k_{eff} = k_{foam} \frac{(1-\varepsilon)}{3} \quad \text{(Lemlich, 1978)} \tag{5.18}$$

$$k_{eff} = k_{foam} - \left(\frac{k_{eff}}{k_{PCM}}\right)^{1/3} (k_{foam} - k_{PCM})\varepsilon \quad \text{(Weaver and Viskanta, 1986)} \tag{5.19}$$

$$k_{eff} = \left(\left(\frac{2}{\sqrt{3}}\right)\left(\frac{r\left(\frac{b}{L_f}\right)}{k_{PCM}+\left(1+\frac{b}{L_f}\right)\left(\frac{k_{foam}-k_{PCM}}{3}\right)} + \frac{(1-r)\left(\frac{b}{L_f}\right)}{k_{PCM}+\frac{2b(k_{foam}-k_{PCM})}{3L_f}} + \frac{\frac{\sqrt{3}}{2}-\frac{b}{L_f}}{k_{PCM}+\left(\frac{4rb}{3\sqrt{3L_f}}\right)(k_{foam}-k_{PCM})}\right)\right)^{-1}$$

(Mesalhy et al., 2005) \qquad (5.20)

$$\frac{b}{L_f} = \frac{-r+\sqrt{r^2+\frac{2}{\sqrt{3}}(1-\varepsilon)\left(2-r\left(1+\frac{4}{\sqrt{3}}\right)\right)}}{\frac{2}{3}\left(2-r\left(1+\frac{4}{\sqrt{3}}\right)\right)} \text{(Mesalhy et al., 2005)} (5.21)$$

$$k_{eff} = \frac{\left(k_{PCM} + \pi\left(\sqrt{\frac{1-\varepsilon}{3\pi}} - \frac{1-\varepsilon}{3\pi}\right)\left(k_{foam} - k_{PCM}\right)\right)\left(k_{PCM} + \frac{1-\varepsilon}{3}\left(k_{foam} - k_{PCM}\right)\right)}{k_{PCM} + \left[\frac{4}{3}\sqrt{\frac{1-\varepsilon}{3\pi}}(1-\varepsilon) + \pi\sqrt{\frac{1-\varepsilon}{3\pi}} - (1-\varepsilon)\right]\left(k_{foam} - k_{PCM}\right)}$$

(Warzoha et al., 2015)                                    (5.22)

Equations 5.13–5.22 all assume a one-temperature model where the PCM and the foam are in thermal equilibrium. However, if the thermal diffusivity of the foam is much higher than that of the PCM then the system is more accurately represented by a two-temperature model. In a two-temperature model, the foam and the PCM are assumed to be at different temperatures and heat transfer between the foam and the PCM must be accounted for. For these models, the conductivity is not developed as an effective conductivity of the combined material but is instead modeled separately for the foam and the PCM. It is common in this case to use the effective conductivity models with either $k_{PCM}$ or $k_{foam}$ set to zero to develop a model for the thermal conductivity as a function of porosity. An example of this is shown in eqs. 5.23 and 5.24 where eq. 5.22 is used as the starting basis (Mesalhy et al., 2005).

$$k_{eff-foam-only} = \frac{\left(\pi\left(\sqrt{\frac{1-\varepsilon}{3\pi}} - \frac{1-\varepsilon}{3\pi}\right)\right)\left(\frac{1-\varepsilon}{3\pi}\right)k_{foam}}{\left[\frac{4}{3}\sqrt{\frac{1-\varepsilon}{3\pi}}(1-\varepsilon) + \pi\sqrt{\frac{1-\varepsilon}{3\pi}} - (1-\varepsilon)\right]}$$

(Warzoha et al., 2015)                                    (5.23)

$$k_{eff,PCM} = \frac{\left(k_{PCM} - \pi\left(\sqrt{\frac{(1-\varepsilon)}{3\pi}} - \frac{(1-\varepsilon)}{3\pi}\right)(k_{PCM})\right)\left(k_{PCM} - \frac{(1-\varepsilon)}{3}k_{PCM}\right)}{k_{PCM} + \left[\frac{4}{3}\sqrt{\frac{(1-\varepsilon)}{3}}(1-\varepsilon) + \pi\sqrt{\frac{(1-\varepsilon)}{3\pi}} - (1-\varepsilon)\right]k_{PCM}}$$

(Warzoha et al., 2015).                                    (5.24)

When dealing with nanoparticles-enhanced *PCMs* rather than foams, many of the same basic models can be used with slight modifications. For instance, the Maxwell-Garnett effective medium theory can be used, but in this case, the PCM will play the role of the matrix, the nanoparticles are the inclusions and the volume percent loading level of nanoparticles in the composite is represented by $\phi$.

$$k_{eff} = k_{PCM} \frac{k_{NP} + 2k_{PCM} - 2\phi(k_{PCM} - k_{NP})}{k_{NP} + 2k_{PCM} + \phi(k_{PCM} - k_{NP})} \tag{5.25}$$

Despite the complexity of these formulations, the accuracy of the results is quite high. The development of models such as these to predict the transient response and energy storage performance of PCMs in the application is a great asset to the thermal engineer wishing to use PCMs in their designs.

## 5.5 HEAT TRANSFER ANALYSIS OF PCM FOR DIFFERENT APPLICATIONS

In the recent past, various authors have been conducting the study of various PCMs for different practical applications such as photovoltaic (PV) panel cooling, solar drying, and solar stills.

### 5.5.1 PCMs FOR PV PANEL COOLING

The efficiency of PV panel decreased with increment in the operating temperature of PV panels. Therefore, cooling of PV panel is needed to enhance the electrical efficiency. The PCMs are the promising technique to enhance the electrical efficiency by incorporating it at the PV panel back surface. Several authors studied numerical and experimental investigation to optimize cost and size of container for thermal management of PV panel. The numerical simulation had been conducted by Kant et al. (2016) to see the effect of PCM incorporation on the PV panel temperature (Kant et al., 2016b). The study has been conducted using material RT-25 as PCM, aluminum as container material. The time-dependent study was conducted for the whole day and it was found that the operating temperature of PV panel with PCM was lower as compared to PV panel without application of PCM. In this study, the natural convection was also considered during the melting of PCM, which affects charging and discharging time of PCM. Kibra et al. (2016) developed thermal model of PV panel with the application of PCMs without considering the natural convection in the PCM during its melting stage (Kibria et al., 2016). Ezan et al. conducted heat transfer analysis of

PV with PCM to see the effect of natural convection during the melting of PCM (Ezan et al., 2018). The authors have been developing a code in C++, and a survey is conducted under constant and variable boundary conditions. Results reveal that the conduction- and convection-dominated models diverge from each other as the PCM thickness is increased. For 10 cm of PCM, the difference between the conduction and convection-dominated models is nearly 50% regarding the maximum PV panel temperature. Emam et al. (2017) evaluated the inclined concentrated PV-PCM system (Emam et al., 2017). The optimum tilt angle for concentrated solar PV PCMs system was 45°. Emam et al. (2018) developed four different configuration of hybrid concentrator PV-PCM heat sink (Emam and Ahmed, 2018). The objective of the study was to attain rapid thermal dissipation by enhancing the typically low thermal conductivity of PCMs. These findings can help identify the optimal configuration of heat sinks and pattern arrangements of PCMs in order to achieve higher performance with concentrator PV systems. Khanna et al. (2017) evaluated the performance of tilted PV panel coupled with phase change materials (Khanna et al., 2017). From the obtained results, it was found that as tilt-angle increases from 0° to 90°, the PV temperature (in PV-PCM system) decreases from 43.4°C to 34.5°C which leads to increase in PV efficiency from 18.1% to 19%. The comparison of PV-PCM with only PV is also carried out and it is found that PV temperature can be reduced by 19°C by using PCM and efficiency can be improved from 17.1% to 19%. Huang et al. (2004) elevated operating temperatures of PV panel with the application of PCMs which enhanced the electrical efficiency (Huang et al., 2004). The parametric study of a design application is also testified. The temperatures, velocity fields, and vortex formation within the system were predicted for a variety of configurations using the experimentally validated numerical model.

### 5.5.2 *HEAT TRANSFER STUDY OF PCM IN DIFFERENT CONTAINER*

The heat transfer analysis of PCM had been conducted by several authors for different container shape and size. Kant et al. (2016) evaluated performance of different fatty acid and a paraffin wax at different wall temperature conditions (Kant et al., 2016a). The melting and solidification study of PCM has been conducted using COMSOL Multiphysics software which is based on finite element analysis. The aluminum square container of 60 mm × 60 mm had been selected for the study. On the basis of obtained results, it was found that the capric acid had best thermal energy storage material as compared to other fatty acids. Ebrahimi et al. (2015) conducted a study of nanoparticle dispersed PCMs in square container with two heat source–sink pairs (Ebrahimi and Dadvand, 2015). Four different cases had been studied: Case I where the sources and sinks are separately placed on two vertical sidewalls; Case II where the sources and sinks are alternately placed on two vertical sidewalls; Case III where the sources are placed below the sinks on the vertical sidewalls; and Case IV where the sources are placed above the sinks on the vertical sidewalls. It was found that, Case II has the highest liquid fraction and Case IV possesses the lowest liquid fraction at the final stages of the melting process. Kant et al. (2017) developed thermal model for nano-enhanced PCMs (Kant et al., 2017). The numerical simulation had been conducted for three PCM, that is, capric acid, salt hydrates, and paraffin in aluminum square cavity with graphene nanoparticle. The results were obtained in the form of melting front variation, melt fraction, and velocity profile. Based on the this study, it can also be said that the effective thermal conductivity of all three latent heat storage media can be significantly increased by using smaller volumetric concentrations of graphene nanoparticles although the convection heat transfer gets hampered by the same additives. Dhaidan et al. (2013) conducted numerical simulation of nano-enhanced PCMs (NePCM) inside an annular cavity formed between two circular cylinders (Dhaidan et al., 2013). The thermal model had been developed and validated experimentally. The results reveal that there was an enhancement in melting characteristics with the emulsion of nanoparticles in PCM (intensifying the effective thermal conductivity) and raising the wall heat flux and the corresponding Rayleigh number (augmenting the role of natural convection).

### 5.5.3 PCMs FOR SOLAR GREEN HOUSE

A thermal model had been developed by Berroug et al. (2011) considering different components of the greenhouse (cover, plants, inside air, and north wall PCM) and based on the greenhouse heat and mass balance (Berroug et al., 2011). The developed thermal model investigates the impact of the PCM on greenhouse temperature and humidity. Calculations were done for typical decade climate of January in Marrakesh ($31.62°N$, $8.03°W$). Results show that with an equivalent to 32.4 kg of PCM per square meter of the greenhouse ground surface area, temperature of plants and inside air were found to be 6°C–12°C more at night time in winter period with less fluctuations. Relative humidity was found to be on average 10–15% lower at night time. A mathematical model is developed for the storage material and for the greenhouse by Najjar and Hasan (2008). The developed models are solved using numerical methods and Java code program. The effect of different parameters on the inside greenhouse temperature is investigated. The temperature swing between maximum and minimum values during 24 h can be reduced by 3°C–5°C using the PCM storage. This can be improved further by enhancing the heat transfer between the PCM storage and the air inside the greenhouse.

## 5.6  CONCLUSIONS

This chapter gives a comprehensive overview of mathematical modeling of PCMs and its application. The mathematical modeling is based on in the thermal properties enhancement of PCMs. The discussion had been made on the mathematical modeling of thermos-physical properties of PCM, that is, specific heat, density, thermal conductivity, and latent heat of fusion.

The chapter also described the recent studies on mathematical modeling of PCM-based application in PV cooling, solar green house, and optimization of different container size. The studies had been conducted to optimize cost and quantity of PCMs for the efficient operation of the system. The various researchers are working on the container size optimize, the container shape, and size to effective utilization of PCM. These types of studies are very helpful in the optimization of overall cost before performing of experimental study.

## KEYWORDS

- optimization
- thermal modeling
- heat transfer
- numerical simulations
- cost efficiency

## REFERENCES

1. Berroug, F.; Lakhal, E. K.; El Omari, M.; et al. Thermal Performance of a Greenhouse with a Phase Change Material North Wall. *Energy Build.* **2011**, *43*, 3027–3035. DOI: 10.1016/j.enbuild.2011.07.020.
2. Biwole, P. H.; Eclache, P.; Kuznik, F. Phase-Change Materials to Improve Solar Panel's Performance. *Energy Build.* **2013**, *62*, 59–67. DOI: 10.1016/j.enbuild.2013.02.059.
3. Brent, A. D.; Voller, V. R.; Reid, K. J. Enthalpy-Porosity Technique for Modeling Convection-Diffusion Phase Change: Application to the Melting of a Pure Metal. *Numer. Heat Transf. Part A Appl.* **1988**, *13*, 297–318.
4. Brillouin, M. Sur quelques problèmes non résolus de la Physique Mathématique classique Propagation de la fusion. In *Annales de l'institut Henri Poincaré*; 1930; pp 285–308.
5. Calmidi, V. V.; Mahajan, R. L. The Effective Thermal Conductivity of High Porosity Fibrous Metal Foams. *J. Heat Transf.* **1999**, *121*, 466–471.
6. Carslaw, H. S.; Jaeger, J. C. *Conduction of Heat in Solids*, 2nd ed.; Clarendon Press: Oxford, 1959.
7. Crank, J. Free and Moving Boundary Problems. Oxford University Press, 1987.
8. Dhaidan, N. S.; Khodadadi, J. M.; Al-Hattab, T. A.; Al-Mashat, S. M. Experimental and Numerical Investigation of Melting of NePCM Inside an Annular Container under a Constant Heat Flux Including the Effect of Eccentricity. *Int. J. Heat Mass Transf.* **2013**, *67*, 455–468. DOI: 10.1016/j.ijheatmasstransfer.2013.08.002.
9. Douglas, J. A Uniqueness Theorem for the Solution of a Stefan Problem, Proceedings of the American Mathematical Society **1957**, *8*, 402–408.
10. Ebrahimi, A.; Dadvand, A. Simulation of Melting of a Nano-Enhanced Phase Change Material (NePCM) in a Square Cavity with Two Heat Source–Sink Pairs. *Alexandria Eng. J.* **2015**, *54*, 1003–1017. DOI: 10.1016/j.aej.2015.09.007.
11. Emam, M.; Ahmed, M. Cooling Concentrator Photovoltaic Systems using Various Configurations of Phase-Change Material Heat Sinks. *Energy Convers. Manag.* **2018**, *158*, 298–314. DOI: 10.1016/j.enconman.2017.12.077.
12. Emam, M.; Ookawara, S.; Ahmed, M. Performance Study and Analysis of an Inclined Concentrated Photovoltaic-Phase Change Material System. *Sol. Energy* **2017**, *150*, 229–245. DOI: 10.1016/j.solener.2017.04.050.
13. Evans, G. W. A Note on the Existence of a Solution to a Problem of Stefan. *Q. Appl. Math.* **1951**, *9*, 185–193.
14. Ezan, M. A.; Yüksel, C.; Alptekin, E.; Yılancı, A. Importance of Natural Convection on Numerical Modelling of the Building Integrated PVP/PCM Systems. *Sol. Energy* **2018**, *159*, 616–627. DOI: 10.1016/j.solener.2017.11.022.
15. Galione, P.; Lehmkuhl, O.; Rigola, J.; et al. In *Numerical Simulations of Thermal Energy Storage Systems with Phase Change Materials*, Proceedings of the ISES Solar World Congress 2011. International Solar Energy Society: Freiburg, Germany, 2011; pp 1–12.
16. Hill, J. M. One-Dimensional Stefan Problems: An Introduction. Longman Sc & Tech, 1987.
17. Huang, M. J.; Eames, P. C.; Norton, B. Thermal Regulation of Building-Integrated Photovoltaics using Phase Change Materials. *Int. J. Heat Mass Transf.* **2004**, *47*, 2715–2733. DOI: 10.1016/j.ijheatmasstransfer.2003.11.015.

18. Kant, K.; Shukla, A.; Sharma, A. Performance Evaluation of Fatty Acids as Phase Change Material for Thermal Energy Storage. *J. Energy Storage* **2016a**, *6*, DOI: 10.1016/j.est.2016.04.002.

19. Kant, K.; Shukla, A.; Sharma, A.; Biwole, P. H. Heat Transfer Studies of Photovoltaic Panel Coupled with Phase Change Material. *Sol. Energy* **2016b**, *140*, 151–161. DOI: 10.1016/j.solener.2016.11.006.

20. Kant, K.; Shukla, A.; Sharma, A.; Biwole, P. H. Heat Transfer Study of Phase Change Materials with Graphene Nano Particle for Thermal Energy Storage. *Sol. Energy* **2017**, *146*, 453–463. DOI: 10.1016/j.solener.2017.03.013.

21. Khanna, S.; Reddy, K. S.; Mallick, T. K. Performance Analysis of Tilted Photovoltaic System Integrated with Phase Change Material under Varying Operating Conditions. *Energy* **2017**, *133*, 887–899. DOI: 10.1016/J.ENERGY.2017.05.150.

22. Kibria, M. A.; Saidur, R.; Al-Sulaiman, F. A.; Aziz, M. M. A. Development of a Thermal Model for a Hybrid Photovoltaic Module and Phase Change Materials Storage Integrated in Buildings. *Sol. Energy* **2016**, *124*, 114–123. DOI: 10.1016/j.solener.2015.11.027.

23. Lamé, G.; Clapeyron, B. P. Mémoire sur la solidification par refroidissement d'un globe liquide. In *Annales Chimie Physique*; 1831; pp 250–256.

24. Lemlich, R. A Theory for the Limiting Conductivity of Polyhedral Foam at Low Density. *J. Colloid Interface Sci.* **1978**, *64*, 0–3.

25. Lunardini, V. J. Heat Transfer in Cold Climates. Van Nostrand Reinhold Company, 1981.

26. Mesalhy, O.; Lafdi, K.; Elgafy, A.; Bowman, K. Numerical Study for Enhancing the Thermal Conductivity of Phase Change Material (PCM) Storage using High Thermal Conductivity Porous Matrix. *Energy Convers. Manag.* **2005**, *46*, 847–867. DOI: 10.1016/j.enconman.2004.06.010.

27. Najjar, A.; Hasan, A. Modeling of Greenhouse with PCM Energy Storage. *Energy Convers. Manag.* **2008**, *49*, 3338–3342. DOI: 10.1016/J.ENCONMAN.2008.04.015.

28. Sharma, A.; Tyagi, V. V.; Chen, C. R.; Buddhi, D. Review on Thermal Energy Storage with Phase Change Materials and Applications. *Renew. Sustain. Energy Rev.* **2009**, *13*, 318–345. DOI: 10.1016/j.rser.2007.10.005.

29. Stefan, I. Uber Einige Probleme der Theorie der Warmeleitung, Sitzber. *Kais. Akad. Wiss. Wien Math. Naturw Kl* **1898**, *98*, 437–438.

30. Viswanath, R.; Jaluria, Y. A Comparison of Different Solution Methodologies for Melting and Solidification Problems in Enclosures. *Numer. Heat Transf. Part B* **1999**, *24*, 77–105. DOI: 10.1080/104077999275857.

31. Warzoha, R. J.; Weigand, R. M.; Fleischer, A. S. Temperature-Dependent Thermal Properties of a Paraffin Phase Change Material Embedded with Herringbone Style Graphite Nanofibers. *Appl. Energy* **2015**, *137*, 716–725. DOI: 10.1016/j.apenergy.2014.03.091.

32. Weaver, J. A.; Viskanta, R. Freezing of Liquid-Saturated Porous Media. *J. Heat Transf.* **1986**, *108*, 654–659.

# CHAPTER 6

# Application of PCMs for Enhancing Energy Efficiency and Indoor Comfort in Buildings

FARAH SOUAYFANE[1,2*], PASCAL HENRY BIWOLÉ[3,4], and
FAROUK FARDOUN[2,5]

[1]*Université Cote d'Azur, J.A. Dieudonné Laboratory,
UMR CNRS 7351, 06108 Nice, France*

[2]*Lebanese University, Centre de Modélisation,
Ecole Doctorale des Sciences et Technologie, Beirut, Lebanon*

[3]*Université Clermont Auvergne, CNRS, SIGMA Clermont,
Institut Pascal, F-63000 Clermont–Ferrand, France*

[4]*MINES Paris Tech, PSL Research University, PERSEE-Center for
Processes, Renewable Energies and Energy Systems,
CS 10207, 06904 Sophia Antipolis, France*

[5]*University Institute of Technology, Department GIM,
Lebanese University, Saida, Lebanon*

*Corresponding author. E-mail: souayfane.farah@gmail.com*

## ABSTRACT

In recent years, applying phase change materials (PCMs) has attracted growing attention due to its potential to enhance energy efficiency and to provide thermal comfort in buildings. The proper use of PCMs can reduce the energy demand significantly, keeping the indoor temperature within the comfort range. This chapter provides an update on recent developments in PCMs used to optimize building envelope and equipment. First, a review of PCMs applications to improve building envelope is presented. This is followed by reviewing articles on building equipment optimized with PCMs to reduce regular energy consumption. Finally, conclusions and existing gaps related to the research works on energy performance improvement with PCMs are identified.

## 6.1  INTRODUCTION

More than one-third of the global energy consumption comes from the building sector (IEA, 2013) which accounts for about 8% of direct energy-related $CO_2$ emissions from final energy consumers (Pérez et al., 2008). In Europe, 50% of this energy use and related gas emissions correspond to space heating and cooling systems (EU, 2014), and this value will increase if no actions are considered, as a significant rise in cooling energy demand (150% globally and 300–600% in developing countries) is expected by 2050 (IEA, 2013).

In recent years, different ways of thermal energy storage (TES) in the buildings are arousing the interest of many researchers as a solution of the energy crisis. Particularly, using phase change materials (PCMs) in buildings to improve the indoor thermal environment and reducing the space heating and cooling energy consumption, by balancing the environment temperature, has attracted more and more attention. In fact, PCMs store and release thermal energy at nearly constant temperature as they undergo phase change and have large TES density. The exothermic and endothermic phase transition of the PCMs can be utilized effectively by incorporating them in TES systems and thermal loads. However, PCMs have several issues such as leakage problem, low thermal conductivity, poor thermal stability, high flammability, supercooling, corrosiveness, and volume and pressure variations during phase change process, limiting the commercial viability of such materials. More researchers are focusing extensively on improving the thermophysical properties of the PCMs toward commercialization.

The integration of PCMs in building envelope and building equipment is a way to enhance the energy storage capacity of enclosures maintaining occupants' thermal comfort. Research has been carried out to incorporate PCM into building envelopes as microencapsulated inside the material, such as concrete (Cabeza et al., 2007), gypsum (Schossig et al., 2005), or fiber insulations (Kosney et al., 2012), or as a new layer inside the constructive system, as impregnated into gypsum plasterboard (Scalat et al., 1996), or installed in macro-encapsulated panels (Zhang et al., 2005). The thermal performance of buildings is associated with complex physical phenomena, thus the effectiveness of the incorporation of PCM in buildings for thermal inertia improvement depends mainly on indoor and outdoor conditions (climatic characteristics and diurnal temperature swing). As such, the PCM properties and especially melting temperature must be optimized specific to the climate (Zhang et al., 2006). Therefore, numerical and experimental methods have been widely used to analyze and optimize the performance of PCM systems in buildings (Saffari et al., 2017). Numerical research has emphasized on the benefits of using PCM technology, the influence of the melting point, and the performance under different weather conditions, concluding that PCM systems should be designed specifically for each building, use, and climate conditions (Saffari et al., 2017). Moreover, experimental research also demonstrated the benefits and limitations of using this technology in building envelopes and building equipment.

This chapter provides an update on recent applications of PCMs to optimize energy efficiency and thermal comfort in buildings, where human body exists most time in a day. The chapter contains two main sections: the first one presents a review of building envelope optimization methods with PCMs. The second one deals with building service equipment optimized with PCMs to reduce regular energy consumption. Finally, the conclusions and research gaps are given.

## 6.2 PCMs APPLICATIONS IN BUILDING ENVELOPE

The heat transfer mechanism in a building is quite complex, where the latter is submitted to internal and external solicitations. External solicitations are due to the local weather conditions while internal ones come from solar radiative flux entering the building and internal loads. Building envelope is an important element affecting the heat exchange between interior and exterior environments. It acts as a filter, reducing heat exchange between the outside and the indoor environment, and protects the occupants from wind, rain, and other conditions. One way to ensure thermal comfort of occupants with a minimum system energy requirement is to provide an energy-efficient envelope (Mao et al., 2017). This can be done using TES strategies in the building envelope based on PCMs applications (Mao et al., 2017). The integration of PCMs into building walls, roofs, and floors can potentially increase the thermal mass of these components, decrease heat transfer rates during peak hours, and reduce the interior temperature fluctuations. Therefore, more and more studies focused on the building envelope optimization by using PCMs.

Saffari et al. (2017) presented a simulation-based single-objective optimization methodology to define the optimum PCM peak melting temperature of a wallboard integrated into a residential building envelope under various climate conditions based on Köppen-Geiger classification. PCM gypsum boards were installed on the inner surface of the exterior walls and roof of the building. The results showed that the use of PCM-enhanced gypsum technology as integrated passive system into the building envelopes, with optimized peak melting temperature in each climate zone, can yield energy savings and ensure indoor thermal comfort, in both heating and cooling dominant climates. In addition, it was found that in a cooling dominant climate, the best PCM melting temperature to reduce the annual energy consumption is close to 26°C (melting range of 24°C–28°C), whereas in heating dominant climates PCM with lower melting temperature of 20°C (melting range of 18°C–22°C) yields higher annual energy benefits. Figure 6.1 shows the worldwide distribution of the optimum PCM melting temperature in different climates. Furthermore, in almost all high-altitude regions, considerable energy savings due to the use of PCM were obtained. Finally, when designing a passive building with PCM technology not only the climate classification should be considered, but also other factors such as elevation from sea level, solar irradiance, and wind profile.

De Gracia (2019) presented a novel technology based on the dynamic use of PCMs in building envelopes. The concept behind the proposed dynamic PCM system is based on the possibility to vary the position of the PCM layer with respect to the insulating layer of a building envelope. A possible application of the dynamic PCM system in a typical Mediterranean vertical construction system is shown in Figure 6.2. The potential of this technology in reducing cooling load was demonstrated numerically when implemented in different constructive systems. The numerical results indicate that the dynamic system facilitates the PCM solidification process even when using a PCM with a peak melting temperature below the indoor set point. This ability allows the technology to be used not only as a thermal barrier but also as a cooling supplier system. In addition, results showed that the system with optimum design (PCM melting temperature of 22°C) and optimum control achieves a cooling load reduction of 379% in comparison to a system without PCM.

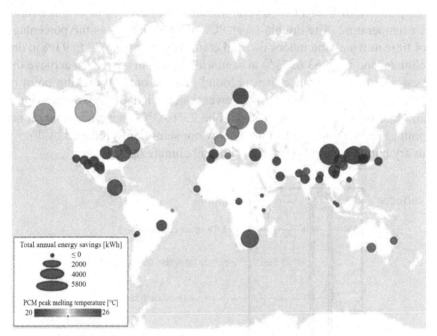

**FIGURE 6.1** Global energy savings due to use of PCM passive system in building enve-lopes. (Reprinted with permission from Saffari et al., 2017. © Elsevier.)

Ahangari and Maerefat (2019) introduced in their work a novel double-layer PCM system in the aim of improving building energy efficiency and indoor thermal comfort in five different climatic zones. The first PCM layer was employed for reducing energy consumption and improving thermal comfort conditions in cold months while the second layer, having higher melting temperature, was employed for hot months. PCM wallboards made of shape-stabilized PCM (SSPCM) were used and were attached to all surfaces of the four walls, ceiling, and the floor. Figure 6.3 shows the location of first and second layers of double-layer PCM system in exterior wall. The Fanger's comfort model and the frequency of thermal comfort (FTC) index were used to assess indoor thermal comfort conditions.

Investigations were also done to choose the most suitable PCMs melting temperature corresponding to each climate. The results showed that the proposed PCM system performs well in improving the thermal comfort and saving building energy in the most dominant climates in Iran (arid and semiarid climates). To ensure indoor thermal comfort, the optimal melting temperature of both PCM-layers system must be close to average room air temperature. The double-layer PCM system improves the percentage of time in which the indoor thermal comfort is met from 73 to 93% in dry climate, and from 63 to 75% in semiarid climate in winter. To achieve the best energy performance, it was found that the optimal melting point of first layer of PCM should be 1°C lower than the indoor set point in winter, and that of the second layer should be 23°C–3°C above the summer set point. Furthermore, the heating energy consumption is reduced by 17.5% in dry climate, and by 10.4% in semiarid climate in Iran.

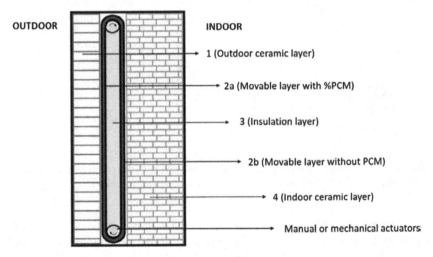

**FIGURE 6.2** Sketch of possible implementation of the dynamic PCM system in a ceramic-based vertical building envelope. (Reprinted with permission from De Gracia, 2019. © Elsevier.)

Stritih et al. (2018) developed and analyzed numerically a composite wall filled with different PCMs with the purpose of integration into passive near zero building applications. In addition, the decrease of energy use in building was investigated for different percentages of microencapsulated PCM in the wall. The results showed that the PCMs in walls can reduce building energy use on daily basis and help achieving the goals of a net zero energy building (NZEB) in future. Further, it was found that to properly select the type of PCM, it is necessary to consider the meteorological conditions in which the PCM wall is located.

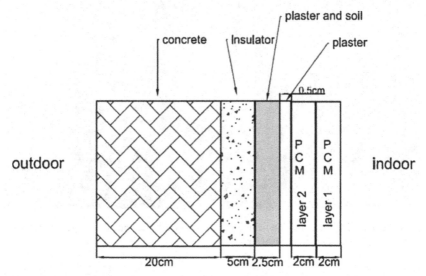

**FIGURE 6.3** Place of first and second layers of double-layer system in exterior wall. (Reprinted with permission from Ahangari and Maerefat, 2019. © Elsevier.)

Nghana and Tariku (2016) assessed, experimentally and numerically, the applicability of PCM in residential construction. This study was carried out for a mild costal climate with focus on thermal comfort and energy consumption improvements, with consideration to occupancy effects and changing building characteristics such as orientation and window to wall ratio. A BioPCM of melting point 23°C was selected in this research, and it was installed in the exterior walls of the building (Fig. 6.4). Experimental results showed that PCMs are effective in stabilizing the indoor air; in fact, the fluctuations of the indoor temperature and wall temperature were reduced by 1.4°C and 2.7°C, respectively. In addition, numerical results showed that, in winter conditions, PCMs are effective in decreasing the heating demand by up to 57% and that, with the PCM, the potential for heat loss across the exterior wall is reduced. In summer, PCMs were found effective in moderating the operative temperature but did not improve the thermal comfort when evaluated over a night time occupancy regime.

**FIGURE 6.4**    PCM installation behind gypsum drywall. (Reprinted with permission from Nghana and Tariku, 2016. © Elsevier.)

Figueiredo et al. (2017) investigated the overheating and heating demand reduction issue for a university department building. Experimentally, a PCM panel was incorporated into the partition wall and the ceiling in one room of the building. Two rooms (one with PCM and the second without PCM, Fig. 6.5) of the building were monitored, and the experimental results were collected during a whole year. The thermal comfort and the PCM influence were quantified using the standard EN 15251. In the numerical work, different PCM solutions were applied and optimized to attain the optimum indoor comfort conditions and energy efficiency. The results showed that the use of PCM reduces overheating by 7.23% representing a PCM efficiency of 35.49%. After the optimization process, an overheating reduction of about 34% was attained. Similarly to other researchers, they found that the selection of PCM melting point is crucial to fully take advantage of the PCM and that the ventilation rate is essential to assure the discharging process of the PCM. Finally, regarding the economic analysis of the use of the PCM for reducing cooling demand, a payback time of 18 years was found.

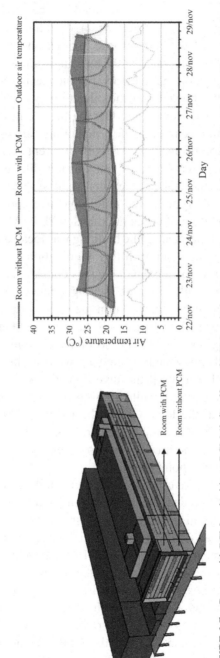

**FIGURE 6.5** Room with PCM and without PCM (*left*), effect of PCM on indoor temperature for a week (*right*). (Reprinted with permission from Figueiredo et al., 2017. © Elsevier.)

Berardi and Manca (2017) investigated the benefits of the adoption of PCMs in lightweight constructions in the climates of Toronto and Vancouver for the cold season and the hot season. They examined the contribution of PCMs, integrated in the building walls, ceiling and floor, in reducing the building cooling demand, and increasing the indoor thermal comfort. The results showed that the climate plays an important role in the charge and discharge cycle and that the best contribution of PCM is when climate conditions and outside temperatures are moderate. In addition, the combination of many PCMs with different melting points could cover a wider range of temperatures, but the main difficulty to this was the decrease of the total thermal conductivity of the system.

Cascone et al. (2018) presented multi-objective optimization analyses for the energy retrofitting of office buildings, with PCM-enhanced opaque building envelope components. The building was located in the climates of Palermo and Turin in Italy. A retrofitting intervention on either the external or internal side of the opaque envelope was considered, and a maximum of two PCM layers with different melting temperatures were selected and placed in different positions within the wall as shown in Figure 6.6. The optimization problem aimed firstly to minimize the primary energy consumption and global costs, and secondly to minimize the building energy needs for heating and cooling as well as the investment cost. The results showed that in all the investigated cases, retrofitting on the internal side of the external walls allowed the best energy performance to be achieved, with the lowest costs. In addition, in the two-objective investigations, the PCM having a greater peak melting temperature than 23°C was preferred in both climates. However, the PCM was found not feasible from economic point of view.

Jamil et al. (2016) investigated, experimentally and numerically, the feasibility of PCM to reduce the zone air temperature and improve occupant thermal comfort, in an existing house in Melbourne, Australia. In the experimental work, a PCM was installed in the ceiling of one of the rooms of the house. Then, a simulation model was developed and validated experimentally. The results showed that the indoor air temperature during daytime was reduced by 1.1°C and the discomfort hours were reduced by 34% following installation of PCM on the ceilings. Moreover, the analysis showed that effectiveness of PCM in reducing peak indoor temperature and improving thermal comfort depends on the behavior of the occupants. For example, the efficiency of PCM was found more important when the windows were kept open on the previous nights and internal doors remained closed all the time.

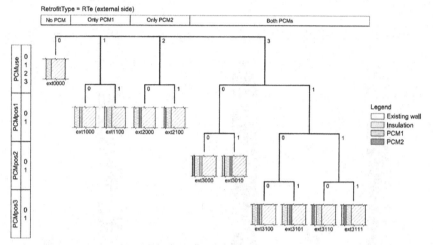

**FIGURE 6.6** Wall configurations for retrofitting on the external side. (Reprinted with permission from Cascone et al., 2018. © Elsevier.)

Wijesuriya et al. (2018) performed a parametric analysis related to the integration of a PCM drywall into a home envelope located in Arizona, the United States of America. The analysis included the optimization of PCM location, PCM properties, and precooling strategy with two different interior convection modes. This study showed that forced convection improves the heat transfer between the walls and indoor air, keeping the indoor environment cooler and with less temperature fluctuations during the peak time. In addition, the maximum electricity cost savings (up to 29.4%) were observed when the PCMs are optimized for all locations under the forced convection mode using high-efficiency fans.

Ramakrishnan et al. (2017) investigated the potential of building refurbishment with PCMs to reduce heat stress risks in residential buildings during heat wave periods. A commercially available BioPCM with a melting temperature of 27°C and layer thickness of 12.5 mm was installed as inner linings of walls and ceilings. Discomfort index (DI) was used as heat stress indicator. The results showed that PCM refurbishment can effectively reduce the indoor heat stress risks, indicating a significant advantage in improving the occupant health and comfort. In addition, it was shown that the appropriate selection of PCM (suitable melting temperature and amount) with better ventilation design could reduce the severe discomfort period by 65% during extreme heat wave conditions (Fig. 6.7).

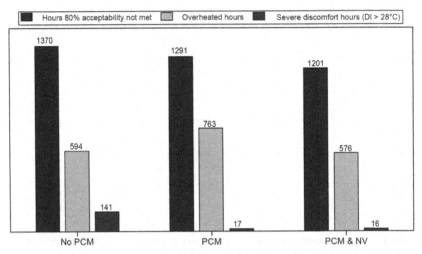

**FIGURE 6.7** Effect of PCM refurbishment in reducing heat stress risks and thermal discomfort (total period of 2160 h). (Reprinted with permission from Ramakrishnan et al., 2017. © Elsevier.)

Guichard et al. investigated the thermal performance of a dedicated test cell with integrated PCMs in the roof with a nonventilated air layer (2017). The experimental equipment was set up under tropical and humid climatic conditions, and a mathematical model was used to predict the actual impact of PCMs on energy consumption as well as thermal comfort. The main result from the study is that the use of PCMs (of melting temperature 23.4°C) in buildings can improve the thermal comfort conditions. For the chosen simulated period, approximately 71.5% persons were dissatisfied for the non-PCM building, whereas the percentage of dissatisfied for the PCM building was 58.9%.

Li et al. investigated experimentally the thermal impact of a specific kind of SSPCM bricks integrated in building walls (2017), for both mid-season and summer conditions. Appearance of testing room is shown in Figure 6.8. In the study, the surface temperature and heat flux of both external and internal walls were monitored. With a PCM melting temperature of 20.93°C, the experimental results showed that the PCM walls were effective for the mid-season, but not for summer. An insufficiency of liquid to solid phase transition has been observed in the days approaching April 30. Moreover, after April 30, the PCM did not experience phase change any more.

Fateh et al. (2017) investigated the thermal performance of a wallboard integrated with Energain® PCM, placed at various locations inside a light wall of a building envelope. They presented the way in which the position of the PCM layer and the insulation layer, in a typical wallboard, affects the temperature and heat flux inside each layer. The results showed that the layers with embedded PCMs are effective for shaving peaks of thermal loads in lightweight structures. In addition, the maximum reduction of heat consumption (about 15%) was obtained when PCMs were in positions three and four (Fig. 6.9).

**FIGURE 6.8** Appearance of the tested room: (a) south view and (b) tested walls. (Reprinted with permission from Li et al., 2017. © Elsevier.)

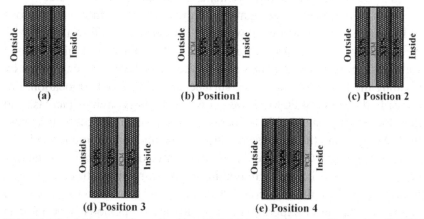

**FIGURE 6.9** (a) Layout of sample without PCM, (b) with PCM in position 1, (c) with PCM in position 2, (d) with PCM in position 3, and (e) with PCM in position 4. (Reprinted with permission from Fateh et al., 2017. © Elsevier.)

Marin et al. (2016) investigated the possibility of PCM application in lightweight relocatable buildings as a passive alternative to save energy under different weather conditions. To investigate the effects of PCM on the heating and cooling energy performances and the thermal comfort, a reference building without PCM was compared to another one with PCM plasterboard installed on the interior surfaces of the exterior walls and the roof. The PCM had a melting point of 25°C. The numerical results highlighted the potential of using PCM-enhanced gypsum boards in lightweight buildings, in order to increase the energy performance during both heating and cooling seasons in arid and warm temperate main climate areas. Kuznik et al. (2011) investigated a renovation project in the south of Lyon, France using PCM wallboards. By testing a room in the same building that was renovated without PCM and then comparing it to the room with PCM, they concluded that the PCM increased the indoor thermal comfort, but it appeared unable to use its latent heat storage capacity for a certain duration due to the incomplete discharge overnight.

Ascione et al. (2014) installed gypsum wallboard PCM on the inner faces of the external building envelope. The results showed that the application of the PCM plaster on the whole vertical envelope leads to the highest energy savings in different climatic zones. In addition, it was found that the cooling demand is reduced with the increment of the thickness of PCM plaster. However, authors investigated until a maximum thickness of 3 cm of PCM plaster because additional increment of PCM thickness did not provide considerable improvements of the indoor temperature.

Alam et al. (2014) investigated energy savings for different locations of PCM (east wall, north wall, west wall, south wall, north wall and roof, west wall and roof, south wall and roof, east wall and roof, all walls, all walls and roof) where the thickness of the PCM layer was calculated by dividing the PCM volume by the surface area of the applied location. For a specified amount of PCM, it was shown that energy savings increase and therefore effectiveness of PCM increase with the decrease of the thickness of PCM layer and the increase of surface area until an optimum level.

In summary, PCMs used in building envelope improve the energy performance of building by regulating indoor air temperature, as well as improving indoor thermal comfort level. From all the above, it can be concluded that both experimental and numerical methods were used in the PCMs investigations. For building envelope thermal performance optimization, paraffin and organic acids were the main choices for PCMs.

It is also found that in most studies on PCM-integrated building envelope, the energy performance is tested. However, more characteristic of PCMs should be analyzed, including their toxicity, thermal stability and durability, and influence on structure safety. In addition, economic analysis should be conducted during designing process. Finally, choosing an appropriate PCMs melting temperature suitable for the considered climate conditions is the key factor to improve the energy performance in buildings.

## 6.3 PCMs APPLICATIONS IN BUILDING EQUIPMENT

Similar to the PCM applications in building envelopes, PCM storage systems can also be used in air cooling, heating, and ventilation systems to store thermal energy from the evaporator or condenser, and consequently improve indoor thermal level and optimize building energy performance. Some researchers investigated series of PCMs for different air cooling systems, such as TES coupled to air conditioning (AC) system, solar cooling systems, and heat pump units, etc. In addition, the purchase of additional equipment for heating can also be deferred and the equipment sizing in new facilities can be reduced coupling PCM to air heating systems such as heat pump and air source heat pump (ASHP) units.

### 6.3.1 PCM-AC SYSTEMS

Integrating PCM in AC system could significantly reduce the cooling load, where AC with smaller power size could be used. Fang et al. (2010) tested the performance of an AC system incorporated with PCM spherical capsules packed bed. They investigated different parameters, mainly the cool storage rate and capacity, the condensation and the evaporation pressures of the refrigeration system, the coefficient of performance (COP) of the system, the inlet and outlet coolant temperatures during charging and discharging periods and others. They concluded that the AC system incorporated with PCM showed better performances. Figure 6.10 shows their investigated experimental system. Parameshwaran and Kalaiselvam (2014) aimed in their work to improve the thermal performance and energy efficiency of chilled water-based variable air volume AC system, integrated with a silver nanoparticles embedded latent TES system. Compared to a conventional AC system, the experimental results showed that the proposed AC system achieved an on-peak and per day average energy savings potential of 36–58% and 24–51%, respectively, for year-round operation. Moreover, the combined AC system was beneficial in terms of accomplishing good thermal comfort, acceptable indoor air quality, and energy redistribution needs in buildings without sacrificing energy efficiency.

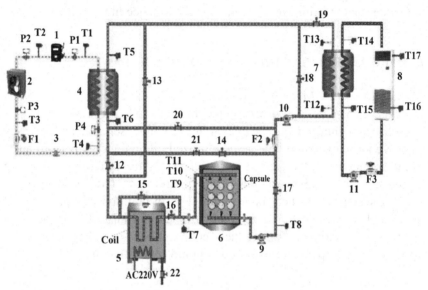

**FIGURE 6.10** Schematic diagram of the AC experimental system with PCM. (Reprinted with permission from Fang et al., 2010. © Elsevier.)

### 6.3.2 SOLAR COOLING SYSTEMS WITH PCM

Solar cooling system attracts more and more attentions, due to the clean characteristics of solar energy. The use of PCM with solar absorption cooling, help significantly to meet cooling demand when the solar energy is not available. For example, Cheng et al. (2016) designed a packed bed cold storage unit, using composite PCM capsules, for high temperature solar cooling application. Thermal performance analysis of the unit under different operating conditions was conducted experimentally and numerically. Effects of inlet heat transfer fluid temperature and flow rate on thermal performance of the unit were discussed in the cold charge process, showing that the unit is feasible in practical high temperature solar cooling applications. As well, Belmonte et al. (2014) investigated the integration of PCMs in the heat rejection loops of absorption solar cooling systems for residential applications in Spain. The first simulation case corresponds to a conventional configuration including an open wet cooling tower, while the second case includes dry coolers in combination with a thermal energy storage system (TES$_{pcm}$) as shown in Figure 6.11. The results showed that the heat rejection loop with a TES$_{pcm}$ of 1 m³ can improve the mean overall system performance coefficient by almost one unit in locations with temperate and humid summers. Helm et al. (2009) examined a solar-driven absorption cooling system coupled with PCM and a dry air cooler instead of a traditional wet cooling tower. They concluded that by integrating PCM in heat rejection circuit of the chiller, a quantity of required power could be shifted to the off-peak hours with minor rise in total electric consumption of the absorption cooling system.

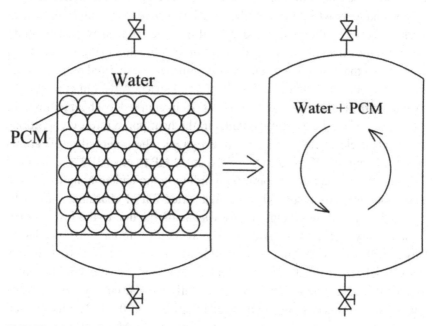

**FIGURE 6.11** Schematic representation of the well-mixed configuration for the TES$_{pcm}$ system. (Reprinted with permission from Belmonte et al., 2014. © Elsevier.)

### 6.3.3 PCM STORAGE IN HEAT PUMP UNITS

Combining a PCM-TES system with heat pump units was also shown to be a promising technology which can lead to energy savings and environmental benefits. A heat pump coupled to TES tanks was experimentally tested under summer conditions by Moreno et al. (2014). Two different configurations of the tanks were compared: a water tank and a PCM tank (PCM phase change temperature of 10°C). The results showed that the PCM tank can supply 14.5% more cold and can maintain the indoor temperature within the comfort range 20.65% longer than the water tank. In addition, Hamada and Fukai (2005) analyzed an experimental setup, as shown in Figure 6.12, where the condenser of a heat pump was connected to two different PCM storage tanks and the evaporator was connected to an ice storage tank, so that the system provided both heating and cooling. The study focused on the heat transfer improvement in the TES tanks. Paraffin wax with a melting temperature of 49°C and latent heat of 180 kJ/kg was used in both tanks to store heat. The tanks were shell and tube heat exchangers where carbon fiber brushes were used to enhance the heat transfer between the heat transfer fluid and the paraffin wax. The study analyzed the effect of including the carbon fiber brushes in the TES tanks. It was shown that the brushes remarkably improve the thermal outputs of the TES tanks, resulting in reduction in cost and space. Agyenim et al. (2010) evaluated the behavior of a PCM-TES system coupled to a heat pump unit. The effect of the PCM storage tanks in the heat pump unit on conditions of off-peak electricity tariff was investigated. The potential implications of integrating the PCM storage unit to an ASHP to meet 100% residential heating energy load, for a common building in the United Kingdom, were also assessed. The results showed that, after 7 h of charging, only 52% of the theoretical maximum energy was available due to the low thermal conductivity of the PCM, and that the average store tank size needed to cover 100% daily heating of a semidetached house was 1116 l. It was also concluded that an improvement on the heat transfer could lead to reduce the store size by 30%. More recently, a solar energy-assisted heat pump system was designed and tested by Qv et al. (2015). Results showed that the PCM-based solar air source heat pump (PCM-SAHP) system can correct the mismatch between thermal energy and electricity supply and demand. Besides, cooling COP of the PCM-SAHP system was enhanced by 17% when the ambient temperature higher than 38°C, and heating COP rose by 65% when outdoor air temperature was below −10°C. These conclusions encourage more heat pump systems considered to be coupled with PCM-TES.

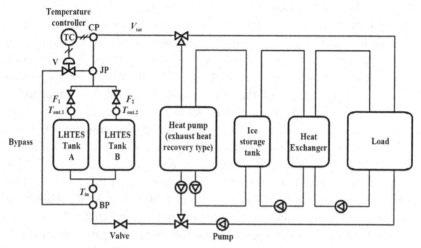

**FIGURE 6.12** Flow diagram of the experimental setup studied by Hamada and Fukai (2005). (Reprinted with permission. © Elsevier.)

### 6.3.4 FREE COOLING SYSTEMS WITH PCM

In the free cooling technique, a separate storage unit is utilized to provide the cold into the room, whenever it is required, by circulating room air through the storage unit. The difference between the natural night ventilation and free cooling is that fans or other mechanical equipment are used to charge or discharge the heat from the storage unit, which improves the cooling potential. The effectiveness of PCM-based free cooling application depends on the diurnal temperature range that should be between 12°C and 15°C (Raj and Velraj, 2010). If the air temperature swing between day and night is relatively small, then other parameters should be accurately considered in the design of free cooling systems coupled with PCM, that is, selection of an appropriate PCM with suitable encapsulation (Zalba et al. 2004). An experimental installation was designed by Zalba et al. (2004) to investigate the performance of PCM in free cooling systems. The principal parameters affecting the melting and solidification processes were discussed. They concluded that the designed installation is technically and economically beneficial, considering further enhancements such as increasing the heat transfer coefficient and the use of a more appropriate PCM. Mosaffa et al. (2013) studied numerically, using the heat capacity method, the performance of a multiple PCM-TES unit for free cooling.

They investigated the impact of some parameters mainly the thickness and the length of PCM slabs and the thickness of air channels, using an energy-based optimization method. Another optimization method, based on energy storage effectiveness, was proposed by the same authors (Mosaffa et al., 2013) to enhance the performance of a multiple PCM free cooling system (Fig. 6.13). They concluded that the suggested method is not appropriate for free cooling system optimization, but the model may be advantageous to design an optimum free cooling system at different climates. Osterman et al. (2015) examined numerically and experimentally the performance of a proposed TES system on a yearly basis, and they discussed its viability for space cooling and space heating. They concluded that the maximum quantity of cold is accumulated in August and July in the climate of Ljubljana, Slovenia, due to larger diurnal temperature fluctuations. Yanbing et al. (2003) investigated the performance of night ventilation with PCM (NVP) storage system. They found that the NVP with coupled PCM system is efficient and can decrease the room energy consumption. Takeda et al. (2004) investigated the potential of a PCM-packed bed storage unit (positioned in the building ventilation system), in decreasing the ventilation load for different Japanese climates. They found that the use of PCM storage unit can decrease the building ventilation load up to 62% in the different considered Japanese cities.

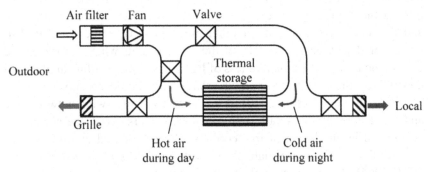

**FIGURE 6.13**    Installation of thermal storage for free cooling. (Reprinted with permission from Mosaffa et al., 2013. © Elsevier.)

### 6.3.5   RADIATIVE COOLING SYSTEM WITH PCM

Another attractive strategy is the night radiative cooling combined with PCM system. Zhang and Niu (2012) investigated a hybrid system

(Fig. 6.14) which is a combination of a nocturnal radiative cooling coupled with microencapsulated PCM slurry storage tank ($T_m$ = 18°C), in order to evaluate its cooling performance in buildings. The investigations were carried out under different climatic conditions in five cities in China (from north to south: Urumqi, Beijing, Lanzhou, Shanghai, and Hong Kong). The results showed that the energy savings in Lanzhou and Urumqi are, respectively, up to 77% and 62% for low-rise buildings, while Hong Kong, under hot and humid climate, showed the weakest performance. Authors recommended using this hybrid system in cities where the temperature is low at night and the weather is dry (north and central china). In addition, Wang et al. (2008) proposed a hybrid system consisting of cooled ceiling, microencapsulated PCM slurry storage (hexadecane $C_{16}H_{34}$ particles and pure water, $T_m$ = 18.1°C) and evaporative cooling technique as shown in Fig. 6.15. They evaluated the system in five cities in China under different climatic conditions. The cooling energy produced by the evaporative cooling system was stored by the MPCM slurry storage. The results showed that energy savings reached 80% under northwestern Chinese climate (Urumqi), about 10% under southeastern Chinese climate (Hong Kong) and between these two values for the other three cities (Shanghai, Beijing, Lanzhou). This hybrid system is suitable for cities under dry climate with high diurnal temperature difference.

In summary, PCMs used in building equipment could improve indoor thermal comfort, and effectively increase the energy efficiency of air cooling, heating, and ventilation systems. Compared to the building envelopment, experiments about building equipment are much easier to carry out since only the system itself needs to be considered. In addition, phase change temperature of PCMs applied in building envelopes should be near the human body comfort level, but the phase change temperature range of PCMs becomes much wider for equipment TES applications. It is noted that for air cooling systems, most researchers focused on energy saving, while the indoor thermal comfort was neglected.

## 6.4 FINAL REMARKS

This chapter has focused on PCMs applications for enhancing energy efficiency and indoor comfort in buildings. Various experimental and numerical studies were reviewed; the first section deals with incorporating

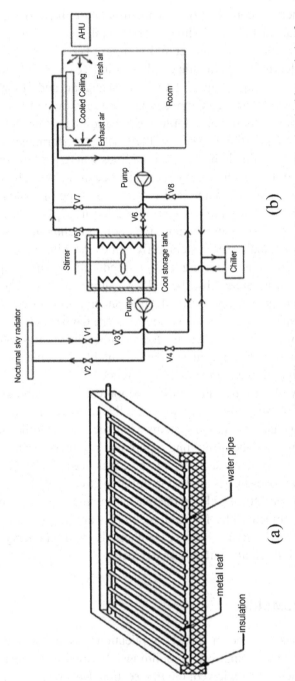

**FIGURE 6.14** (a) Construction of the nocturnal sky radiator, (b) schematic diagram of the hybrid system. (Reprinted with permission from Zhang and Niu, 2014. © Elsevier.)

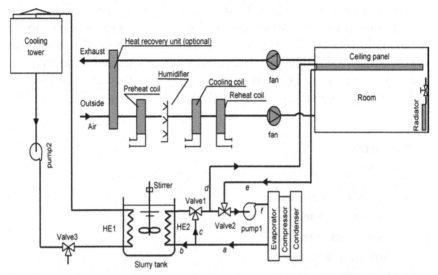

**FIGURE 6.15** Schematic diagram of the hybrid system. (Reprinted with permission from Wang et al., 2008. © Elsevier.)

PCMs into building walls, roofs, and floors and the second one pertains to PCMs applications in building equipment. PCM applications in building envelope are mostly considered as passive systems that do not require additional energy. They depend completely on the outdoor temperature and the variable weather conditions. If night temperatures do not drop considerably below the PCM phase change temperature, the PCM will not fully solidify which delays the pursuit of its operation. Therefore, it is necessary to choose an appropriate PCMs melting temperature, suitable for each considered climate. Paraffin and binary organic acids are the mainly used PCMs in envelopes. PCMs integrated in the envelope components, increasing their thermal mass, can decrease heat transfer rates during peak hours, and reduce large interior temperature fluctuations. Consequently, energy savings and indoor thermal comfort are both expected to be improved. Furthermore, air cooling, heating, and ventilation systems coupled with PCMs-TES system have been presented, such as integration of PCMs in solar cooling systems, heat pumps, etc. It was noted that phase change temperature of PCMs used in the building envelope ranges from 10°C to 39°C, while it ranges from −15.4°C to 77°C for equipment. However, according to the recent publications presented in this chapter, it is found that large-scale commercial application of latent heat storage through PCMs is still limited, and the durability is lower than that of sensible heat storage materials. More attention should be paid to the location of PCM plates when their effects on indoor thermal environment are investigated. In addition, more studies are expected to be conducted in the aim of enhancing PCMs properties, especially their thermal conductivity and their cyclability.

## KEYWORDS

- **PCMs applications**
- **thermal comfort**
- **energy efficiency**
- **building envelope**
- **building equipment**

## REFERENCES

1. Ahangari, M.; Maerefat, M. An Innovative PCM System for Thermal Comfort Improvement and Energy Demand Reduction in Building under Different Climate Conditions. *Sustain. Cities Soc.* **2019,** *44*, 120–129.
2. Agyenim, F.; Hewitt, N. The Development of a Finned Phase Change Material (PCM) Storage System to Take Advantage of Off-Peak Electricity Tariff for Improvement in Cost of Heat Pump Operation. *Energy Build.* **2010,** *42* (9), 1552–1560.
3. Alam, M.; Jamil, H.; Sanjayan, J.; Wilson, J. Energy Saving Potential of Phase Change Materials in Major Australian Cities. *Energy Build.* **2014,** *78*, 192–201.
4. Ascione, F.; Bianco, N.; De Masi, R. F.; de'Rossi, F.; Vanoli, G. P. Energy Refurbishment of Existing Buildings Through the Use of Phase Change Materials: Energy Savings and Indoor Comfort in the Cooling Season. *Appl. Energy* **2014,** *113*, 990–1007.
5. Belmonte, J. F.; Izquierdo-Barrientos, M. A.; Eguía, P.; Molina, A. E.; Almendros-Ibáñez, J. A. PCM in the Heat Rejection Loops of Absorption Chillers. A Feasibility Study for the Residential Sector in Spain. *Energy Build.* **2014,** *80*, 331–351.
6. Berardi, U.; Manca, M. The Energy Saving and Indoor Comfort Improvements with Latent Thermal Energy Storage in Building Retrofits in Canada. *Energy Procedia.* **2017,** *111*, 462–471.
7. Cabeza, L. F.; Castellón, C.; Nogués, M.; Medrano, M.; Leppers, R.; Zubillaga, O. Use of Microencapsulated PCM in Concrete Walls for Energy Savings. *Energy Build.* **2007,** *39*, 113–119.

8. Cascone, Y.; Capozzoli, A.; Perino, M. Optimisation Analysis of PCM-Enhanced Opaque Building Envelope Components for the Energy Retrofitting of Office Buildings in Mediterranean Climates. *Appl. Energy* **2018,** *211,* 929–953.

9. Cheng, X.; Zhai, X.; Wang, R. Thermal Performance Analysis of a Packed Bed Cold Storage Unit using Composite PCM Capsules for High Temperature Solar Cooling Application. *Appl. Therm. Engineering.* **2016,** *100,* 247–255.

10. Fang, G.; Shuangmao, Wu; Xu, Liu. "Experimental Study on Cool Storage Air-Conditioning System with Spherical Capsules Packed Bed." *Energy Build.* **2010,** *42* (7), 1056–1062.

11. Fateh, A.; Klinker, F.; Brütting, M.; Weinläder, H.; Devia, F. Numerical and Experimental Investigation of an Insulation Layer with Phase Change Materials (PCMs). *Energy Build.* **2017,** *153,* 231–240.

12. Figueiredo, A.; Vicente, R.; Lapa, J.; Cardoso, C.; Rodrigues, F.; Kämpf, J. Indoor Thermal Comfort Assessment using Different Constructive Solutions Incorporating PCM. *Appl. Energy* **2017,** *208,* 1208–1221.

13. Gracia Cuesta, A. D. Dynamic Building Envelope with PCM for Cooling Purposes-Proof of Concept. *Appl. Energy* **2019,** *235,* 1245–1253.

14. Guichard, S.; Miranville, F.; Bigot, D.; Malet-Damour, B.; Beddiar, K.; Boyer, H. A Complex Roof Incorporating Phase Change Material for Improving Thermal Comfort in a Dedicated Test Cell. *Renew. Energy* **2017,** *101,* 450–461.

15. Guichard, S.; Miranville, F.; Bigot, D.; Malet-Damour, B.; Beddiar, K.; Boyer, H. A Complex Roof Incorporating Phase Change Material for Improving Thermal Comfort in a Dedicated Test Cell. *Renew. Energy* **2017,** *101,* 450–461.

16. Hamada, Y.; Fukai, J. Latent Heat Thermal Energy Storage Tanks for Space Heating of Buildings: Comparison Between Calculations and Experiments. *Energy Convers Manag.* **2005,** *46,* 3221–3335.

17. Helm, M.; Keil, C.; Hiebler, S.; Mehling, H.; Schweigler, C. Solar Heating and Cooling System with Absorption Chiller and Low Temperature Latent Heat Storage: Energetic Performance and Operational Experience. *Int. J. Refrig.* **2009,** *32* (4), 596–606.

18. International Energy Agency, Technology Roadmap: Energy Efficient Building Envelopes, OECD, 2013.

19. International Energy Agency, Transition to Sustainable Buildings Strategies and Opportunities to 2050, 2013.

20. Jamil, H.; Alam, M.; Sanjayan, J.; Wilson, J. Investigation of PCM as Retrofitting Option to Enhance Occupant Thermal Comfort in a Modern Residential Building. *Energy Build.* **2016,** *133,* 217–229.

21. Kosny, J.; Kossecka, E.; Brzezinski, A.; Tleoubaev, A.; Yarbrough, D. Dynamic Thermal Performance Analysis of Fiber Insulations Containing Bio-Based Phase Change Materials (PCMs). *Energy Build.* **2012,** *52,* 122–131.

22. Kuznik, F.; Virgone, J.; Johannes, K. In-Situ Study of Thermal Comfort Enhancement in a Renovated Building Equipped with Phase Change Material Wallboard. *Renew. Energy* **2011,** *36* (5), 1458-1462.

23. Li, L.; Yu, H.; Liu, R. Research on Composite-Phase Change Materials (PCMs)-Bricks in the West Wall of Room-Scale Cubicle: Mid-Season and Summer Day Cases. *Build. Environ.* **2017,** *123,* 494–503.

24. Mao, N.; Pan, D.; Song, M.; Li, Z.; Xu, Y.; Deng, S. Operating Optimization for Improved Energy Consumption of a TAC System Affected by Nighttime Thermal Loads of Building Envelopes. *Energy* **2017**, *133*, 491–501.

25. Mao, N.; Song, M.; Pan, D.; Li, Z.; Deng, S. Numerical Investigations on the Effects of Envelope Thermal Loads on Energy Utilization Potential and Thermal Non-Uniformity in Sleeping Environments. *Build. Environ.* **2017**, *124*, 232–244.

26. Marin, P.; Saffari, M.; de Gracia, A.; Zhu, X.; Farid, M. M.; Cabeza, L. F.; Ushak, S. Energy Savings Due to the use of PCM for Relocatable Lightweight Buildings Passive Heating and Cooling in Different Weather Conditions. *Energy Build.* **2016**, *129*, 274–283.

27. Moreno, P.; Castell, A.; Sole, C.; Zsembinszki, G.; Cabeza, L. F. PCM Thermal Energy Storage Tanks in Heat Pump System for Space Cooling. *Energy Build.* **2014**, *82*, 399–405.

28. Mosaffa, A. H.; Ferreira, C. I.; Rosen, M. A.; Talati, F. Thermal Performance Optimization of Free Cooling Systems using Enhanced Latent Heat Thermal Storage Unit. *Appl. Therm. Eng.* **2013**, *59* (1–2), 473–479.

29. Mosaffa, A. H.; Ferreira, C. I.; Talati, F.; Rosen, M. A. Thermal Performance of a Multiple PCM Thermal Storage Unit for Free Cooling. *Energy Convers. Manag.* **2013**, *67*, 1–7.

30. Nghana, B.; Tariku, F. Phase Change Material's (PCM) Impacts on the Energy Performance and Thermal Comfort of Buildings in a Mild Climate. *Build. Environ.* **2016**, *99*, 221–238.

31. Osterman, E.; Butala, V.; Stritih, U. PCM Thermal Storage System for 'Free' Heating and Cooling of Buildings. *Energy Build.* **2015**, *106*, 125–133.

32. Parameshwaran, R.; Kalaiselvam, S. Energy Conservative Air Conditioning System using Silver Nano-Based PCM Thermal Storage for Modern Buildings. *Energy Build.* **2014**, *69*, 202–212.

33. Parameshwaran, R.; Kalaiselvam, S. Energy Conservative Air Conditioning System using Silver Nano-Based PCM Thermal Storage for Modern Buildings. *Energy Build.* **2014**, *69*, 202–212.

34. Pérez-Lombard, L.; Ortiz, J.; Pout, C. A Review on Buildings Energy Consumption Information. *Energy Build.* **2008**, *40* (3), 394–398.

35. Qv, D.; Ni, L.; Yao, Y.; Hu, W. Reliability Verification of a Solar–Air Source Heat Pump System with PCM Energy Storage in Operating Strategy Transition. *Renew. Energy* **2015**, *84*, 46–55.

36. Raj, V. A.; Velraj, R. Review on Free Cooling of Buildings using Phase Change Materials. *Renew. Sustain. Energy Rev.* **2010**, *14* (9), 2819–2829.

37. Ramakrishnan, S.; Wang, X.; Sanjayan, J.; Wilson, J. Thermal Performance of Buildings Integrated with Phase Change Materials to Reduce Heat Stress Risks During Extreme Heatwave Events. *Appl. Energy* **2017**, *194*, 410–421.

38. Saffari, M.; de Gracia, A.; Fernández, C.; Cabeza, L. F. Simulation-Based Optimization of PCM Melting Temperature to Improve the Energy Performance in Buildings. *Appl. Energy* **2017**, *202*, 420–434.

39. Saffari, M.; de Gracia, A.; Ushak, S.; Cabeza, L. F. Passive Cooling of Buildings with Phase Change Materials using Whole-Building Energy Simulation Tools: a Review. *Renew. Sustain. Energy Rev.* **2017**, *80*, 1239–1255.

40. Scalat, S.; Banu, D.; Hawes, D.; Paris, J.; Haghighat, F.; Feldman, D. Full Scale Thermal Testing of Latent Heat Storage in Wallboard. *Sol. Energy Mater. Sol. Cells* **1996,** *44,* 46–61.

41. Schossig, P.; Henning, H. M.; Gschwander, S.; Haussmann, T. Micro-Encapsulated Phase Change Materials Integrated into Construction Materials. *Sol. Energy Mater. Sol. Cells* **2005,** *89,* 297–306.

42. Stritih, U.; Tyagi, V. V.; Stropnik, R.; Paksoy, H.; Haghighat, F.; Joybari, M. M. Integration of Passive PCM Technologies for Net-Zero Energy Buildings. *Sustain. Cities Soc.* **2018,** *41,* 286-295.

43. Takeda, S.; Nagano, K.; Mochida, T.; Shimakura, K. Development of a Ventilation System Utilizing Thermal Energy Storage for Granules Containing Phase Change Material. *Sol. Energy* **2004,** *77* (3), 329–338.

44. The European Union Explained: Energy, Sustainable, Secure and Affordable Energy for Europeans, European Commissioner for Energy, 2014.

45. Wang, X.; Niu, J.; Van Paassen, A. H. Raising Evaporative Cooling Potentials using Combined Cooled Ceiling and MPCM Slurry Storage. *Energy Build.* **2008,** *40* (9), 1691–1698.

46. Wijesuriya, S.; Brandt, M.; Tabares-Velasco, P. C. Parametric Analysis of a Residential Building with Phase Change Material (PCM)-Enhanced Drywall, Precooling, and Variable Electric Rates in a Hot and Dry Climate. *Appl. Energy* **2018,** *222,* 497–514.

47. Yanbing, K.; Yi, J.; Yinping, Z. Modeling and Experimental Study on an Innovative Passive Cooling System—NVP System. *Energy Build.* **2003,** *35* (4), 417–425.

48. Zalba, B.; Marín, J. M.; Cabeza, L. F.; Mehling, H. Free-Cooling of Buildings with Phase Change Materials. *Int. J. Refrig.* **2004,** *27* (8), 839–849.

49. Zhang, M.; Medina, M. A.; King, J. B. Development of a Thermally Enhanced Frame Wall with Phase Change Materials for On-Peak Air Conditioning Demand Reduction and Energy Savings in Residential Buildings. *Int. J. Energy Res.* **2005,** *29,* 795–809.

50. Zhang, S.; Niu, J. Cooling Performance of Nocturnal Radiative Cooling Combined with Microencapsulated Phase Change Material (MPCM) Slurry Storage. *Energy Build.* **2012,** *54,* 122–130.

51. Zhang, Y. P.; Lin, K. P.; Yang, R.; Di, H. F.; Jiang, Y. Preparation, Thermal Performance and Application of Shape-Stabilized PCM in Energy Efficient Buildings. *Energy Build.* **2006,** *38* (10), 1262–1269.

52. Zhu, N.; Hu, P.; Lei, Y.; Jiang, Z.; Lei, F. Numerical Study on Ground Source Heat Pump Integrated with Phase Change Material Cooling Storage System in Office Building. *Appl. Therm. Eng.* **2015,** *87,* 615–623.

# CHAPTER 7

# Phase Change Materials for Temperature Regulation of Photovoltaic Cells

KARUNESH KANT[1,2*], AMRITANSHU SHUKLA[1], and ATUL SHARMA[1]

[1]*Non-Conventional Energy Laboratory, Rajiv Gandhi Institute of Petroleum Technology, Jais, Amethi 229304, India*

[2]*Department of Mechanical Engineering, Eindhoven University of Technology, 5600 MB-Eindhoven, Netherlands*

*Corresponding author. E-mail: k1091kant@gmail.com*

## ABSTRACT

Strong solar radiation and high ambient temperature can induce an elevated photovoltaic (PV) cell operating temperature, which is normally negative for its life span and power output. There are many techniques to maintain the temperature of PV cell at a level consistent with higher efficiency of solar PV cell. Phase change materials (PCMs) are one of the most effective tools to maintain the temperature of solar panel at level consistent with its higher efficiency. The uses of PCMs for cooling of PV cell have been studied within the past decade. There are large numbers of PCMs that melt and solidify at a wide range of temperatures, making them attractive in a number of applications. This paper also summarizes the investigation and analysis of the available temperature regulation of PV cell with PCMs for different applications.

## 7.1 INTRODUCTION

Photovoltaic (PV) are a method of transforming sunlight into direct current electricity using semiconducting materials that exhibit the PV effect. PV system is composed of a number of solar cells to supply usable solar power. Power generation from solar PV has long been seen as a clean sustainable energy technology, which appeals upon the planets' most ample and widely spread renewable energy source: the sun. The straight conversion of sunlight to electricity takes place without any moving parts or ecological emanations during operation. It is well proven, as PV systems have now been used for 50 years in focused applications, and grid-connected PV systems have been in use for over several years. It is well known that the efficiency of PV solar cells decreases with an increase of temperature, this decrease is determined first of all by the drop of open circuit cell voltage; therefore, an efficient performance of PV cells in conditions, for example, of concentrated sunlight, demands cooling. The earlier theoretical studies have shown that this decrease is inevitable, but the actual temperature coefficient of the efficiency depends on the carrier transport mechanism, and, in different cases (ideal current or recombination one), could differ by several times; the effects of recombination normally enhance the efficiency deviation with temperature.

The efficiency of PV panel can be enhanced by retaining the operating temperature of PV panel at level of standard test conditions (solar radiation ~1000 and operating temperature ~25°C). There are several techniques that can be utilized for this purpose. The merits and demerits of these cooling techniques are mentioned in Table 7.1. These cooling techniques are as follows:

i.    Air cooling
ii.   Water cooling
iii.  Heat pipe cooling
iv.   Thermo electric cooling
v.    Nanofluid cooling
vi    Phase change material (PCM) cooling

In the aforementioned cooling techniques, the PCM is the most suitable among all due to its favorable properties. This technique has been utilized by several authors (Huang et al., 2006; Hasan et al., 2010; Jun Huang, 2011; Biswas et al., 2012; Ciulla et al., 2012; Lo Brano et al., 2014; Park et al., 2014; Atkin and Farid, 2015; Stropnik and Stritih, 2016; Kant et al., 2016c). This chapter is mainly focused on the temperature regulation of PV panel with PCM. In this chapter, authors has discussed about application of various PCMs and their advantages toward the application of PV panels.

**TABLE 7.1** Merits and Drawbacks of Different Thermal Management Techniques.

| | Merits | Drawbacks |
|---|---|---|
| **Natural air circulation** | Low initial cost, no maintenance, easy to integrate, longer life, no noise, no electricity consumption, passive heat exchange | Low heat transfer rates, accumulation of dust in inlet grating further reducing heat transfer, dependent on wind direction and speed, low thermal conductivity and heat capacity of air, low mass flow rates of air, limited temperature reduction, not suited for roof-integrated solar panels |
| **Forced air circulation** | Higher heat transfer rates compared to natural circulation of air, independent of wind direction and speed, higher mass flow rates than natural air circulation, higher temperature reduction compared to natural air circulation | High initial cost for fans, ducts to handle large mass flow rates, high electrical consumption, high maintenance costs, noisy system, difficult to integrate compared to natural air circulation system |
| **Hydraulic cooling** | Higher heat transfer rate compared to natural and forced circulation of air, higher mass flow rates compared to natural and forced circulation of air, higher thermal conductivity and heat capacity of water compared to air, higher temperature reduction | Higher initial cost due to pumps, higher maintenance cost compared to forced air circulation, higher electricity consumption compared to forced air circulation, a lower lifetime compared to forced air circulation due to corrosion |
| **Heat pipes** | Passive heat exchange, low cost, easy to integrate | Low heat transfer rates, dust accumulation on the inlet grating, dependent on the wind speed and direction |
| **PCM thermal management** | Higher heat transfer rates compared to both forced air circulation and forced water circulation, higher heat absorption due to latent heating, isothermal heat removal, no electricity consumption, passive heat exchange, no noise, no maintenance cost, heat delivery on demand | Higher PCM cost compared to natural and forced air circulation, some PCMs are toxic, some PCMs have fire safety issues, some PCMs are strongly corrosive, PCMs may have disposal problem after their life cycle is complete |
| **Thermo-electric (Peltier) cooling** | No moving parts, noise free, small size, easy to integrate, low maintenance costs, solid state heat transfer | Heat transfer depends on ambient conditions, active systems, require electricity, reliability issues, costly for PV cooling, no heat storage capacity, requires efficient heat removal from the warmer side for effective cooling |

## 7.2  PHASE CHANGE MATERIALS

There are many promising ongoing developments in the field of PCM applications for heating and cooling. PCMs are "latent" thermal storage materials possessing a large amount of heat energy stored during its phase change stage. The energy required to change the phase of a substance is known as latent heat. The storage capacity of LHTES devices is given by Kant et al. (2016a).

$$Q = \int_{T_i}^{T_m} mC_p dT + ma_m \Delta h_m + \int_{T_m}^{T_f} mC_p dT \tag{7.1}$$

$$Q = m[C_{sp}(T_m - T_i) + a_m \Delta h_m + C_{lp}(T_f - T_m)] \tag{7.2}$$

where $Q$ is the storage capacity, $Cp$ specific heat, $T_i$, $T_m$, and $T_f$ are initial, melting, and freezing temperature, respectively, and $h$ is the enthalpy. The desirable properties of PCMs are presented in Table 7.2. There are several authors who had developed novel PCMs that can be utilized for the thermal management (Sharma et al., 2009, 2013, 2014; Sharma and Shukla, 2015; Shukla et al., 2015; Kant et al., 2016b).

**TABLE 7.2**  Main Desirable Properties of Phase Change Materials.

| | |
|---|---|
| **Thermal properties** | Suitable phase-transition temperature, high latent heat of transition, high thermal conductivity in both liquid and solid phases, good heat transfer |
| **Physical properties** | Favorable phase equilibrium, high density, small volume change, low vapor pressure |
| **Kinetic properties** | No super cooling, sufficient crystallization rate |
| **Chemical properties** | Long-term chemical stability, compatibility with materials of construction, no toxicity, no fire hazard |
| **Economic properties** | Abundant, available, cost effective |

The operating temperature of PV panel reaches around 60°C to 80°C during field operations; however, the temperature at standard test condition is 25°C. Therefore, we need the PCMs that have melting temperature in the range of 25 ± 5°C. These PCMs should have high thermal conductivity, latent heat of fusion, and density to make it container compact and effective. Several PCMs had been investigated for temperature control of PV panels, which is presented in Table 7.3.

**TABLE 7.3** The PCMs That Can Be Used for Regulating the Temperature of the PV Panel.

| Phase change materials | Melting point (°C) | Latent heat (kJ/kg) | Density (kg/m3) | | Thermal conductivity (W/mK) | | Specific heat (kJ/kgK) | | Type |
|---|---|---|---|---|---|---|---|---|---|
| | | | Solid | Liquid | Solid | Liquid | Solid | Liquid | |
| Polyethylene glycol 600(H(OC$_2$H$_2$)n·OH) (Sharma et al., 2009) | 20–25 | 146 | – | – | – | – | – | – | Inorganic |
| RT21 (2016) | 21 | 134 | 760 | 840 | 0.2 | – | – | – | Organic |
| C$_{14}$H$_{28}$O$_2$ + C$_{10}$H$_{20}$O$_2$ (Sharma et al., 2009) | 24 | 147.7 | – | – | – | – | – | – | Eutectic |
| Commercial blend (SP22) (Sharma et al., 2016) | 24.6 | 182 | 1490 | 1430 | 0.6 | – | 2.5 | – | Inorganic |
| Eutectic mixture of capric-lauric acid (C–L) (Sharma et al., 2013; Sharma and Shukla, 2015) | 24.66 | 171.98 | 880 | 863 | 0.139 | – | – | – | Eutectic |
| CaCl$_2$ + MgCl$_2$·6H$_2$O (Sharma et al., 2009) | 25 | 95 | – | – | – | – | – | – | Eutectic |
| Mn(NO$_3$)$_2$·6H$_2$O (Sharma et al., 2009) | 25.5 | 148 | – | – | – | – | – | – | Inorganic |
| Paraffin wax (RT20) (He et al., 2004) | 25.73 | 140.3 | 880 | 770 | 0.2 | – | 1.8 | 2.4 | Organic |
| d-Lattic acid (Sharma et al., 2009) | 26 | 184 | – | – | – | – | – | – | Organic |
| Eutectic mixture of capric–palmitic acid (C–P) (Sharma et al., 2013) | 26.4 | 196.07 | 883 | 840 | 0.143 | – | – | – | Eutectic |
| RT25 (Lu and Tassou, 2012) | 26.6 | 232 | 785 | 749 | 0.19 | 0.18 | 1.413 | 1797.6 | Organic |
| Paraffin (PCM$_1$) (He et al., 2004) | 26–28 | 179 | 870 | 750 | 0.2 | – | 1.8 | 2.4 | Organic |
| RT27 (2016) | 27 | 184 | 750 | 840 | 0.2 | – | – | – | Organic |
| FeBr$_3$·6H$_2$O (Sharma et al., 2009) | 27 | 105 | – | – | – | – | – | – | Inorganic |
| CH$_3$CONH$_2$ + NH$_2$CONH$_2$ (Sharma et al., 2009) | 27 | 163 | – | – | – | – | – | – | Eutectic |

**TABLE 7.3** *(Continued)*

| Phase change materials | Melting point (°C) | Latent heat (kJ/kg) | Density (kg/m3) | | Thermal conductivity (W/mK) | | Specific heat (kJ/kgK) | | Type |
|---|---|---|---|---|---|---|---|---|---|
| | | | Solid | Liquid | Solid | Liquid | Solid | Liquid | |
| Paraffin (PCM$_3$) (He et al., 2004) | 28–32 | 157 | 880 | 760 | 0.2 | – | 1.8 | 2.4 | Organic |
| RT31 (2016) | 29 | 196 | 770 | 890 | 0.2 | – | – | – | Organic |
| Methyl palmitate (Sharma et al., 2009) | 29 | 205 | – | – | – | – | – | – | Organic |
| Pure salt hydrate (CaCl$_2$,6H$_2$O) (Kapsalis and Karamanis, 2016; Cabeza et al., 2011) | 29.66 | 213.12 | 1710 | – | 1.09 | – | 1.4 | – | Inorganic |
| CaCl$_2 \cdot$12H$_2$O (Sharma et al., 2009) | 29.8 | 174 | – | – | – | – | – | – | Inorganic |
| Triethylolethane + urea (Sharma et al. 2009) | 29.8 | 218 | – | – | – | – | – | – | Eutectic |
| Paraffin (PCM$_2$) (He et al., 2004) | 31 | 169 | 870 | 760 | 0.2 | – | 1.8 | 2.4 | Organic |

## 7.3 CLASSIFICATIONS OF PCM

Latent heat storage materials are considered PCMs. Numerous researchers have presented the classifications of PCMs (Fig. 7.1) (Sharma et al., 2009; Oró et al., 2012). The latent heat thermal energy storage materials are mainly classified in organics, inorganic, and eutectic. The organics are paraffin and nonparaffin, inorganics are metallic and salt-hydrate, and eutectic are combination of organic–organic, inorganic–inorganic, and organic–inorganic.

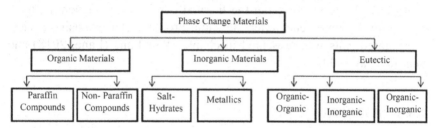

**FIGURE 7.1**   Classification of PCMs (Kant et al., 2016a).

### 7.3.1   ORGANIC PCMs

Organic PCMs such as paraffin wax consist of straight n-alkanes chain ($CH_3$–$(CH_2)$–$CH_3$) and fatty acids that are made up of straight-chain hydrocarbons and are relatively expensive and possess combustible nature. Organic materials possess the capability of congruent melting without phase separation. These compounds are available in a wide range of melting points (Khudhair and Farid, 2004). Paraffin is safe, reliable, predictable, inexpensive, noncorrosive and chemically inert and stable below 500°C but possesses extremely low thermal conductivity (0.1–0.3 W/m K) and is not suitable for encapsulation in plastic containers.

### 7.3.2   INORGANIC PCMs

Inorganic materials are generally hydrated salts and metallic and have a large number of applications in solar energy (Dincer and Rosen, 2002; Lane, 2018). As PCM, these materials are capable of maintaining the heat of fusion (~350 MJ/m$^3$) even after a large number of cycles and relatively higher thermal conductivity (~0.5 W/m°C), but they melt incongruently. One of the cheapest inorganic materials that is suitable to be used as thermal energy storage is Glauber salt (Na2SO4·H2O), which contains 44% Na2SO4 and 56% H2O in weight and was studied by Telkes (1952). This salt has high latent heat (254 kJ/kg) and melting point of about 32.4°C but it is highly prone to phase segregation and subcooling. The corrosion of salt on metal container is also a concern (Farrell et al., 2006). The use of thickening agents, for example, Bentonite clay, and nucleating agent, for example, Borax, help to overcome the subcooling but they reduce the heat transfer rate by lowering the thermal conductivity. Generally, most of the salt hydrates encounter the same problem. An extensive review of hydrated salts was presented in chapter 1 of Lane (Lane, 2018) and Sharma et al. (2004).

### 7.3.3 EUTECTICS

The eutectics are the composition of two or more components such as organic–organic, organic–inorganic, and inorganic–inorganic; each of them change their phase congruently and form a mixture of component crystal during crystallization (George, 1989). Eutectics generally melt and freeze congruently and leave no chances of separation of components (Sharma et al., 2009).

### 7.4 THERMAL REGULATION OF PV PANEL WITH PCM

Several authors had been conducted extensive research and review on thermal management of PV panel with PCMs. The experimental and numerical study has been conducted to investigate the performance of different PV panel with different PCMs. In this section, the authors has described the various methods, PCMs, that can be used for thermal management of PV panels. List of previous studies is presented in Table 7.4.

Nada and El-Nagar (2018) conducted experimental study to investigate the performance of four different PV modules: free stand, building integrated, PCM integrated, and Al2O3 nanoparticles-enhanced PCM integrated. The obtained results revealed that (i) the temperature of the building integrated PV module is substantially high and its efficiency is low compared to those of the free stand PV module, (ii) integrating PCM

**TABLE 7.4** List of Previous Studies on Thermal Management of PV Panel with PCM.

| Study type | PCM | System description | Remarks | Ref. |
|---|---|---|---|---|
| Experimental | RT55, RT55 +2% Al2O3 nanoparticle | Four type of systems had been constructed; standalone, building integrated, PCM integrated, and Al2O3 nanoparticles-enhanced PCM integrated | PCM and PCM enhanced with nanoparticle enhanced the PV panel efficiency considerably. | Nada and El-Nagar (2018) |
| Experimental and numerical | RT-30 | Three type of different PV panel systems, i.e., conventional PV panel, water-based PV thermal collector with double absorber plate, water-based photovoltaic thermal collector with PCM | Water-based photovoltaic thermal collector with phase change materials shows higher efficiency. | Preet et al. (2017) |
| Numerical | RT-44 | The movable shutter had been used to ensure complete melting | The analysis indicated that if the melting temperature of PCM is approximately equal to average of the maximum ambient temperature of all summer months, the maximum electrical efficiency can be achieved. | Waqas and Ji (2017) |
| Experimental | Capric acid | PV/T and PV/T-PCM | Electrical efficiency as well as thermal efficiency has been investigated. | Yang et al. (2018) |
| Numerical | RT-28 HC | PV/PCM | The electricity production of PV-PCM panel for a city of Ljubljana was higher for 7.3% in a period of one year. | Stropnik and Stritih (2016) |

or PCM enhanced with nanoparticles additions reduce the temperature of the module and increase its efficiency, (iii) integrating PCM to the free stand module adversely affect its temperature and performance, and (iv) adding nanoparticles metals to the PCM integrated to the PV modules enhance its effectiveness of thermal regulation and performance enhancement of the PV modules. Preet et al. (2017) conducted experimental and numerical investigation of water-based PV thermal collector with and without PCMs. Three different types of PV system had been constructed, that is, conventional PV panel, water-based PV thermal collector with double absorber plate, and water-based PV thermal collector with PCMs. In the water-based PV/T system, double absorber plate had been used and first absorber plate had been connected to the PV systems and the second is connected to the cooper pipe of the same shape of profile as that of piping arrangement. The RT-30 had been used as PCM in the third system. The experiments were performed at three different mass flow rates, that is, 0.013 kg/s, 0.023 kg/s, and 0.031 kg/s. The maximum electrical efficiency has been obtained with the system that is integrated with the PCM. Waqas and Ji (2017) conducted a numerical study of conventional PV panel with movable shutter. The function of the movable shutter is insuring to complete solidification of PCM during off-sunshine hours. The study has also been conducted at different melting point of PCM. The modeling results stated that the operating temperature of conventional PV panel reaches to the 64°C and that can be reduced to 42°C with the PCM having the melting temperature of 35°C. The analysis had been suggested that to achieve maximum electrical efficiency, the melting temperature of PCM should be approximately equal to the average of the maximum ambient temperature of all summer months. Yang et al. (2018) conducted experimental study of PV thermal collectors and PV with PCM. The operating temperature of PV panel had been reduced to 15.8°C with the application of PCM. The power production of PV/T-PCM is higher as compared to conventional PV/T system and it was 7.4 W. The solar electrical efficiency of PV/T and PV/T-PCM were recorded at 6.98% and 8.16%, respectively. At the end of the experiment, the temperature of water collected in the water tank were 42.4°C and 40.3°C for PV/T and PV/T-PCM, respectively. The total thermal efficiencies of the system were achieved at 58.35% and 70.34% for PV/T and PV/T-PCM system, respectively. The total solar energy conversion efficiencies of the PV/T and PV/T-PCM modules were determined to be 63.93% and 76.87%, while their primary energy-saving efficiencies were 73% and 87.5%, respectively.

Stropnik and Stritih (2016) focused on the development of experimental setup and simulation of heat extraction from the PV panel with the use of TRNSYS software. A modification of PV panel Canadian Solar CS6P-M was made with a PCM RT28HC. The experimental results show that the maximum temperature difference on the surface of PV panel without PCM was 35.6°C higher than on a panel with PCM in a period of one day. The final results of simulation shows that the electricity production of PV-PCM panel for a city of Ljubljana was higher for 7.3% in a period of one year.

## 7.5  SUMMARY AND CONCLUSION

This chapter gives an overview of the thermal management of PV panels with application of PCMs. The various PCMs that can be utilized for the thermal management of PV panel is presented. It has been observed that PCM can effectively reduce the operating temperature of the PV panel to certain degrees resulting in better and improved electrical conversion efficiency of the PV panels. Suitability of the PCM for PV cooling depends on geographical location and the year-round climatic conditions. It can be expected that these systems will have a relatively long payback period taking into account the initial cost of PV panels and the additional cost of the cooling arrangements.

## KEYWORDS

- **temperature regulation**
- **phase change material**
- **photovoltaic (PV) cell**
- **latent heat**
- **storage systems**

# REFERENCES

1. Atkin, P.; Farid, M. M. Improving the Efficiency of Photovoltaic Cells Using PCM Infused Graphite and Aluminium Fins. *Solar Energy* **2015**, *114*, 217–228, doi: 10.1016/j.solener.2015.01.037.
2. Biswas, K.; Miller, W.; Kriner, S.; Kos, J. Field Thermal Performance of Naturally Ventilated Solar Roof with PCM Heat Sink. *Sol. Energy* **2012**, *86*, 2504–2514, doi: 10.1016/j.solener.2012.05.020.
3. Cabeza, L. F. F.; Castell, A.; Barreneche, C.; et al. Materials Used as PCM in Thermal Energy Storage in Buildings: A Review. *Renew. Sustain. Energy Rev.* **2011**, *15*, 1675–1695, doi: 10.1016/j.rser.2010.11.018.
4. Ciulla, G.; Lo, Brano V.; Cellura, M.; et al. A Finite Difference Model of a PV-PCM System. *Energy Procedia* **2012**, *30*, 198–206, doi: 10.1016/j.egypro.2012.11.024.
5. Dincer, I.; Rosen, M. *Thermal Energy Storage: Systems and Applications.* John Wiley & Sons: West Sussex, UK, 2002.
6. Farrell, A. J.; Norton, B.; Kennedy, D. M. Corrosive Effects of Salt Hydrate Phase Change Materials Used with Aluminium and Copper. *J. Mat. Proc. Technol.* **2006**, *175*, 198–205, doi: 10.1016/j.jmatprotec.2005.04.058.
7. George, A. *Hand Book of Thermal Design. Phase Change Thermal Storage Materials.* McGraw Hill Book Co, New York, 1989.
8. Hasan, A.; McCormack, S. J.; Huang, M. J.; Norton, B. Evaluation of Phase Change Materials for Thermal Regulation Enhancement of Building Integrated Photovoltaics. *Solar Energy* **2010**, *84*, 1601–1612, doi: 10.1016/j.solener.2010.06.010.
9. He, B.; Martin, V.; Setterwall, F. Phase Transition Temperature Ranges and Storage Density of Paraffin Wax Phase Change Materials. *Energy* **2004**, *29*, 1785–1804, doi: 10.1016/j.energy.2004.03.002.
10. Huang, M.; Eames, P.; Norton, B. Comparison of a Small-Scale 3D PCM Thermal Control Model with a Validated 2D PCM Thermal Control Model. *Solar Energy Mat. Solar Cells* **2006**, *90*, 1961–1972, doi: 10.1016/j.solmat.2006.02.001.
11. Jun Huang, M. The Effect of Using Two PCMs on the Thermal Regulation Performance of BIPV Systems. *Solar Energy Mat. Solar Cells* **2011**, *95*, 957–963, doi: 10.1016/j.solmat.2010.11.032.
12. Kant, K.; Shukla, A.; Sharma, A.; et al. Thermal Energy Storage Based Solar Drying Systems: A Review. *Innovat. Food Sci. Emerg. Technol.* **2016**, *34*, 86–99, doi: 10.1016/j.ifset.2016.01.007.
13. Kant, K.; Shukla, A.; Sharma, A. Ternary Mixture of Fatty Acids as Phase Change Materials for Thermal Energy Storage Applications. *Energy Rep.* **2016**, *2*, 274–279, doi: 10.1016/j.est.2016.04.002.
14. Kant, K.; Shukla, A.; Sharma, A.; Biwole, P. H. Heat Transfer Studies of Photovoltaic Panel Coupled with Phase Change Material. *Solar Energy* **2016**, *140*, 151–161, doi: 10.1016/j.solener.2016.11.006.
15. Kapsalis, V.; Karamanis, D. Solar Thermal Energy Storage and Heat Pumps with Phase Change Materials. *Appl. Therm. Engineer.* **2016**, *99*, 1212–1224. doi: 10.1016/j.applthermaleng.2016.01.071.

16. Khudhair, A. M.; Farid, M. M. A Review on Energy Conservation in Building Applications with Thermal Storage by Latent Heat Using Phase Change Materials. *Energy Conver. Manage.* **2004**, *45*, 263–275, doi: 10.1016/S0196-8904(03)00131-6.

17. Lane, G. A. *Solar Heat Storage: Volume II: Latent Heat Material.* CRC Press: Boca Raton, 2018.

18. Lo Brano, V.; Ciulla, G.; Piacentino, A.; Cardona, F. Finite Difference Thermal Model of a Latent Heat Storage System Coupled with a Photovoltaic Device: Description and Experimental Validation. *Renew. Energy* **2014**, *68*, 181–193, doi: 10.1016/j.renene.2014.01.043.

19. Lu, W.; Tassou, S. A. Experimental Study of the Thermal Characteristics of Phase Change Slurries for Active Cooling. *Appl. Energy* **2012**, *91*, 366–374, doi: 10.1016/j.apenergy.2011.10.004.

20. Nada, S. A.; El-Nagar, D. H. Possibility of Using PCMs in Temperature Control and Performance Enhancements of Free Stand and Building Integrated PV Modules. *Renew. Energy* **2018**, *127*, 630–641, doi: 10.1016/J.RENENE.2018.05.010.

21. Oró, E.; de Gracia, A.; Castell, A.; et al. Review on Phase Change Materials (PCMs) for Cold Thermal Energy Storage Applications. *Appl. Energy* **2012**, *99*, 513–533, doi: 10.1016/j.apenergy.2012.03.058.

22. Park, J.; Kim, T.; Leigh, S-BB. Application of a Phase-Change Material to Improve the Electrical Performance of Vertical-Building-Added Photovoltaics Considering the Annual Weather Conditions. *Solar Energy* **2014**, *105*, 561–574, doi: 10.1016/j.solener.2014.04.020.

23. Preet, S.; Bhushan, B.; Mahajan, T. Experimental Investigation of Water Based Photovoltaic/Thermal (PV/T) System with and without Phase Change Material (PCM). *Solar Energy* **2017**, *155*, 1104–1120, doi: 10.1016/j.solener.2017.07.040.

24. Sharma, A.; Shukla, A. Thermal Cycle Test of Binary Mixtures of Some Fatty Acids as Phase Change Materials for Building Applications. *Energy Build.* **2015**, *99*, 196–203, doi: 10.1016/j.enbuild.2015.04.028.

25. Sharma, A.; Shukla, A.; Chen, C. R.; Wu, T-N. Development of Phase Change Materials (PCMs) for Low Temperature Energy Storage Applications. *Sustain. Energy Technol. Assess.* **2014**, *7*, 17–21, doi: 10.1016/j.seta.2014.02.009.

26. Sharma, A.; Shukla, A.; Chen, C. R. R.; Dwivedi, S. Development of Phase Change Materials for Building Applications. *Energy Build.* **2013**, *64*, 403–407, doi: 10.1016/j.enbuild.2013.05.029.

27. Sharma, A.; Tyagi, V. V.; Chen, C. R.; Buddhi, D. Review on Thermal Energy Storage with Phase Change Materials and Applications. *Renew. Sustain. Energy Rev.* **2009**, *13*, 318–345, doi: 10.1016/j.rser.2007.10.005.

28. Sharma, S.; Tahir, A.; Reddy, K. S. S.; Mallick, T. K. Performance Enhancement of a Building-Integrated Concentrating Photovoltaic system Using Phase Change Material. *Solar Energy Mat. Solar Cells* **2016**, *149*, 29–39, doi: 10.1016/j.solmat.2015.12.035.

29. Sharma, S. D.; Kitano, H.; Sagara, K. Phase Change Materials for Low Temperature Solar Thermal Applications. *Res. Rep. Fac. Eng. Mie. Univ.* **2004**, *29*, 31–64.

30. Shukla, A.; Sharma, A.; Shukla, M.; Chen, C. R. Development of Thermal Energy Storage Materials for Biomedical Applications. *J. Med. Engineer. Technol.* **2015**, *39*, 363–368, doi: 10.3109/03091902.2015.1054523.

31. Stropnik, R.; Stritih, U. Increasing the Efficiency of PV Panel with the Use of PCM. *Renew. Energy* **2016,** *97*, 671–679, doi: 10.1016/j.renene.2016.06.011.
32. Telkes, M. Nucleation of Supersaturated Inorganic Salt Solutions. *Indust. Engineer. Chem.* **1952,** *44*, 1308–1310.
33. Waqas, A.; Ji, J. Thermal Management of Conventional PV Panel Using PCM with Movable Shutters—A Numerical Study. *Solar Energy* **2017,** *158*, 797–807, doi: 10.1016/J.SOLENER.2017.10.050.
34. Yang, X.; Sun, L.; Yuan, Y.; et al. Experimental Investigation on Performance Comparison of PV/T-PCM System and PV/T System. *Renew. Energy* **2018,** *119*, 152–159, doi: 10.1016/J.RENENE.2017.11.094.
35. Rubitherm GmbH, 2016. Available at http://www.rubitherm.de.

# CHAPTER 8

# Solar Cooking Applications Through Phase Change Materials

ATUL SHARMA*, ABHISHEK ANAND, SHAILENDRA SINGH, and
AMRITANSHU SHUKLA

*Non-Conventional Energy Laboratory,
Rajiv Gandhi Institute of Petroleum Technology, Jais, Amethi, India*

*Corresponding author. E-mail: asharma@rgipt.ac.in*

## ABSTRACT

Cooking through solar energy has become a popular option in the whole world and also has an abundant additional potential for fuel-wood in food preparation, especially in much of the developing world. The rural, urban, and semi-urban population depends mainly on noncommercial fuels to meet their energy needs. Cooking through solar energy is a promising response; however, its acceptance has been limited moderately due to a few issues. The device, that is, solar cooker cannot cook the meal in late evening hours. This particular problem can be resolved by an additional storage unit, which can be linked within the solar cooker and will also provide additional heating support to the device so that food can be cooked at late evening hours. The authors have summarized in this chapter the investigation on solar cooking through phase change materials.

## 8.1 INTRODUCTION

India is blessed with good sunshine. Maximum portions of the country receive mean daily solar radiation in the range of 5–7 kWh/m² and have more than 275 bright days in a year.[1] Energy intake for cooking in developing countries is a main factor of the total energy consumption, with commercial and noncommercial energy sources. Energy requirements for cooking account for 36% of total primary energy consumption in India. However, the maximum portion of the energy demand for cooking is met by noncommercial fuels, such as firewood, agricultural waste, and cow dung in rural areas and kerosene and liquid petroleum gas (LPG) in urban areas.[2]

Solar cooking has been a part of India since last several decades. Overall, solar cooking has a vast potential with respect to cooking in the domestic sector, the commercial sector, and the industrial sector, which also saves a lot of energy as well as energy resources. It is also identified as a suitable technology for Indian masses and has several advantages, such as no regular costs, likely to decrease drudgery, highly nutritious food, longer stability, etc. Solar cookers do not damage the environment and cause breathing infections like the traditional cookstoves in most Indian villages. Additionally, LPG and other clean energy sources are not certainly available to many rural communities in remote areas.

Solar energy is freely available on the globe, environmentally clean too, and hence is accepted as one of the favorable alternative energy recourses possibilities. Solar energy has grown outstanding importance in the existing worldwide discussions on energy and environment. As the whole world becomes environmentally conscious, there is a growing fear regarding deforestation and finding renewable energy sources other than the fossil fuels. At present, solar energy is resolving the need for energy supplies for a large share of the entire world's population, particularly in the developing countries.[3] In near future, several large-scale energy systems, directly converting solar radiation into thermal energy (heat), can be looked forward to.

In 1982, a new department, that is, Department of Nonconventional Energy Sources (DNES) had been set up by Government of India, and later on in 1992, DNES became the Ministry of Non-conventional Energy Sources (MNES), the world's first ministry committed to renewable energy. Thereafter, in October 2006, the Ministry was re-christened as the Ministry of New and Renewable Energy (MNRE).[4,5] MNRE supports the solar cooking program in the country and due to MNRE's effort, the country has a good number of the box-type solar cookers in an operative mode. Sharma et al. reported that up to 2009, overall 6,00,000 box-type solar cookers already sold so far in India; however, this number is now too much larger then as quoted by the authors in India.[6]

The use of solar energy potential for thermal applications, that is, cooking, heating, and drying, is well known in tropical and semitropical regions. The different types of solar cookers developed for cooking are (i) concentrator type (ii) box type, and (iii) indirect type.[7,8] Out of this, the box-type solar cooker is too common and easily available in the local commercial market in the whole world. Solar cookers are used to cook meals such as rice, vegetables, eggs, meat, cakes, etc. Among the various types of solar cookers, the box-type design is becoming very popular. In spite of these advantages, the main hurdles in its dissemination are a reluctance to accept as it is a novel technology, intermittent nature of solar radiation, there is no sun at night, limited space availability in urban areas, higher initial costs, and suitability concerns. Its total available value is seasonal and also reliant on the meteorological conditions of the site. Unpredictability is also the major factor for extensive use of solar energy.

In addition to this, these box-type solar cookers cannot cook the meal in partial cloudy days or in late evening hours due to the unavailability of the storage section in it. The reliability of a solar cooker can be managed by incorporating a suitable energy storage device in the solar cooker. During load condition, the storage device will store the thermal energy and can be utilized whenever needed. Therefore, energy storage is playing a good role in solar thermal systems and adjusts the time-based mismatches between the supply and demand. Energy storage also improves the performance and reliability of energy systems and plays an important role in conserving the energy. If in a solar cooker the storage concept applied, then it will be more reliable and there is a possibility of cooking food during partial clouds and/or in late evening hours.[9]

Phase change materials (PCM) can store the thermal energy during the charging of the system and release the stored amount of energy whenever it's required. These materials absorb energy during the heating process as phase change takes place and release energy to the environment in the phase change range during a reverse cooling process. Basically, there are three methods of storing thermal energy: sensible, latent, and thermo-chemical heat or cold storage. Thermal energy storage in solid-to-liquid phase change retaining through PCM has already attracted a lot of attention in solar thermal systems due to the following advantages:

i)    high latent heat storage capacity,
ii)   melt and solidify at a nearly constant temperature, and
iii)  a small volume is required for a latent heat storage system, thereby the heat losses from the system maintain at a reasonable level during the charging and discharging of heat.

A large number of solid–liquid PCMs have been investigated for heating and cooling applications.[9,10–13] In recent times, the use of PCM in several applications has shown good attention to the researchers. Many review articles are available on PCMs for various applications but it is quite rare for solar cookers. Therefore, in this chapter, an attempt has been taken to summarize the research of the solar cooking system using PCMs. This chapter is a collection of considerable practical information on few selected PCMs used in a box-type solar cooker and concentrator solar cooker. This chapter will help to find the design and development of suitable PCM storage unit for solar cookers.

## 8.2  SOLAR COOKERS WITH LATENT HEAT STORAGE MATERIALS: A REVIEW

Ramadan et al.[14] designed a simple flat-plate solar cooker with focusing plane mirrors with energy storage capabilities. In this proposed design, authors used sand as a sensible heat storage medium. A jacket of sand (0.5 cm) around the cooking unit was used during the course of experiments and due to this arrangement, the cooker performance tremendously improved. Authors also suggested that $Ba(OH)_2 \cdot 8H_2O$ can be a suitable PCM for the solar cooking applications; however, they also recommend to used paraffin in similar type of applications.

David L. Bushnell[15] used 27.3 kg pentaerythritol (solid–solid phase change at 182°C) as PCM to store the solar energy during daytime through a heat exchanger concept, which was connected to a concentrating array of CPC cylindrical troughs. The author suggested that the presented solar cooking concept is feasible for domestic application. Bushnell and Sohi[16] also designed a modular phase change heat exchanger with pentaerythritol used as a PCM for thermal storage. The system was tested in an oven by circulating heat transfer oil, which was heated electrically to simulate a concentrating solar collector. The modular configuration provides a highly improved extraction rate for cooking due to its wrap-around character and its increased surface-to-volume ratio.

Domanski et al.[17] investigated the possibility of cooking during late-evening hours using PCM (Fig. 8.1). Two concentric cylindrical vessels (0.0015 m thick), made from aluminum, are connected together at their tops using four screws to form a double-wall vessel with a 2 cm gap between the outer and inner walls. This gap was approximately fulfilled by the stearic acid or magnesium nitrate hexahydrate as a PCM. A small portion of the gap in the vessel left sufficient space for expansion of the PCM during melting. The cooker performance was evaluated during the course of experiments. The overall efficiency of the cooker during discharge was found to be 3–4 times greater than that for steam and heat-pipe solar cookers, which can be used for indoor cooking.

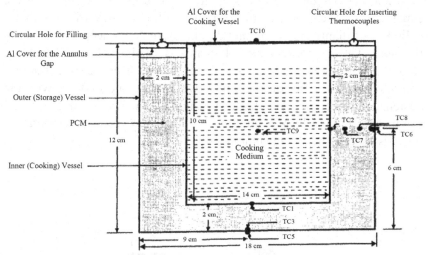

**FIGURE 8.1** A schematic diagram of the storage cooking vessel.[17] Reprinted with permission from Ref. 17. © 1995 Elsevier.

Buddhi and Sahoo[18] designed and fabricated a box-type solar cooker with latent heat storage concept. Authors used commercial-grade stearic acid (melting point 55.1°C, latent heat of fusion 160 kJ/kg) as a PCM for this research work. Authors also suggested that PCM is a good option for the solar cooking systems; however, they reported that high-temperature thermal energy storage (around 80°C) would be a more suitable option for the similar type of research work.

Buddhi and Sahoo[18] reported that storage material should have a high melting temperature to cook food properly through storage concept. Experiments with solar cookers also indicated that foods can be well cooked at temperatures between 95 and 97°C. However, there is no appropriate and promising PCM available between 95 and 105°C melting temperature range in the literature. Therefore, Sharma et al.[19] designed and developed a thermal storage unit for a box-type solar cooker to store solar energy during sunshine hours. The stored energy can be utilized later to cook the food in the late evening hours. Authors used commercial grade acetamide (melting point 82°C, latent heat of fusion 263 kJ/kg) as a latent heat storage material in this research work. The authors claimed that evening cooking is possible with a solar cooker having the PCM storage unit, while it is not possible in a standard box-type solar cooker. Figure 8.2 shows the layout if storage unit is used in the course of the experiments. The storage unit had two hollow concentric aluminum cylinders of diameters 18 cm and 25 cm and is 8 cm deep with a 2 mm thickness. The gap between the cylinders was filled with 2.0 kg acetamide as a PCM. Authors reported that three batches a day during summer and two batches a day during winter were made successfully with the designed storage unit. Authors also conclude that late evening cooking can be possible if a PCM within the 105 and 110°C melting temperature range is used in the storage unit to cook the food.

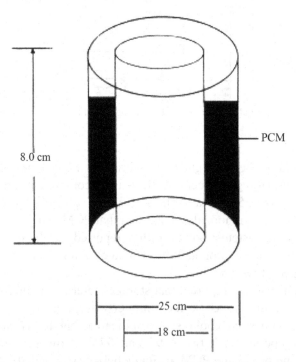

**FIGURE 8.2** Schematic diagram of a latent heat storage unit. Reprinted with permission from Ref. 19. © 2000 Elsevier.

Sharma et al.[19] recommended that the melting temperature of a PCM should be between 105 and 110°C for evening cooking. Therefore, there was a need to find out a storage material that can fulfill the requirement. Later, Buddhi et al.[20] used acetanilide (melting point 118.9°C, latent heat of fusion 222 kJ/kg) as a PCM for night cooking. Authors increased the size of the storage unit as reported by Sharma et al.[19] The gap between the aluminum cylinders was first filled with 2.25 kg acetanilide used as the PCM. At first, the experiments were conducted with box-type solar cooker with a single reflector. To increase the heat transfer rate between the PCM and the inner wall of the PCM container, eight fins were fabricated at the inner wall of the PCM container. Later, authors filled 4.0 kg of the PCM in the storage unit, which could not melt by the single reflector. Therefore, three reflectors were used in this research work to concentrate more solar radiation on the glass window of the solar cooker. The other two reflectors were fixed by a ball and socket mechanism in the left and right sides of the middle reflector. Authors concluded that late evening cooking is possible in a box-type solar cooker having three reflectors through a storage unit.

Sharma and Sagara[21] and Sharma et al.[22] designed and developed a cylindrical storage unit (two hollow concentric aluminum cylinders) that were connected through evacuated tube solar collector (ETSC). The space between the hollow cylinders was filled with 45 kg erythritol (melting point 118°C, latent heat of fusion 339.8 kJ/kg) used as the PCM. Solar energy was stored in the PCM storage unit during sunshine hours and was utilized for cooking in late evening/night time. The schematic layout of the experimental setup is shown in Figure 8.3. Authors reported that such type of a system is expensive; however, it also showed good potential for community applications.

Murty and Kanthed[23] presented the thermal performance of a box-type solar cooker, which was filled in the glass cover to use as transparent insulation during low solar radiation or off-sunshine hours; it would not allow thermal loss from the solar cooker. Commercial grade Lauric Acid (melting point 42.2°C, latent heat of fusion 181 kJ/kg) was used as a latent heat storage material in this research work. Authors reported that during day time as PCM melts, it allowed the solar radiation into the solar cooker and improves the cooking performance and during discharging process; as PCM solidifies, the thickness of the insulating medium increased in the glass cover with time and the rate of heat loss from the box-type solar cooker also decreased and, henceforth, the heat energy retains for a long time inside the PCM solar cooker and, therefore, it could be used as a hot case. Authors also recommend evaluating the heat loss coefficient and the thermal resistance of the PCM medium of a required thickness to evaluate the thermal performance of the PCM solar cooker during charging and discharging processes.

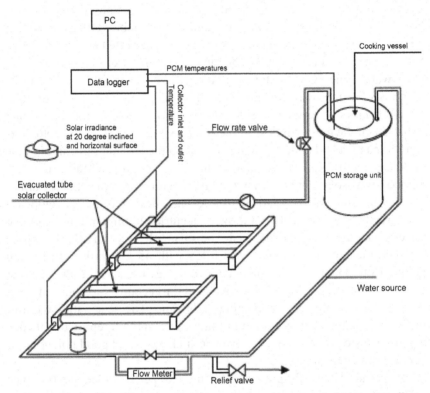

**FIGURE 8.3**  Outline of the prototype solar cooker based on evacuated tube solar collector with PCM storage unit. (Reprinted with permission from Ref. 9. © 2009 Elsevier.)

Parabolic solar cooker (PSC) generally provide high temperature up to 250°C, which can be fruitful for several thermal applications such as making chapattis, baking of food material, and distillation; however, it creates an inconvenience to the user due to its high amount of glare. Murty et al.[24] designed and developed an inclined heat exchanger unit for SK-14 PSC for off-place cooking. Authors used heat transfer fluid (HTF) to transfer the heat to the off-place cooking unit. Later, Murty et al.[25,26] have also optimized the angle of inclination of the HTF column experimentally. Commercial grade sodium acetate (melting point 104°C, latent heat of fusion 230 kJ/kg) and acetanilide (melting point 115.42°C, latent heat of fusion 189.4 kJ/kg) were used as PCM in this study. Thermal energy obtained from a PSC can be transported to a comfortable place for the efficient utilization of solar energy by using an inclined heat exchanger unit. The method can be applied for various domestic and industrial purposes. The authors reported that inclined cylindrical heat exchanger unit along with the PCM can be used for off-place cooking and night cooking. Authors also reported that cooking time was least for the inclination angle of 45° for HTF column.

Chen et al.[27] presented a theoretical investigation on the PCMs, which generally used PCM for box-type solar cookers. Authors identified magnesium nitrate hexahydrate, stearic acid, acetamide, acetanilide, and erythritol as a suitable PCM for the solar cooking application. Authors presented a two-dimensional simulation model that was based on the enthalpy approach, and calculations have been made for the melt fraction with conduction only. Many heat exchanger container materials, that is, glass, stainless steel, tin, aluminum mixed, aluminum, and copper are used in the numerical calculations. Authors claimed that stearic acid and acetamide were found to be good compatibility within the system. It was also reported that the initial temperature of PCM does not have very important effects on the melting time, however, the boundary wall temperature showed an important role during the melting and has a strong effect on the melt fraction. Authors claimed that the effect of thickness of container material on the melt fraction is insignificant.

Hussein et al.[28] designed, developed, and tested a novel indirect solar cooker with a storage unit. Magnesium nitrate hexahydrate (melting point 89°C, latent heat of fusion 134 kJ/kg) is used as the PCM inside the indoor cooking unit of the cooker. Two plane reflectors were used to increase the insolation falling on the cooker's collector. Authors reported that the designed cooker can be cooked different kinds of food at noon, afternoon, and evening hours. Simultaneously, the same unit can also be used as a hot case; therefore, the food temperature can be maintained the whole night and until early morning too for breakfast of the next day. Figure 8.4 shows the cross-sectional view of the solar cooking system with a storage unit.

Dash et al.[29] presented a cooking unit (steam cooking mechanism), which is designed to cook various Indian foods (Fig. 8.5). The steam comes from the solar thermal battery (PCM) through the copper pipe to the heat exchanger of the cooking vessel and supplies heat to the material to be cooked. This solar cooker can be used for baking, making of chapattis, and frying, too. The system uses solar radiation during day time and also stores the energy in the PCM, which can be utilized later as per the requirement. Indoor cooking can be also possible through such type of designed solar cooker.

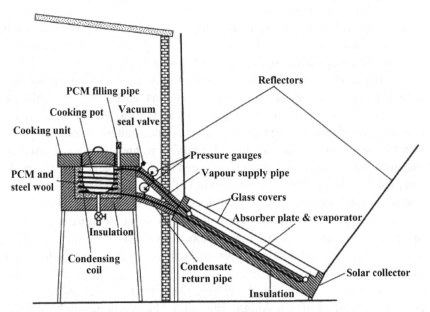

**FIGURE 8.4** Cross-sectional side view of the present solar cooker shows its main components. (Reprinted with permission from Ref. 28. © 2008 Elsevier.)

N. Kumar et al.[30] developed a solar cooker with a PCM storage unit, which was supported by the evacuated tube collector to provide solar energy to the storage unit. The cooking device can be installed in the kitchen area, too. Commercial grade acetamide is used as a PCM for this research work. Authors concluded that the storage device can be used for cooking food during late evening hours due to the available stored energy within the PCM unit.

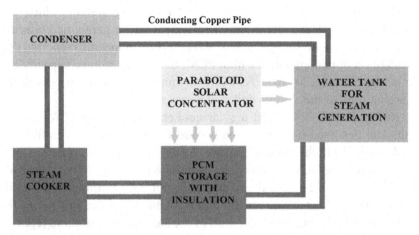

**FIGURE 8.5** Block diagram for the design. (Reprinted from Ref. 29. Open access.)

Veremachi et al.[31] presented a double-reflector setup with storage unit. The binary mixture of sodium nitrate and potassium nitrate ($NaNO_3$–$KNO_3$) at a mixing ratio of 60:40 (mol %) was used as a PCM in this experiment. The melting temperature of the PCM was about 220°C, which can be a good option for cooking applications. Authors presented a secondary reflector, which was positioned above the focal point of the primary reflector that directs the rays onto heat storage positioned below a hole in the primary reflector. Figure 8.6 shows the camera photograph of the experimental setup.

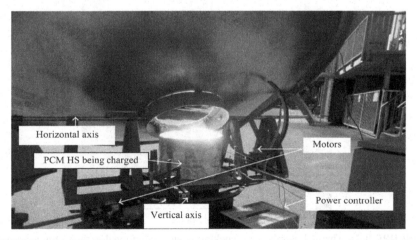

**FIGURE 8.6** PCM heat storage and the tracking system. (Reprinted with permission from Ref. 31. https://creativecommons.org/licenses/by/4.0/)

Bamane and Papade[32] proposed a solar cooking system, which can cook food in late evening hours. The system had a parabolic collector, heat exchanger with storage media, and pumps. Authors also reported that by addition of nanoparticles within PCM, the energy storage capacity can be increased up to 20–30%. As PCM thermal conductivity increased, the charging time of the storage unit also decreased.

Reddy et al.[33] developed a solar cooker along with a PCM storage unit. The designed cooker can be cooked for 2–4 persons. Paraffin was used as a PCM in this research work. The payback of the cooker was 7.87 years, thereby reducing a carbon footprint of 80.541 kg $CO_2$/year. Mahale and Bhangale[34] used a parabolic dish collector with a storage unit in this research study. Acetanilide was used as PCM and was filled in between the hollow space of inner and outer wall of the storage unit. Authors reported that a storage unit with a parabolic dish collector showed better results than box-type solar cooker with storage unit.

Bhave and Thakare[35] designed a concentrating-type solar cooker and magnesium chloride hexahydrate (melting point 118°C) used as PCM for this research work. The thermal energy would be achieved by a combination of a thermic oil (HTF) and sealed aluminum tubes containing the PCM in the lower part of the vessel (Fig. 8.7). The HTF would act as heat storage

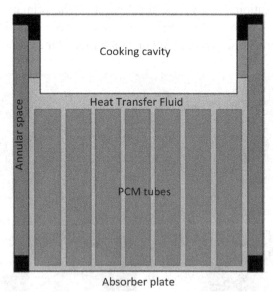

**FIGURE 8.7**  Thermal storage cum cooking device. (Reprinted with permission from Ref. 35. © 2018 Elsevier.)

and heat transfer medium from the absorber to PCM by convection during the solar energy collection phase when the device would be placed at the focus of the dish during day time. The device's absorbing surface can then be covered with an insulating disk when it is stored in the kitchen for later use. The developed device was able to store a charge of heat in about 50 min and cooked about 140 gm of rice in 30 min from the stored heat.

Coccia et al.[36] proposed a storage unit, which is a double-walled vessel composed of two stainless steel cylindrical pots assembled concentrically (Fig. 8.8). The annular gap between both cylinders was filled with a ternary mixture of nitrite and nitrate salts (solar salt: 53 wt% $KNO_3$, 40 wt% $NaNO_2$, 7 wt% $NaNO_3$). The mixture, having a melting point of about 145°C, allowed to keep the testing fluid (silicone oil) temperature stabilized within a high-temperature range. With the help of this storage, food was cooked when solar radiation was not available.

Rekha and Sukchai[37] designed and developed solar concentric parabolic cooker with a storage unit. The storage unit is a hollow concentric cylinder with inner and outer radii of 0.09 m and 0.1 m, respectively. The annular gap between the two layers of the hollow concentric cylinder is 0.01 m and is filled with heat transfer oil. The outer layer of the cooking unit is surrounded by the vertical cylindrical PCM tubes of diameter 0.025 m. The cooking power of the solar cooker with PCM storage unit is 125.3 W, which is 65.6 W more than that of the cooking power without PCM storage unit.

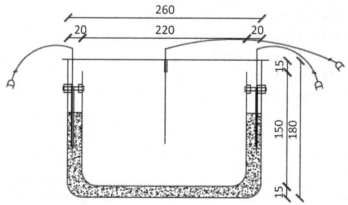

**FIGURE 8.8** Thermal storage unit. (Reprinted with permission from Ref. 36. © 2018 Elsevier.)

## 8.3 CONCLUSION

The box-type solar cookers are matching the energy demand of the urban and rural areas in India as well as in the whole world. Such type of the cookers is successfully commercialized in many parts of the world. Solar cooking energy plays an important role to manage the energy resources in Indian households as well as worldwide. This chapter is focused on the available thermal energy storage technology for solar cookers. Several authors reported that with the support of the storage unit, food can be cooked at late evening hours; however, late evening cooking was not possible with a normal type of solar cooker. Therefore, solar cooker with storage unit is too much essential for the manufactures, industries, as well as for energy conservation.

## ACKNOWLEDGMENT

Authors are highly thankful to the Council of Science and Technology, UP (Reference No. CST 3012-dt.26-12-2016) for providing the research grants to carry out the work at the institute.

## KEYWORDS

- **solar cooker**
- **thermal energy storage systems**
- **phase change material**
- **solar energy**
- **latent heat**

## REFERENCES

1. Mani, A. R. S. *Solar Radiation Over India*. Allied Publishers, Delhi, 1982.
2. Pohekar, S. D.; Kumar, D.; Ramachandran, M. Dissemination of Cooking Energy Alternatives in India—A Review. *Renew. Sustain. Energy Rev.* **2005,** *9,* 379–393, doi:10.1016/j.rser.2004.05.001.

3. Muthusivagami, R. M.; Velraj, R.; Sethumadhavan, R. Solar Cookers with and without Thermal Storage—A Review. *Renew. Sustain. Energy Rev.* **2010,** *14,* 691–701, doi: 10.1016/j.rser.2008.08.018.

4. Sharma, A.; Srivastava, J.; Kar, S. K.; Kumar, A. Wind Energy Status in India: A Short Review. *Renew. Sustain. Energy Rev.* **2012,** *16,* 1157–1164, doi:10.1016/j.rser. 2011.11.018.

5. Ministry of New and Renewable Energy. Https://mnre.gov.in/ n.d.

6. Sharma, A.; Chen, C. R.; Murty, V. V. S.; Shukla, A. Solar Cooker with Latent Heat Storage Systems: A Review. *Renew. Sustain. Energy Rev.* **2009,** *13,* 1599–1605, doi: 10.1016/j.rser.2008.09.020.

7. Telkes, M. Solar Cooking Ovens. *Sol Energy* **1959,** *3,* 1–11, doi:10.1016/0038-092X(59)90053-2.

8. Garg, H. P.; Mann, H. S.; Thanvi, K. P. Performance Evaluation of Five Solar Cookers. *Int. Sol. Energy Soc.* **1978,** doi:10.1016/B978-1-4832-8407-1.50286-5.

9. Sharma, A.; Tyagi, V. V.; Chen, C. R.; Buddhi, D. Review on Thermal Energy Storage with Phase Change Materials and Applications. *Renew. Sustain. Energy Rev.* **2009,** *13,* 318–345, doi:10.1016/j.rser.2007.10.005.

10. Zalba, B.; Marin, J. S.; Cabeza, L. F.; Mehling, H. Review on Thermal Energy Storage with Phase Change Materials, Heat Transfer Analysis and Applications. *Appl. Therm. Eng.* **2003,** *23,* 251–283, doi:10.1016/S1359-4311(02)00192-8.

11. Farid, M. M.; Khudhair, A. M.; Razack, S. A. K.; Al-Hallaj, S. A Review on Phase Change Energy Storage: Materials and Applications. *Energy Convers. Manage.* **2004,** *45,* 1597–1615, doi:10.1016/j.enconman.2003.09.015.

12. Kenisarin, M. M. High-Temperature Phase Change Materials for Thermal Energy Storage. *Renew. Sustain. Energy Rev.* **2010,** *14,* 955–970, doi:10.1016/j.rser.2009.11.011.

13. Buddhi, D. Review on Coolness Storage Technologies Used In Refrigeration and Air-Conditioning System with Various Mode of Operating Strategies. *Renew. Sustain. Energy Rev.* **2011,** *1,* 27–35.

14. Ramadan, M. R. I.; Aboul-Enein, S.; El-Sebaii, A. A. A Model of an Improved Low Cost-Indoor-Solar-Cooker in Tanta. *Sol. Wind Technol.* **1988,** *5,* 387–393, doi: 10.1016/0741-983X(88)90005-7.

15. Bushnell, D. L. Performance Studies of a Solar Energy Storing Heat Exchanger. *Sol. Energy* **1988,** *41,* 503–512, doi:10.1016/0038-092X(88)90053-9.

16. Bushnell, D. L.; Sohi, M. A Modular Phase Change Heat Exchanger for a Solar Oven. *Sol. Energy* **1992,** *49,* 235–244, doi:10.1016/0038-092X(92)90002-R.

17. Domanski, R.; El-Sebaii, A. A.; Jaworski, M. Cooking During Off-Sunshine Hours Using PCMs as Storage Media. *Energy* **1995,** *20,* 607–616, doi:10.1016/ 0360-5442 (95)00012-6.

18. Buddhi, D.; Sahoo, L. K. Solar Cooker with Latent Heat Storage: Design and Experimental Testing. *Energy Convers. Manage.* **1997,** *38,* 493–498.

19. Sharma, S. D.; Buddhi, D.; Sawhney, R. L.; Sharma, A. Design, Development and Performance Evaluation of a Latent Heat Storage Unit for Evening Cooking in a Solar Cooker. *Energy Convers. Manag.* **2000,** *41,* 1497–1508, doi:10.1016/S0196-8904 (99)00193-4.

20. Buddhi, D.; Sharma, S. D.; Sharma, A. Thermal Performance Evaluation of a Latent Heat Storage Unit for Late Evening Cooking in a Solar Cooker Having Three

Reflectors. *Energy Convers. Manag.* **2003,** *44,* 809–817, doi:10.1016/S0196-8904 (02)00106-1.

21. Sharma, S. D.; Iwata, T.; Kitano, H.; Kakuichi, H. S. K. Experimental Results of Evacuated Tube Solar Collector for Use in Solar Cooking with Latent Heat Storage. *Annex* **2004,** *17;* Proceeding.

22. Sharma, S. D.; Iwata, T.; Kitano, H.; Sagara, K. Thermal Performance of a Solar Cooker Based on an Evacuated Tube Solar Collector with a PCM Storage Unit. *Sol. Energy* **2005,** *78,* 416–426, doi:10.1016/j.solener.2004.08.001.

23. Murty, V. V. S.; Kanthed, P. Thermal Performance of a Solar Cooker Having Phase Change Material as Transparent Insulation. *IEA, ECESIA Annex 17,* 2003.

24. Murty, V. V. S.; Gupta, A.; Shukla, A. Design, Development and Thermal Performance Evaluation of an Inclined Heat Exchanger Unit for SK-14 Parabolic Solar Cooker for Off-Place Cooking. *Int. Green Energy Conf.,* 2006.

25. Murty, V. V. S.; Gupta, A.; Patel, K.; Patel, N.; Shukla, A. *Design, Development and Thermal Performance Evaluation of an Inclined Heat Exchanger Unit Assisted SK-14 Parabolic Solar Cooker for Off-Place Cooking with and without Phase Change Material.* 3rd International Conference on Solar Radiation. Day Light (SOLARIS 2007), 2007, pp. 8–15.

26. Murty, V. V. S.; Sharma, A.; Shukla, A. *Effect of Variation of Viscosity of HTF in an Inclined Heat Exchanger Unit Assisted SK-14 Parabolic Solar Cooker for Offplace Cooking with and without Phase Change Material.* Kun Shan University of Tainan, Taiwan, ROC., 2007.

27. Chen, C. R.; Sharma, A.; Tyagi, S. K.; Buddhi, D. Numerical Heat Transfer Studies of PCMs Used in a Box-Type Solar Cooker. *Renew. Energy* **2008,** *33,* 1121–1129, doi: 10.1016/j.renene.2007.06.014.

28. Hussein, H. M. S.; El-Ghetany, H. H.; Nada, S. A. Experimental Investigation of Novel Indirect Solar Cooker with Indoor PCM Thermal Storage and Cooking Unit. *J. Hazard Mater.* **2008,** *157,* 480–489, doi:10.1016/j.jhazmat.2008.01.012.

29. Dash, J. K.; Parida, O.; Tripathy, S.; Dutta, S. Molten Salt Solar Cooker. *Int. J. Innov. Technol. Res.* **2011,** *14,* 201–206.

30. Kumar, N.; Budhiraja, A.; Rohilla, S. Feasibility of a Solar Cooker in Off Sunshine Hours Using PCM as the Source of Heat. *Adv. Eng. Int. J.* **2016,** *1,* 33–39.

31. Veremachi, A.; Cuamba, B. C.; Zia, A.; Lovseth, J.; Nydal, O. J. PCM Heat Storage Charged with a Double-Reflector Solar System. *J. Sol. Energy* **2016,** *2016,* 1–8, doi: 10.1155/2016/9075349.

32. Bamane, V. V.; Papade, C. V. A Review Paper on Nano Mixed Phase Change Material for Indoor and Outdoor Solar Cooker Application. *IJETT* **2017,** 393–397.

33. Mallikarjuna, R. S.; Sandeep, V.; Sreekanth, M.; Daniel, J. Development and Testing of a Solar Cooker with Thermal Energy Storage System. *Int. Energy J.* **2017,** *17,* 185–192.

34. Bhangale, J. H. Study of Solar Energy Storage with Phase Change Materials and Improving its Performance. *SEST* **2017,** *3,* 56–64.

35. Bhave, A. G.; Thakare, K. A. Development of a Solar Thermal Storage Cum Cooking Device Using Salt Hydrate. *Sol. Energy* **2018,** *171,* 784–789, doi:10.1016/j.solener.2018.07.018.

36. Coccia, G.; Di Nicola, G.; Tomassetti, S.; Pierantozzi, M.; Chieruzzi, M.; Torre, L. Experimental Validation of a High-Temperature Solar Box Cooker with a Solar-Salt-Based Thermal Storage Unit. *Sol. Energy* **2018,** *170*, 1016–1025, doi:10.1016/j. solener.2018.06.021.

37. Santhi Rekha, S. M.; Sukchai, S. Design of Phase Change Material Based Domestic Solar Cooking System for Both Indoor and Outdoor Cooking Applications. *J. Sol. Energy Eng. Trans. ASME* **2018,** *140*, 1–8, doi:10.1115/1.4039605.

## CHAPTER 9

# Role of Phase Change Materials in Solar Water Heating Applications

SHAILENDRA SINGH*, ABHISHEK ANAND, AMRITANSHU SHUKLA, and ATUL SHARMA

*Non-Conventional Energy Laboratory, Rajiv Gandhi Institute of Petroleum Technology, Jais, Amethi 229304, India*

*Corresponding author. E-mail: pre18001@rgipt.ac.in*

## ABSTRACT

Solar water heating systems (SWHSs) are generally used in the residential, commercial, and industrial sectors in the whole world community. In such type of systems, thermal energy storage is playing a critical role to maintain the temperature of stored water. As solar radiation is available only in day time, however, several users would like to use the hot water during off sunshine hours. Therefore, storage concept is needed to sustain the hot water temperature during late evening hours. To resolve such type of issue, phase change materials (PCMs) can be a good option. PCM can store the energy in the form of latent heat of fusion during the charging mode and stored heat can be released later to heat water as per the requirement. It has been seen that for a good thermal performance of solar water heater a PCM with the high latent heat of fusion and with large surface area for heat transfer is required. This chapter summarizes the role of the PCM in SWHSs.

## 9.1 INTRODUCTION

Today's world requirement is to identify new or nonconventional energy sources to fulfill the energy requirement around the globe. We are seeing that global energy demand is increasing day by day and this demand for energy cannot be fulfilled by fossil fuel sources. It is expected that fossil fuel must be replaced by a clean and abundantly available source of energy. India receives 5000 trillion kWh per annum of solar radiation with a daily average of 4–7 KWhm$^{-2}$. The maximum solar radiation is received at Jodhpur, that is, 20.97 MJm$^{-2}$ day$^{-1}$ and the minimum value obtained at Shillong, that is, 15.90 MJm$^{-2}$ day$^{-1}$. The aridest parts of the country receive maximum radiation, that is, 7200–7600 MJm$^{-2}$ per annum, semi-arid areas receive 6000–7200 MJm$^{-2}$ per annum and the amount received in the mountainous region is around 6000 MJm$^{-2}$ per annum. The ambient temperature in northern and central parts of the country varies between 2°C (minimum) and 25°C (maximum) in winters.[1] However, the use of solar energy still remains far less prevalent than the equivalent in Japan and Israel. In this context, it is also important to utilize proper energy storage systems. So that energy storage method becoming a big challenge to everybody for minimizing waste of energy and to maximize effective method of utilization of stored energy. A phase change material (PCM) plays an important role in the storage of thermal energy in different technological types of equipment. We can say that PCMs are capable of storing plenty amount of latent heat in the fusion state. The biggest advantage of PCM is that it stores energy and deliver it in a very narrow temperature range while keeping its thermal and physical properties at the constant condition. In this way, PCMs are treated as Energy reservoirs in different types of equipment. With these properties of PCM, it is advantageous that it should be incorporated with the storage tank of solar water heater systems for storing latent heat during sunshine hours and must be delivered during night time of requirement. This chapter will describe the analysis of the incorporation of PCM with a storage tank of solar water heating system (SWHS).

## 9.2   SWHS

SWHS is a clean energy technology for heating of water for domestic and industrial purposes. It is studied and developed since the last decades and still under the process of enhancing the performance to a more advantageous stage. Present day's research work is focused over to, how maximum attainable water temperature can be achieved for a long duration of time even during off sunshine periods. It is also an effective way of reducing $CO_2$ emissions by reducing the conventional way of water heating in which coal-fired systems are used. It is expected that about 3.43% of $CO_2$ emissions can be reduced by the utilization of suitable PCMs with different thermal energy systems by the year 2020.[2] SWHS has a wide range of applications like domestic hot water requirement, hotels and guesthouse, hospitals, hostels, industries such as food processing, rice mill, textile processing, pharmaceutical, pulp & paper, chemical, and auto component industries, etc. It has many advantages among which some are listed here:

- SWHS is simple in design as well as in installation also.
- It has almost no maintenance and running cost.
- Clean energy technology and results in no environmental pollution.
- Existing houses are capable of installing of equipment easily.
- Economically competitive with electric water heaters.
- The desired temperature can be easily achieved with simple equipment.

Small SWHS is suitable for domestic purposes while larger systems are used in industrial processes. Majorly there are two types of water heating systems are available: (1) Natural circulation or passive solar system (thermosyphon): This type of solar water heaters is comparatively simple in design and also has low cost. In this type of system, the tank is located above the collector and water is circulated due to density difference when heat is added through collectors. However, their applications are usually limited to nonfreezing climates although they may also be designed with heat exchangers for mild freezing climates if required; (2) Forced circulation or active solar system: Forced circulation water heaters are used in freezing climates and for industrial process heating purposes. It contains a pump for circulation of water and it is operated through differential thermostat for maintaining the required level of water at the desired temperature difference of inlet and outlet water. Here a check valve is also needed to prevent reverse circulation.[3]

Solar water heaters are manufactured and installed throughout the world for different climatic zones so that it is necessary to do some changes in the layout of design for getting desired results. In this sequence, auxiliary energy is incorporated with storage tank, these basically heat exchangers. Auxiliary energy can also be provided by standard electric, oil, or gas water heaters.

## 9.3 PCMs

PCMs are capable of storing latent heat during its phase change from solid state to the liquid state and this stored heat is delivered back when needed.

During its phase change from solid to liquid it gains energy and during its again change of phase from liquid to solid it delivers the almost the same amount of energy without any loss. In this way, PCMs have cycles of charging and discharging of energy. It is also important that PCMs must have constant thermal properties within a certain temperature limit and for a specified duration of time. In this context, researchers and scientists are trying to develop suitable PCMs for different applications so that the efficiency of thermal systems can be increased by incorporating suitable PCMs to different types of equipment. Some desirable properties of PCMs are accepted worldwide are listed here:[4]

i)   Thermal properties: Suitable phase-transition temperature, high latent heat of transition, good heat transfer rate.
ii)  Physical properties: favorable phase equilibrium, high density, small volume change, low vapor pressure.
iii) Kinetic properties: No supercooling, sufficient crystallization rate.
iv)  Chemical properties: Long-term chemical stability, compatibility with materials of construction, no toxicity, no fire hazard.

Above listed properties of PCMs cannot be found in the single composition of materials so that for achieving the desired range of properties combinations of more than single material is required. PCMs are classified into three categories namely organic, inorganic, and eutectic (Fig. 9.1). Further, organic is divided into paraffin compounds and non-paraffin compounds; inorganic is divided into salt hydrates and metallics; lastly, eutectic is divided into three components namely organic-organic, inorganic-inorganic, and inorganic-organic. Flow diagram of such classification is also drawn here.

*Paraffin compounds* are basically wax containing a mixture of a straight chain of n-alkanes $CH_3-(CH_2)-CH_3$ as length increases its capacity of containing latent heat is increases. *Non-paraffin compounds* are materials having properties desired by PCM, that is, formic acid, caprilic acid, etc. *Salt hydrates* are alloys of inorganic salts and water, it forms typical crystalline solid of general formula $AB.nH_2O$. *Metallics* include low-melting materials, such as Gallium-gallium antimony eutectic, cerrolow eutectic, etc. However, due to a weight problem, it is generally not frequently used as PCM. Organic-organic eutectic, inorganic-inorganic eutectic, and inorganic-organic eutectic are a mixture of two or more components they melt and solidify simultaneously.[4] Thermal properties of PCM like latent heat of fusion and melting point temperatures are presently measured through differential scanning calorimeter (DSC) and differential thermal analysis (DTA).

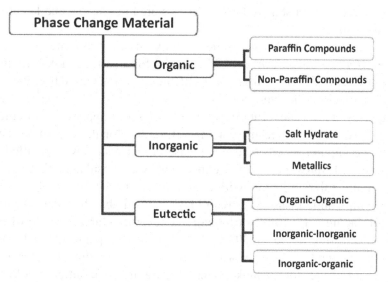

**FIGURE 9.1** Flow diagram of PCM classification.[4] Reprinted with permission from Ref. 4. © 2009 Elsevier.

## 9.4 APPLICATION OF PCMs TO STORAGE TANK OF SOLAR WATER HEATER

Solar energy is available only day time so that storage of energy in the form of heat in a medium is necessary for utilizing energy during off sunshine time. There is a requirement of some material that has to be capable of storing energy in the day time and get back it during off sunshine hours and this requirement is fulfilled by PCM. During sunshine hours energy is stored in the form of sensible heat in any medium like water, rocks, and soils, etc. So that in solar water heaters energy is stored in the form of sensible heat by using a flat plate collector. Now, this stored sensible heat in water of storage tank can be further stored in suitable PCMs, if incorporated in a storage tank. When the temperature of hot water inside of storage tank increases beyond the transition temperature of PCM, it gets melted and absorbs the energy in the form of latent heat and when the temperature of water decreases the energy is provided back to the water through solidification of PCM.[5] In this way, incorporated PCMs create a mechanism of storing energy in the form of latent heat. Without the use of PCMs in storage tank heat is stored in the form of sensible heat only while incorporated PCM storage tanks, heat is stored in sensible and latent heat forms both.

Canbazoglu et al.[5] has done a study of incorporating of *sodium thiosulfate pentahydrate* as PCM material for a storage tank for natural circulation SWHSs. In their study of the experimental setup, they use solar collectors, hot and cold water tanks with measurement and data logger systems. The general layout of the setup apparatus is depicted in Figure 9.2. For measuring temperature and for the accumulation of data, a computer-aided electronic measurement system was installed. All pipes and fittings used in the experiment were insulated. Flat parallel aluminum solar collectors with a fixed plate of 1.94 m × 0.94 m × 0.10 m having single glass cover with black painted absorber plate are used. South facing solar collector having net absorption area 1.65 m² of each one and tilting angle of 30° with the horizontal plane is used. Three major temperatures are recorded here, the temperature of water at the midpoint of the storage tank, water temperature at an outlet of collector, and ambient temperature. Other temperatures were also recorded here for a comparison of study purpose like water temperature above and below the midpoint of the storage tank, water temperature at the inlet of the collector, and at the outlet of the heat storage tank, at the midpoint of the cold water tank. All measured temperatures are shown in the figure along with the location of the point.

After observation, it was found that the midpoint temperature of the heat storage tank is almost equivalent to the mean temperature of the upper and lower temperature of the heat storage tank. All temperature is recorded with the help of sensors having a sensitivity of ± 0.1°C after calibrated with a water-ice mixture before the experiment. Here heat storage tank is filled with PCM and cross-sectional view of the heat storage tank is given in the Figure 9.3. Tank having a volume of 190 L and made of galvanized steel in shape of a cylindrical vessel with well-insulated glass wool is used. Sodium thiosulfate pentahydrate as PCM material is filled in polyethylene bottles having volume 0.44 L of each bottle is used. Every bottle has a weight of 0.7347 kg and all the bottles are arranged in three rows as shown in the figure. The volumes of PCM and water in the heat storage tank are almost equal to 107.8 L and 82.2 L, respectively.

**Temparature Measurement Points**

1- Ambient temperature
2- Water temperature at the upper point of heat storage tank
3- Water temperature at the midpoint of heat storage tank
4- Water temperature at the lower point of heat storage tank
5- Collector outlet temperature
6- Collector inlet temperature
7- Water temperature at the midpoint of cold water tank
8- Outlet temperature of water from heat storage tank

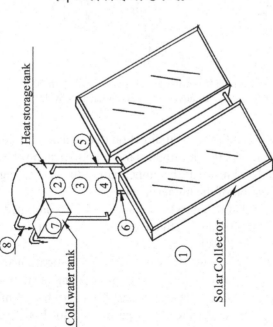

**FIGURE 9.2** Experimental setup of natural circulation SWHS. (Reprinted with permission from Ref. 31. https://creativecommons.org/licenses/by/4.0/)

The total mass of PCM used in the tank is 180 kg and having a density of 1666 kg/m³ in the solid state is used. Bottles are arranged so classily, in that manner, it allows only vertical movement of water and preventing horizontal movement.

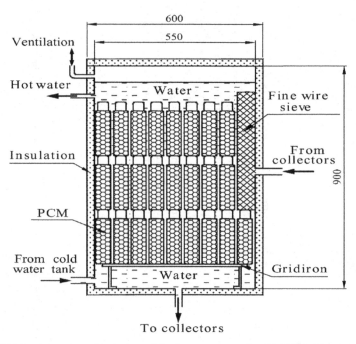

**FIGURE 9.3**   A cross-sectional view of the heat storage tank with PCM. (Reprinted with permission from Ref. 31. https://creativecommons.org/licenses/by/4.0/)

Water is flowing in the vertical direction due to available cavities only in that direction only so that it provides more homogeneous heat distribution horizontally. It is known as that upper part of the heat storage tank had a thermal stratification of the higher level.

It was observed that in natural circulation SWHS combined with PCM increased the thermal storage effectiveness of solar radiation even not increased in the cost of SWHS.

Reddy et al.[6] carried out an experimental investigation for different PCM such as paraffin and stearic acid by varying heat transfer flow (HTF) rates and size of the spherical capsules (68 mm, 58 mm, and 38 mm in diameter). The experiment was performed for both charging and discharging processes. The results proved that 38 mm diameter spherical capsules show better performances over other spherical size capsules. The experimental setup is shown in Figure 9.4a and Figure 9.4b. The objective of the present work is to predict the optimum spherical-sized capsules among three different diameters (68 mm, 58 mm, and 38 mm) for better efficiency of a sensible and latent heat thermal energy storage unit integrated with a varying (solar) heat source. Different diameters of the high density polyethylene (HDPE) spherical capsules were used and surrounded by a solar heating systems (SHS) material (water). Parametric studies are carried out to examine the effects of the diameter, PCM, and HTF flow rates on the performance of the storage unit for varying inlet fluid temperature.

1. Solar flat plate collector
2. Pump
3. Flow control valve
4. Flow control valve
5. Flow meter
6. Thermal energy storage tank
7. PCM capsules
8. Temperature indicator
9. Tp & Tf Temperature sensors

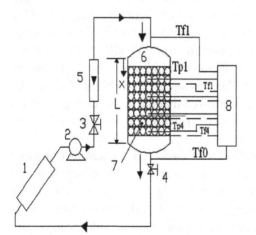

1-Solar flat plate collector;
2-Pump;
3-Flow control valve;
4-Flow control valve;
5-Flow meter;
6-Thermal energy storage tank;
7-PCM capsules;
8-Temperature indicator
9- Tp & Tf Temperature sensors.

**FIGURE 4a** Experimental setup outline. (Reprinted with permission from Ref. 6. Open access.)

**FIGURE 4b** The photographic view of the experimental setup. (Reprinted with permission from Ref. 6. Open access.)

Padmaraju et al.[7] carried out an experiment to investigate the utilization of PCM to heat water for domestic application during night time. This experiment will ensure the availability of hot water throughout the day. The system consisted of two simultaneously functioning heat-absorbing unit. One was a solar water heater and other was heat storage unit consisted of PCM. The solar water heater would function normally and provide hot water during the day time. The storage unit would supply hot water during night time and overcast condition. The storage unit was filled with paraffin in aluminum containers. The result showed that the above configuration is effective for storing 48 L of water which is sufficient for a family of four persons.

Regin et al.[8] analyzed the behavior of packed bed latent heat thermal storage system numerically. The bed was composed of spherical capsules which were filled with paraffin wax as PCM (Fig. 9.5). The layout is shown in Figure 9.6. The result was used for thermal performance analysis for both charging and discharging process. The effects of the inlet heat transfer fluid temperature (Stefan number), mass flow rate, and phase change temperature range on the thermal performance of the capsules of various radii had been investigated. The results indicate that forth proper modeling of the performance of the system the phase change temperature range of the PCM must be accurately known, and should be taken into account. It was further concluded that the complete solidification time was too long compared to the melting time. This was due to the very low heat transfer coefficient during solidification. Higher the Stefan number (i.e., higher inlet heat transfer fluid temperature) the shorter was the time for complete charging. Similarly, for higher the mass flow rate of heat transfer fluid Shorter was the time for complete charging. The charging and discharging rates were significantly higher forth capsule of smaller radius compared those of larger radius. The phase transition temperature range reduced the complete melting time; a difference of 31.6% was observed for the case when the PCM had melting in the temperature range as compared to that for a PCM with melting at a fixed temperature.

Mettawee and Assassa[9] have studied that paraffin wax can be used as PCM for a solar collector. In this collector, the absorber plate–container unit performs the function of both absorbing the solar energy and storing PCM. The collector's effective area was assumed to be 1m² and its total volume was divided into five sectors. The time-wise temperatures of the

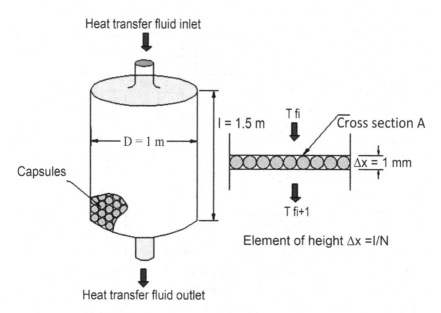

**FIGURE 9.5** Storage system layout and details. (Reprinted with permission from Ref. 8. © 2009 Elsevier.)

PCM were recorded during the processes of charging and discharging. Experiments were conducted for different water flow rates of 8.3–21.7 kg/h. The experimental apparatus was designed to simulate one of the collector's sectors, with an apparatus-absorber effective area of 0.2 m². Outdoor experiments were carried out to demonstrate the applicability of using a compact solar collector for water heating. The experimental results showed that in the charging process, the average heat transfer coefficient increases sharply with increasing the molten layer thickness, as the natural convection grows strong. In the discharge process, the useful heat gain was found to increase as the water mass flow rate increases, shown in the diagram of Figure 9.6.

The research group of the authors, Sharma et al.[10] also designed, developed and performance evaluate of a latent heat storage unit for evening and morning hot water requirements, using a box type solar collector. Paraffin wax (melting temperature 54°C) was used as PCM and found that the performance of the storage unit in the system was very good to get the hot water in the desirable temperature range (Fig. 9.7). Authors also reported that to get the hot water in the desirable temperature range more fins may be provided to increase the effectiveness of the storage unit.

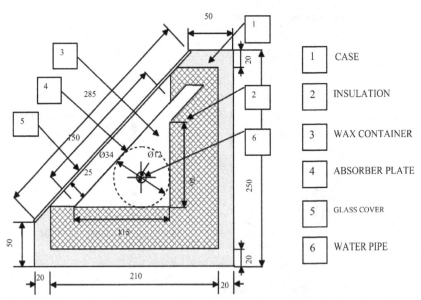

| | |
|---|---|
| 1 | CASE |
| 2 | INSULATION |
| 3 | WAX CONTAINER |
| 4 | ABSORBER PLATE |
| 5 | GLASS COVER |
| 6 | WATER PIPE |

**FIGURE 9.6** Cross-section view of the experimental setup. (Reprinted with permission from Ref. 9. © 2006 Elsevier.)

**FIGURE 9.7** Camera photograph of the box type solar collector with heat exchangers. (Reprinted from Ref. 10.)

## 9.5 CONCLUSION

As we all know fossils fuel resources are depleting day by day. A major form of this energy is utilized for hot water in cold climatic countries. There is a need felt all over the world for the requirement of some innovative technology that can reduce fossils fuel consumption for getting hot water for residential, commercial and as well as industrial purposes. This book chapter has given a brief outline for this type of technology. Some of the recent works in these areas are also reviewed. Overall, this work provides light to further research in this field.

## ACKNOWLEDGMENT

The author (Abhishek Anand) is highly obliged to the University Grants Commission (UGC) and Ministry of Human Resource Development (MHRD), Government of India, New Delhi, for providing the Junior Research Fellowship (JRF). Furthermore, authors are also thankful to the Council of Science and Technology, UP (Reference No. CST 3012-dt., December 26, 2016) for providing research grants to carry out the work at the institute.

## KEYWORDS

- **phase change materials**
- **thermal energy storage**
- **solar water heating system**

## REFERENCES

1. Shukla, A.; Buddhi, D.; Sawhney, R. L. Solar Water Heaters with phase Change material Thermal Energy Storage Medium : A Review, Vol. 13; 2009; pp 2119–2125.
2. Anisur, M. R.; Mahfuz, M. H.; Kibria, M. A.; Saidur, R.; Metselaar, I. H. S. C.; Mahlia, T. M. I. Curbing Global Warming with Phase Change Materials for Energy Storage. *Renew. Sustain. Energy Rev.* **2013**, *18*, 23–30.
3. Deceased, J. A. D.; Beckman, W. A. *Solar Engineering of Thermal Processes.*
4. Sharma, A.; Tyagi, V. V.; Chen, C. R.; Buddhi, D. Review on Thermal Energy Storage with Phase Change Materials and Applications, Vol. 13; 2009; pp 318–345.
5. Canbazoğlu, S., Şahinaslan, A., Ekmekyapar, A., Aksoy, Ý. G., Akarsu, F. Enhancement of Solar Thermal Energy Storage Performance using Sodium Thiosulfate Pentahydrate of a Conventional Solar Water-Heating System, Vol. 37; 2005; pp 235–242.
6. Reddy, R. M.; Nallusamy, N.; Reddy, K. H. Experimental Studies on Phase Change Material-Based Thermal Energy Storage System for Solar Water Heating Applications, Vol. 2; 2012.
7. Padmaraju, S. A. V.; Viginesh, M.; Nallusamy, N. Comparative Study Of Sensible And Latent Heat Storage Systems Integrated With Solar Water Heating Unit, Renewable Energy & Power Quality Journal, Vol. 1, No.6, March 2008, pp 55–60.
8. Regin, A. F.; Solanki, S. C.; Saini, J. S., An Analysis of a Packed Bed Latent Heat Thermal Energy Storage System using PCM Capsules: Numerical Investigation. *Renew. Energy* **2009**, *34*, (7), 1765–1773.
9. Mettawee, E.B.S.; Assassa, G.M.R. Experimental Study of a Compact PCM Solar Collector. *Energy* **2006**, *31*, 2958–2968.
10. Sharma, A.; Sharma, A.; Pradhan, N.; Kumar, B. Performance Evaluation of a Solar Water Heater Having Built in Latent Heat Storage Unit, IEA, ECESIA Annex 17. Advanced Thermal Energy Storage through Phase Change Materials and Chemical Reactions—Feasibility Studies and Demonstration Projects. 4th workshop, Indore, India. March 21–24, 2003; pp 109–115.

## CHAPTER 10

# Application of Latent Heat Energy-Based Thermal Energy Storage Materials in Solar Driers and Other Agricultural Applications

G. RAAM DHEEP and A. SREEKUMAR*

*Solar Thermal Energy Laboratory, Department of Green Energy Technology, Pondicherry Central University, Pondicherry 605014, India*

*Corresponding author. E-mail: sreekmra@gmail.com*

### ABSTRACT

Fossil fuel is the major source of energy in many of the industrial and domestic drying systems used for food processing. Solar drying is a simple, low cost, and hygienic method to preserve agriculture and other marine food products. Solar dryer acts as substitute to conventional drying that reduces the extensive usage of conventional fossil fuels. However, the intermittent nature of solar energy necessitates of a backup storage device with drier to ensure continuous operation. Solar thermal energy storage is considered as a most promising option in this direction. Reliability of a solar dryer is increased by storing excess thermal energy available during sun shine hours and using it when solar energy is inadequate. Solar thermal energy storage system bridges the gap between the energy supply and demand and thereby reduces the instability in the amount of energy available. This chapter presents the studies on solar dryers integrated with latent heat storage system and also details about the performance analysis of solar dryers based on energy, exergy, economic feasibility, and quality of solar dried agricultural food products.

## 10.1   SOLAR DRYING

Solar energy is the most abundant source of renewable energy that has enormous potential, not only for electricity generation but also for many industrial and domestic thermal applications. Solar energy improves our energy security by reducing the usage of expensive fossil fuels for applications like water heating, space heating, thermal electricity generation, and industrial process heating systems. Among many industrial thermal applications, drying food materials is an energy intensive process which consumes almost 10–15% of nation's industrial energy consumption.

Drying of agricultural food products is considered to be the most important food preservation technique to reduce postharvest losses. Drying is the method of expulsion of water to the minimum desired level from the food products to minimize storage losses, reduce storage space, transportation cost and enhance the shelf life of the products. Drying is a process of moisture removal through simultaneous complex heat and mass transfer.[1] This method involves heat transfer to the product from the heat source (heat transfer) and moisture removal from the interior of product to the surface and from the surface to atmosphere (mass transfer).

Various techniques are employed to dehydrate the moisture present in food products which has got its own merits and demerits. Among numerous techniques, open sun drying is the oldest and traditional method of drying agricultural food products. But this method suffers serious disadvantages such as low quality, increased drying time, losses due to weather and contaminations through dust, insects, birds and rodents. Moreover, electrical oven and freeze drying methods nullify the disadvantages of open sun drying and improves the quality of the dried products.[2,3] However, these industrial drying techniques which again energy consuming, high capital and operating cost as it relies on conventional fossil fuels. Thus, negative factors involved in open drying and mechanical drying rejuvenated the usage of solar dryers which reduces postharvest losses, price, and also reinstate the quality of the dried products such as color, texture, and taste.

Solar drying of agricultural food products using specially designed solar dryers is one of the most promising methods to overcome food preservation problems. The drying temperature, initial and final moisture content of a few agriculture food products is shown in Table 10.1. Solar dryers generate higher air temperature, enough heat to remove moisture, large dry air to absorb released moisture, adequate air circulation, and lower relative humidity compared to ambient atmospheric conditions.[4] Due to these proficient characteristics, solar dryers offer several advantages such as increased drying rate, significant reduction in drying time, protection from sudden environmental factors, reduce losses, improved food quality, requires smaller area, low capital and running costs, and greater income.

**TABLE 10.1** Drying Temperature, Initial and Final Moisture Content of Agriculture Food Products.[4]

| Food products | Moisture content (%) | | Drying temperature (°C) |
|---|---|---|---|
| | Initial | Final | |
| Potatoes | 75 | 13 | 75 |
| Chilies | 80 | 5 | 65 |
| Apples | 80 | 24 | 70 |
| Apricot | 85 | 18 | 65 |
| Grapes | 80 | 15–20 | 70 |
| Banana | 80 | 15 | 70 |
| Guava | 80 | 7 | 65 |
| Pineapple | 80 | 10 | 65 |
| Tomatoes | 96 | 10 | 60 |
| Brinjal | 95 | 6 | 60 |
| Fish raw | 75 | 15 | 30 |
| Onion ring | 80 | 10 | 55 |
| Peaches | 85 | 18 | 65 |
| Mulberries | 80 | 10 | 65 |
| Figs | 80 | 24 | 70 |
| Yams | 80 | 10 | 65 |
| Nutmeg | 80 | 20 | 65 |
| Sorrel | 80 | 20 | 65 |
| Cotton | 50 | 9 | 75 |
| Cotton seed | 50 | 8 | 75 |
| Copra | 30 | 5 | 80 |
| Ground nut | 40 | 9 | 50 |

## 10.1 1   CLASSIFICATION OF SOLAR DRYER

Solar dryer is classified based on the mode of operation and heat transfer. Classification of solar dryer is illustrated in Figure 10.1. Natural and forced convection are the two main heat transfer modes of solar dryers.[5,6]

Passive solar dryer is also called as natural convection solar dryer in which the air flows through the dryer due to density difference. Passive solar dryer do not require any blower or fan to circulate the air. In passive solar dryer, the drying duration is high due to inadequate air circulation and poor moist air removal rate. Active solar drier is known as forced circulation or forced convection solar drier. The air is forced into the drying chamber by means of external fans or air blowers. In active solar dryer, air flow and air temperature can be controlled, resulting in enhanced heat transfer rate and reduced drying time.[7]

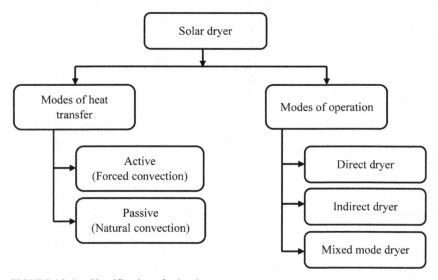

**FIGURE 10.1**   Classification of solar dryer.

Based on operation methodology, solar dryers are classified into three subclasses: direct mode, indirect, and mixed mode. Diagrammatic representation of different types of solar dryer is shown in Figure 10.2. In direct mode, the solar collectors and drying chamber are an integral part of the drying system where the solar radiation passes through the glazing layer reaches the drying chamber thereby increasing the temperature. In an indirect mode solar dryer, the solar collectors and drying chambers are separate units. In this mode, the incident solar radiation is absorbed by solar collectors.[8,9] The air flows through the solar collectors gets heated up through conduction and convection. This warm air flows into the drying chamber using blowers to remove the moisture from food products. In mixed mode solar dryer, both direct and indirect mode of heat transfer techniques is employed to dry agricultural food crops.

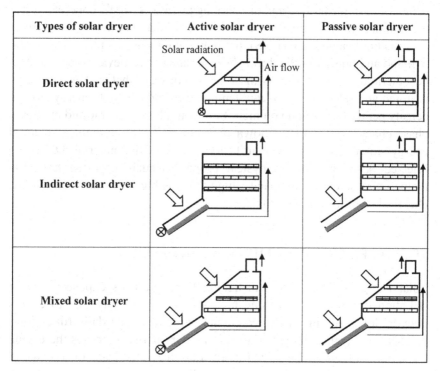

| Types of solar dryer | Active solar dryer | Passive solar dryer |
|---|---|---|
| Direct solar dryer | | |
| Indirect solar dryer | | |
| Mixed solar dryer | | |

**FIGURE 10.2**   Diagrammatic representation on different types of solar dryer.

## 10.2   SOLAR THERMAL ENERGY STORAGE

Solar thermal energy storage balances the peaks and troughs of energy supply, provides energy security, and improves efficiency and feasibility of the system. Solar thermal energy storage system should possess certain characteristics such as large storage capacity per unit mass and volume, low loss, noncorrosive, long life, and less expensive.

Solar thermal energy is stored as a change in internal energy of a material either as sensible heat, latent heat or thermo-chemical reactions. In sensible heat storage (SHS), thermal energy is stored based on the specific heat capacity of the material, change in temperature, and the amount of storage material.[10] The amount of heat stored and released during charging and discharging depends on the heat capacity of the storage material. SHS systems have certain disadvantages such as low heat storage capacity per unit volume, large storage space requirement, and non-isothermal behavior during charging and discharging processes. Rock, concrete, brick, therminol, and water are a few examples of solid and liquid SHS materials.

In latent heat storage (LHS), the material changes its phase from solid to liquid and liquid to solid during heat storage and retrieval process. Latent heat thermal energy storage is an isothermal process, that is, thermal energy is stored and released at constant temperature and has high-energy storage density per unit mass and volume.[11] Few examples of organic and inorganic phase change materials is shown in Table 10.2. The amount of thermal energy stored depends on latent heat capacity of the material. PCMs are selected based on thermophysical properties such as phase transition temperature, latent heat capacity, high density, high thermal conductivity, chemical stability, and less toxic.

## 10.3   APPLICATION OF LHS IN SOLAR DRYER

Solar energy is intermittent in nature. Food products exposed to solar radiation during peak sunshine hour's losses its quality due to excess dehydration resulting in shrinkage and hardening. The daily (day-night) and seasonal (rainy days) variations of solar energy increases the drying duration and reduces the reliability of solar dryer that leads to spoilage of food products due to micro-organisms. These challenges associated with solar dryer are overwhelmed by integrating latent heat thermal energy storage system to provide required thermal energy for drying during inadequate solar energy and nonsolar hours. Several researchers developed solar dryers with different designs integrated with LHS system.

Shalaby[12] designed a novel indirect forced convection solar dryer integrated with LHS unit. Paraffin wax of 12 kg with phase transition temperature of 49°C is used as PCM. Experimental thermal performance of solar dryer is tested without and with storage system for drying *Ocimum Basilicum* and *Thevetia Neriifolia* medicinal herbs. The authors reported that the solar dryer with PCM storage unit maintains the drying air temperature at 2.5°C–7.5°C higher than ambient temperature for 5 h after sunset. LHS system provides drying temperature for approximately 7 consecutive h per day which helps to reach the final moisture content of medicinal herbs in 12 and 18 h, respectively.

**TABLE 10.2**   Examples of Organic and Inorganic Phase Change Materials.

| Phase change material (PCM) | Melting temperature (°C) | Latent heat (kJ/kg) |
|---|---|---|
| Paraffin wax | 5–76 | 170–269 |
| Docosyl bromide | 40 | 201 |
| Caprylone | 40 | 259 |
| Phenol | 41 | 120 |
| Heptadecanone | 41 | 201 |
| 1-Cyclohexyl Octadecane | 41 | 218 |
| $KF.2H_2O$ | 42 | 162 |
| p-Toluidine | 43.3 | 167 |
| Cyanamide | 44 | 209 |
| Methyl eicosanoate | 45 | 230 |
| $Zn(NO_3)_2.4H_2O$ | 45 | 110 |
| $Fe(NO_3)_2.9H_2O$ | 47 | 155 |
| 3-Heptadecanone | 48 | 218 |
| Hydrocinnamic acid | 48 | 118 |
| Lauric acid | 49 | 178 |
| Cetyl alcohol | 49.3 | 141 |
| a-Nepthylamine | 50 | 93 |
| Camphene | 50 | 238 |
| 9-Heptadecanone | 51 | 213 |
| Pentadecanoic acid | 52.5 | 178 |
| Diphenylamine | 52.9 | 107 |
| Oxalate | 54.3 | 178 |
| Palmitic acid | 55 | 163 |
| Hypophosphoric acid | 55 | 213 |
| Chloroacetic acid | 56 | 130 |
| Tristearin | 56 | 191 |
| $MnCl_2.4H_2O$ | 58 | 151 |
| Heptadecanoic acid | 60.6 | 189 |
| Glycolic acid | 63 | 109 |
| $Na_3PO_4.12H_2O$ | 65 | 190 |
| Acrylic acid | 68 | 115 |
| Stearic acid | 69.4 | 199 |
| $Al(NO_3)_2.9H_2O$ | 72 | 155 |
| Phenylacetic acid | 76.7 | 102 |
| Bromocamphor | 77 | 174 |
| Durene | 79.3 | 156 |
| Benzylamine | 78 | 174 |
| $Ba(OH)_2.8H_2O$ | 78 | 265 |
| Acetamide | 81 | 241 |

The feasibility of LHS on the drying kinetics of sweet potato was examined by Devahastin.[13] Experimental setup of solar dryer with LHS is shown in Figure 10.3. Paraffin wax with melting temperature of 35°C–54°C is used as PCM. The investigators reported the effects of different inlet air temperature and velocity on thermal charging and discharging characteristics of PCM. The useful energy extracted from the PCM is 1920 and 1386 kJ min kg$^{-1}$ for the inlet air velocity of 1 and 2 ms$^{-1}$. The use of LHS has saved around 40% and 34% of energy for drying applications.

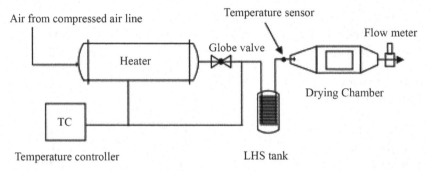

**FIGURE 10.3** Diagrammatic representation of experimental setup. (Reprinted with permission from Ref. 13. © 2006 Elsevier.)

Cakmak[14] designed a solar dryer and experimentally evaluated the drying kinetics of seeded grapes. The solar dryer is shown in Figure 10.4. The experimental system consists of two solar air heaters without and with PCM connected to drying chamber. Solar air collector with expanded surface is used to enhance heat transfer and turbulence during sunshine hours while a solar air collector with PCM is used after the sunset to perform the drying process. Calcium chloride hexahydrate with phase transition temperature of 30°C is used as PCM. The authors performed experiments for three different air velocities and reported that drying time reduces as drying air velocity increases.

Reyes[15] carried out an experimental study on the hybrid solar dryer with PCM for dehydration of mushrooms. The dryer contains solar energy accumulator, solar panel, centrifugal fan, electrical heater, and drying chamber. Accumulator is filled with 14 kg of paraffin wax with melting temperature 58–60°C. Experimental results show that thermal efficiency of solar air heater and the accumulator has varied from 22–67% and 10–21%, respectively. The incorporation of accumulator in solar dryer reduces the

40–70% of electrical energy consumption and also capable to supply the outlet air at 20°C higher than ambient for 2 h after solar hours. The authors also reported the dehydration of kiwi fruit on hybrid solar dryer with 59.5 kg of paraffin wax.[16] The outlet air temperature is maintained at 12°C above the ambient temperature during a total drying period of 9 h.

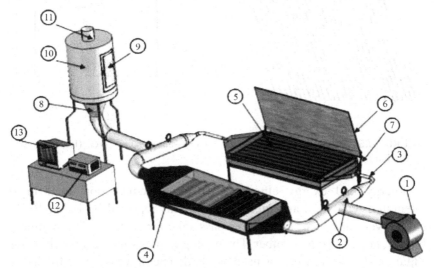

**FIGURE 10.4**   Indirect solar dryer with (1) Fan, (2) valves, (3) connection pipe, (4) expanded-surface solar air collector, (5) collector with PCM, (6) adjustable mirror, (7) adjustable collector tripod, (8) diffuser, (9) observation glass, (10) drying room, (11) air exit chimney, (12) datalogger and (13) PC. (Reprinted with permission from Ref. 14. © 2011 Elsevier.)

Experimental studies on drying kinetics and exergetic analysis of evacuated tube collector (ETC)-based solar dryer with thermal storage is reported by Shringi.[17] The ETC-based solar dryer is shown in Figure 10.5. PCM with the melting temperature of 87°C is used in LHS unit. During the charging process, the heat transfer fluid (60% propylene glycol and 40% water) is circulated through a PCM heat exchanger. The heat energy stored inside the LHS is recovered by circulating the ambient air inside the heat exchanger. The air flowing inside the drying chamber is maintained between 39°C and 69°C for 8 h. The moisture content of garlic cloves reduced from 55.5% (w.b.) to 6.5% (w.b.). The investigators reported that the energy and exergy efficiency of drying chamber varied from 43–83% and 5–55%.

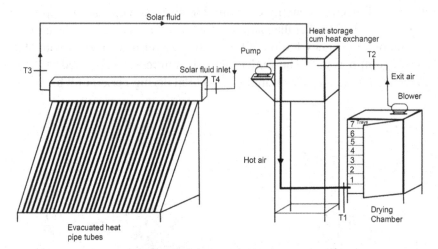

**FIGURE 10.5**  Experimental setup based on ETC-based solar dryer. (Reprinted with permission from Ref. 17. © 2014 Elsevier.)

Energy and exergy efficiency of forced convection solar dryer integrated with shell and tube LHS unit was reported by Rabha.[18] The solar dryer consists of two double pass solar air heaters, LHS module, a blower, and a drying chamber. The photograph of solar dryer is shown in Figure 10.6. Paraffin wax is used as PCM. The dryer is tested by drying 20 kg of red chili operating at outlet air temperature of 36–60°C. The experiments were performed from 8:00 to 18:00 h. The LHS integrated with dryer, smoothens the drying air temperature from 14:30 to 18:00 h. The average energy and exergy efficiency of solar dryer is shown in Table 10.3. The moisture content of the chili is reduced from 73.5% (w.b.) to 9.7% (w.b.) in 4 consecutive days.

Thermal performance and economic analysis of solar crop dryer integrated with thermal energy storage system is studied by Jain.[19] The dryer consists of flat plate solar collector of area 1.5 m², a packed bed LHS unit, and drying chamber. The LHS and drying chamber has a capacity to hold 50 kg paraffin wax PCM and 12 kg of fresh leafy herbs. The photograph of solar crop dryer is shown in Figure 10.7. The experimental results showed that the temperature inside the drying chamber is 6°C higher than ambient temperature for 5–6 h after sunshine. The thermal efficiency of solar dryer is 28.2% with a payback period of 1.5 years.

Baniasadi[20] studied the experimental thermal performance of a forced convection mixed-mode solar dryer integrated with thermal energy storage.

The paraffin wax with melting point of 70°C is used as PCM. The authors observed that PCM absorbs thermal energy until 2 pm and releases till 5 pm. The pickup efficiency and the overall thermal efficiency of the dryer with thermal energy storage is enhanced to 10% and 11%. The drying time reduces to about 50% compared to without storage.

**FIGURE 10.6** Photograph of solar dryer with shell and tube thermal energy storage. (Reprinted with permission from Ref. 18. © 2017 Elsevier.)

**TABLE 10.3** Average Energy and Exergy Efficiency of Solar Dryer.

| Component of solar dryer | Average energy efficiency | Average exergy efficiency |
| --- | --- | --- |
| Solar air heater 1 | 32.4 | 0.9 |
| Solar air heater 2 | 14.1 | 0.8 |
| latent heat storage unit | 43.6–49.8 | 18.3–20.5 |

The feasibility of LHS system integrated with indirect type forced convection solar dryer is studied by Khadraoui.[21] An indirect solar dryer has two solar collectors, solar air panel (without energy storage) and solar energy accumulator (with energy storage) connected with drying chamber as shown in Figure 10.8. During daytime from 6:00 a.m. to 4:00 p.m. the solar air panel provides hot air to the drying chamber whereas the thermal energy is stored simultaneously in solar energy accumulator filled with 60 kg of paraffin wax. At night time from 4:00 p.m. to 6:00 a.m. the thermal energy stored in solar energy accumulator is utilized for drying applications. The authors observed that temperature inside the drying chamber temperature is 4°C–16°C higher than the ambient temperature for 12 h. The results also indicated that relative humidity inside the drying chamber is 17%–34.5% lower than the ambient during nonsolar hours due to thermal energy storage.

**FIGURE 10.7**   Photograph of solar crop dryer. (Reprinted with permission from Ref. 19. © 2015 Elsevier.)

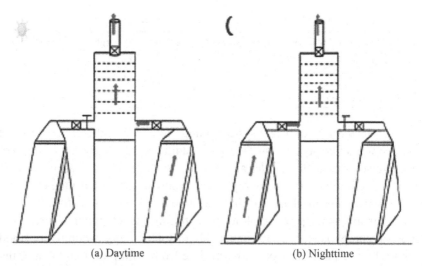

(a) Daytime                                    (b) Nighttime

**FIGURE 10.8**   Schematic representation of indirect solar dryer. (Reprinted with permission from Ref. 21. © 2017 Elsevier.)

## 10.4 PERFORMANCE EVALUATION OF SOLAR DRIER

The performance of solar dryer combined with LHS is evaluated based on physical features of dryer, thermal performance, quality of dried products, and economic analysis. It is necessary to evaluate the performance of developed solar dryer to compare with other dryers and also to assist researchers and manufacturers to improve dryer design, efficiency, optimize testing conditions, and selection of food products. Table 10.4 shows the different types of evaluation and parameters reported on the performance of solar dryer. Literature review on evaluation procedures reveals that there is no specific procedure to evaluate the performance of solar dryer.[22,23] However, the following section describes the parameters generally reported by the researchers to test the performance of solar dryer.

**TABLE 10.4** Parameters used to Evaluate the Performance of Solar Dryer.

| Types of evaluation | Testing parameters |
|---|---|
| Physical features of dryer | Type of solar drier, size and shape, collector material, collector area, drying capacity, tray area and number of layers, nature and dimensions of food products, PCM used and mass of PCM. |
| Thermal performance | Drying time, drying rate, drying air temperature, relative humidity, airflow rate, moisture extraction rate, collector efficiency, dryer efficiency, energy and exergy efficiency of dryer and thermal energy storage |
| Quality of dried product | Sensory quality (color, flavor, taste, texture, aroma), nutritional analysis (sugar content, vitamin C, beta-carotene content, and acidity content), rehydration capacity |
| Economic analysis | Cost of solar dryer, cost of thermal energy storage system, cost of drying, and payback period. |

## 10.5 ENERGY ANALYSIS OF SOLAR DRYER WITH THERMAL ENERGY STORAGE SYSTEM

The effectiveness of solar dryer integrated with LHS depends on thermal performance of the system. Energy and exergy efficiency of a solar dryer is the important parameter to predict the thermal performance of the system. Thermal design evaluated based on the first and second laws of thermodynamics addresses the issues related to the energy and exergy balance of the system. It helps to identify the maximum work potential that can be extracted from the solar dryer during the transfer of energy.[18] The energy analysis measures the unutilized energy while the exergy analysis details about the potential for utilization of unutilized energy in solar drying system. Energy analysis combined with exergy analysis derived from mass and energy conservation equations evaluates the quantity of the energy transferred in and out of the solar dryer. The energy and exergy studies of solar dryer is evaluated by measuring the parameters such as pressure, temperature of the air medium at inlet and outlet of solar collector, storage system and drying chamber, mass of moisture evaporated from the product and heat losses from the system.[24] The energy analysis on the various components of indirect solar drying with thermal energy storage system is evaluated using mass and energy conservation equation expressed as follows:

$$\Sigma \dot{m}_i = \Sigma \dot{m}_o \tag{10.1}$$

$$\dot{Q} - \dot{W} = \Sigma \dot{m}_o \left( h_o + \frac{v_o^2}{2} + X_o g \right) - \Sigma \dot{m}_i \left( h_i + \frac{v_i^2}{2} + X_i g \right)$$

## 10.5.1 ENERGY ANALYSIS OF SOLAR AIR HEATER

The mass and energy conservation equation for solar air heater is expressed by:

$$\dot{m}_i = \dot{m}_o = \dot{m}_a,$$

$$\dot{Q}_{in} - \dot{Q}_{ls} = \dot{m}_a \left( h_o - h_i \right), \tag{10.2}$$

where, $\dot{Q}_{in}$ is the heat energy input to solar air heater and is given by:

$$\dot{Q}_{in} = \alpha_c \tau_c I A_s . \tag{10.3}$$

The useful heat energy gained[25,26] from the solar air heater is expressed as:

$$\dot{Q}_u = \dot{m}_a C_{pa} \left( T_o - T_i \right) \tag{10.4}$$

The instantaneous energy efficiency of solar air heater[27] is defined as the ratio of the useful heat energy gained by air medium to the heat energy input to the system:

$$\eta = \frac{\dot{m}_a C_{pa} \left( T_o - T_i \right)}{\alpha_c \tau_c I A_s}, \tag{10.5}$$

where the absorptivity ($\alpha_c$) and the transmissivity ($\tau_c$) of the absorber and cover plates are 0.95 and 0.85, respectively.

### 10.5.2   ENERGY ANALYSIS OF THERMAL ENERGY STORAGE SYSTEM

The instantaneous thermal energy input to the storage system during the charging process is expressed by:

$$\dot{Q}_{i,ch} = \dot{m}_a C_{pa} \left( T_i - T_o \right). \tag{10.6}$$

The instantaneous thermal energy extracted from the storage system during the discharging process is given by:

$$\dot{Q}_{e,dc} = \dot{m}_a C_{pa} \left( T_o - T_i \right). \tag{10.7}$$

The net heat input and heat recovered of the storage system during the charging and discharging process is given by:

$$Q_{ch} = \int_0^t \dot{Q}_{i,ch} dt \tag{10.8}$$

$$Q_{dc} = \int_0^t \dot{Q}_{e,dc} dt. \tag{10.9}$$

Therefore, the overall energy efficiency of the storage system[28,29] is defined as the ratio of the net heat energy extracted to the net heat energy input of the storage. It is expressed by:

$$\eta_{es} = \frac{Q_{dc}}{Q_{ch}}. \tag{10.10}$$

### 10.5.3   ENERGY ANALYSIS OF THE DRYING CHAMBER

The specific energy consumption (SEC) of the drying system[30] is defined as the ratio of total energy input to the drying system to the amount of moisture evaporated from product.

$$SEC = \frac{P_t}{m_w}. \tag{10.11}$$

The total energy input ($P_t$) is expressed as:

$$P_t = \left[ A_s I + p_b \right] t_d. \tag{10.12}$$

The energy consumed by drying system for the total drying time $(t_d)$ are incident solar radiation $(I)$ and the electrical energy consumed by the blower $(p_b)$.

The mass of the moisture $(m_w)$ evaporated from the product $(m_p)$ with the initial value $(M_i)$ to the final value $(M_f)$ is calculated using the expression below:

$$m_w = \frac{m_p\left(M_i - M_f\right)}{100 - M_f}. \tag{10.13}$$

The overall energy efficiency of the drying chamber is defined as the ratio of energy required to evaporate the moisture from the product to the total energy input into the drying system.

$$\eta_{ds} = \frac{m_w h_f}{P_t}. \tag{10.14}$$

## 10.6 EXERGY ANALYSIS OF SOLAR DRYER WITH THERMAL ENERGY STORAGE SYSTEM

Exergy analysis is useful to design, analyze, and optimize the thermal performance of solar drying system. Exergy analysis based on second law of thermodynamics is more realistic than energy analysis and used to evaluate the useful energy delivered and inefficiencies associated with solar thermal system.[31] The exergy efficiency of each component of solar dryer with thermal energy storage system is evaluated based on the steady flow exergy equation is expressed as:

$$\dot{Ex} = \dot{m}_a\left[C_{pa}\left(T - T_a\right) - T_a\left\{C_{pa}\,ln\left(\frac{T_a}{T}\right) - R\,ln\left(\frac{P}{P_a}\right)\right\}\right]. \tag{10.15}$$

### 10.6.1 EXERGY ANALYSIS OF SOLAR AIR HEATER

The exergy received by the air medium from solar air heater[32,33] is given by:

$$\dot{Ex}_{air} = \dot{m}_a C_{pa}\left[\left(T_{oa} - T_{ia}\right) - T_a\,ln\left(\frac{T_{oa}}{T_{ia}}z\right)\right]. \tag{10.16}$$

The exergy inflow of solar air heater is expressed as:

$$\dot{Ex}_{in,sah} = \left[1 - \left(\frac{T_a}{T_s}\right)\right],$$ (10.17)

where, $T_s$ is the sun temperature and is assumed to 4500 K.
The exergy efficiency of the solar air heater is given by:

$$\eta_{Ex,sah} = 1 - \frac{\dot{Ex}_{dest}}{\dot{Ex}_{in,sah}}$$ (10.18)

or it can also be expressed as follows:

$$\eta_{Ex,sah} = \frac{\dot{Ex}_{air}}{\dot{Ex}_{in,sah}}.$$ (10.19)

The exergy destruction of solar air heater[34] is given by:

$$\dot{Ex}_{dest} = 1 - \left(\frac{T_a}{T_s}\right)\dot{Q}_{in} - \overset{\dot{y}}{m}_a C_{pa}\left[\left(T_{o,sah} - T_{i,sah}\right) - T_a \ln\left(\frac{T_{o,sah}}{T_{i,sah}}\right)\right].$$ (10.20)

### 10.6.2 EXERGY ANALYSIS OF THE ENERGY STORAGE

The exergy input to thermal energy storage during charging process is expressed as:

$$\dot{Ex}_{c,es} = \dot{m}_c C_{pc}\left[\left(T_i - T_o\right) - T_a \ln\left(\frac{T_i}{T_o}\right)\right].$$ (10.21)

The exergy recovered from thermal energy storage during discharging process is expressed as:

$$\dot{Ex}_{dc,es} = \dot{m}_{dc} C_{pdc}\left[\left(T_o - T_i\right) - T_a \ln\left(\frac{T_o}{T_i}\right)\right].$$ (10.22)

The net exergy input and exergy recovered during the charging and discharging of thermal energy[35] is calculated using equations below:

$$Ex_c = \int_0^t \dot{Ex}_{c,es}\, dt$$ (10.23)

$$Ex_{dc} = \int_0^t \dot{Ex}_{dc,es}\, dt.$$ (10.24)

The exergy efficiency of the energy storage system is defined as the ratio of the net exergy recovered from storage unit during the discharging period to the net exergy input to storage unit during charging period. It is calculated using the eq 10.25.

$$\eta_{Ex,es} = \frac{Ex_{dc}}{Ex_c}. \tag{10.25}$$

### 10.6.3   EXERGY ANALYSIS OF THE DRYING CHAMBER

The exergy inflow and outflow of drying chamber is expressed using eqs. 10.26 and 10.27.

$$\dot{Ex}_{id} = \dot{m}_d C_{pd} \left[ (T_{id} - T_a) - T_a ln\left(\frac{T_{id}}{T_a}\right) \right] \tag{10.26}$$

$$\dot{Ex}_{od} = \dot{m}_d C_{pd} \left[ (T_{od} - T_a) - T_a ln\left(\frac{T_{od}}{T_a}\right) \right]. \tag{10.27}$$

The exergy efficiency of the drying chamber is defined as the ratio of the exergy outflow to the exergy inflow of the drying chamber. It is written as follows:

$$\eta_{Ex,d} = \frac{\dot{Ex}_{od}}{\dot{Ex}_{id}}. \tag{10.28}$$

### 10.7   PERFORMANCE ANALYSIS OF SOLAR DRYER

The efficiency of the solar dryer depends on several factors such as solar radiation, inlet air temperature, dryer design, and size and shape of the product to be dried. The effectiveness of the solar dryer is determined by the pickup efficiency, percentage of moisture removed (M%), specific moisture extraction rate (SMER), moisture extraction rate (MER), and saving in drying time ($S_T\%$). The initial mass ($M_i$) and final mass ($M_f$) of the product are measured using electronic weighing balance.[23,36,37] The percentage of moisture removed from the product is calculated by wet basis method using eq 10.29.

$$M\% = \frac{M_i - M_f}{M_i} \times 100. \tag{10.29}$$

SMER signifies the mass of moisture removed from the product to the given amount of energy. It is calculated using the eq 10.30 and is expressed in kg/kWh.

$$SMER = \frac{Moisture\,removed\,from\,product}{Energy\,input}.$$

(10.30)

MER is defined as the mass of moisture removed from the product per unit drying time and is calculated using eq. 10.31. It is expressed in kg/hr.

$$MER = \frac{Moisture\,removed\,from\,product}{Drying\,time}.$$

(10.31)

The drying rate is defined as the difference in moisture content of the material and the equilibrium moisture content.

$$D_r = \frac{dM}{dt} = -k\left(M_t - M_e\right).$$

Saving in drying time is used to determine the percentage of time saved in drying the product using solar dryer compared to other drying techniques. Saving in drying time is calculated using eq 10.32 where $t_o$ is the time taken to dry products in either electric tray dryer or open sun drying and $t_s$ is the time taken to dry products in solar dryer.

$$S_T\% = \frac{t_o - t_s}{t_o} \times 100.$$

(10.32)

The collector efficiency of indirect solar dryer is the ratio of energy absorbed to the amount of solar radiation received on the collector area of the dryer. It is written as:

$$\eta_c = \frac{Q_a}{Q_r}$$

Pick up efficiency is used to determine the efficiency of moisture removed from the product by the dried air. Pickup efficiency decreases with reduction in moisture content of the product. It is expressed as:

$$\eta_p = \frac{W}{\rho V t \left(h_a - h_i\right)},$$

where, $W$ is the mass of water evaporated from the product (kg).

The overall thermal performance of the drying system is calculated using collector efficiency and dryer efficiency. The system efficiency of natural convection solar dryer is:

$$\eta_d = \frac{WL}{IA},$$

where, $W$ is the mass of water evaporated from the product (kg), $L$ latent heat of vaporization of water (J/kg), $I$ is the solar radiation intensity, and $A$ is the area of the solar dryer.

Similarly, for forced convection solar dryer the efficiency is calculated by considering the energy consumed by the fan/blower using the expression below:

$$\eta_d = \frac{WL}{IA + P}.$$

The efficiency of hybrid solar dryer integrated with secondary energy sources such as biomass, LPG etc., is given by:

$$\eta_d = \frac{WL}{(IA + P) + (m_b \times C_f)},$$

where $(m_b \times C_f)$ is the energy supplied by secondary energy source. $m_b$ is mass of fuel and $C_f$ is the calorific value of the fuel. The aforementioned characteristics are used to study about the effectiveness of solar dryer in comparison with open sun drying and electrical tray drying.

## 10.8  ECONOMIC ANALYSIS OF SOLAR DRYER

Solar drying is an energy conserving technique and also a method of producing good quality dried food products. Solar dried product has high selling price which leads to higher marketing income and savings per year. Economic analysis of solar dryer is determined using three consecutive methods namely (1) annualized cost, (2) life cycle savings, and (3) payback period. Economic analysis of solar dryer is calculated based on capital cost of dryer, rate of interest, conventional energy prices, and cost of dried products.[38–40]

Annualized cost method is used to determine the total expenses incurred to operate, maintain, and owning the solar dryer per year. The annualized cost of the dryer ($Ca$) is calculated using eq 10.33.

$$C_a = C_{ac} + C_m - V_a + C_{rf} + C_{re}, \qquad (10.33)$$

where $C_{ac}$ annualized capital cost, $C_m$ annualized maintenance cost (15% of annualized capital cost), $V_a$ annualized salvage value, $C_{rf}$ annual running fuel cost, and $C_{re}$ annual electricity cost for solar dryer.

Annualized cost method also helps to estimate the cost incurred to dry per unit weight of the dried product. The cost of drying ($C_s$) per kilogram is calculated as given in eq 10.34.

$$C_s = \frac{C_a}{M_y} \qquad (10.34)$$

$$M_y = \frac{M_d D}{D_b},$$

where $C_a$ is annualized cost of dryer, $M_y$ mass of product dried per year, $M_d$ mass of dried product per batch, $D$ number of total drying days per year, and $D_b$ number of drying days per batch.

The total savings earned from the entire lifetime of the solar dryer is calculated through life cycle savings method. The annual savings of solar dryer is given by the eq 10.35. The present worth of annual savings obtained from the solar dryer is determined using eq 10.36. The life cycle savings of the solar dryer is obtained by the sum of present worth of annual savings over the lifetime of the system.

$$S_j = S_d * D * (1+i)^{j-1} \qquad (10.35)$$

$$P_j = F_{pj} * S_j \qquad (10.36)$$

$$F_{pj} = \frac{1}{(1+d)^j},$$

where $S_j$ is annual savings in $j^{th}$ year, $S_d$ savings per day, $i$ rate of inflation, $j$ corresponding year, $P_j$ present worth of annual savings in $j^{th}$ year, $F_{pj}$ present worth factor in $j^{th}$ year, and d is the rate of interest.

Payback period method is used to evaluate the time taken for a system to recover its initial cost. A system with short payback period promises the quick inflow of initial annualized cost and reduces the losses due to variable economic conditions. Payback period of the solar dryer is calculated using eq 10.37.

$$N = \frac{In\left(\dfrac{C_{cc}}{S_1}(d-i)\right)}{In\left(\dfrac{1+i}{1+d}\right)},$$
(10.37)

where N is the payback period and $C_{cc}$ capital cost of dryer.

## 10.9   SUMMARY

Solar drying of agricultural food products is one of the potential applications of solar thermal energy. It plays a major role in sustainable energy development by conserving the extensive usage of nonrenewable energy resources. However, the nonperiodic and unpredictable supply of solar energy reduces the reliability of solar dryer. Solar dryers integrated with thermal energy storage system bridges the gap between energy supply and demand and increases the reliability of dryer for uninterrupted drying of agricultural food products. LHS using phase change materials with high latent heat and high energy storage density is the most efficient method of thermal energy storage. Incorporating PCM in solar dryers reduces the drying time and electrical energy consumption and also significantly improves the quality of food product and efficiency of the solar dryer. Only very few researchers has carried out the thermal performance studies on solar dryers combined with LHS. Paraffin wax is the only PCM reported in the literature due to low cost and easy availability. Therefore, the research gap in identifying the new PCMs with good thermophysical properties and development of high efficiency, long life, and economical solar dryer integrated with LHS system requires utmost attention through research studies.

## KEYWORDS

- latent heat storage
- solar drier
- thermal performance
- economic analysis
- energy and exergy studies

## REFERENCES

1. Belessiotis, V.; Delyannis, E. Solar Drying. *Sol. Energy* **2011**, *85*, 1665–1691.
2. El-Sebaii, A. A.; Shalaby, S. M. Solar Drying of Agricultural Products: A Review. *Renew. Sustain. Energy Rev.* **2012**, *16*, 37–43.
3. Raman, S. V. V.; Iniyan, S.; Goic, R. A Review of Solar Drying Technologies. *Renew. Sustain. Energy Rev.* **2012**, *16*, 2652–2670.
4. Kant, K.; Shukla, A.; Sharma, A.; Kumar, A; Jain, A. Thermal Energy Storage Based Solar Drying Systems: A Review. *Innov. Food Sci. Emerg. Technol.* **2016**, *34*, 86–99.
5. Ekechukwu, O. V.; Norton, B. Review of Solar-Energy Drying Systems II: An Overview of Solar Drying Technology. *Energy Convers. Manag.* **1999**, *40*, 615–655.
6. Agrawal, A.; Sarviya, R. M. A Review of Research and Development Work on Solar Dryers with Heat Storage. *Int. J. Sustain. Energy* **2014**, 1–23.
7. Bal, L. M.; Satya, S.; Naik, S. N. Solar Dryer with Thermal Energy Storage Systems for Drying Agricultural Food Products: A Review. *Renew. Sustain. Energy Rev.* **2010**, *14*, 2298–2314.
8. Bal, L. M.; Satya, S.; Naik, S. N.; Meda, V. Review of Solar Dryers with Latent Heat Storage Systems for Agricultural Products. *Renew. Sustain. Energy Rev.* **2011**, *15*, 876–880.
9. Shalaby, S. M.; Bek, M. A.; El-Sebaii, A. A. Solar Dryers with PCM as Energy Storage Medium: A Review. *Renew. Sustain. Energy Rev.* **2014**, *33*, 110–116.
10. Raam Dheep, G.; Sreekumar, A. *Green Energy and Technology, Energy Sustainability through Green Energy: Phase Change Materials—A Sustainable Way of Solar Thermal Energy Storage*; Springer: India, 2015.
11. Raam Dheep, G.; Sreekumar, A. Influence of Nanomaterials on Properties of Latent Heat Solar Thermal Energy Storage Materials—A Review. *Energy Convers. Manag.* **2014**, *83*, 133–148.
12. Shalaby, S. M.; Bek, M. A. Experimental Investigation of a Novel Indirect Solar Dryer Implementing PCM as Energy Storage Medium. *Energy Convers. Manag.* **2014**, *83*, 1–8.
13. Devahastin, S.; Pitaksuriyarat, S. Use of Latent Heat Storage to Conserve Energy During Drying and its Effect on Drying Kinetics of a Food Product. *Appl. Therm. Eng.* **2006**, *26*, 1705–1713.
14. Cakmak, G.; Yildiz, C. The Drying Kinetics of Seeded Grape in Solar Dryer with PCM-Based Solar Integrated Collector. *Food Bioprod. Process* **2011**, *89*, 103–108.
15. Reyes, A.; Mahn, A.; Vásquez, F. Mushrooms Dehydration in a Hybrid-Solar Dryer, Using a Phase Change Material. *Energy Convers. Manag.* **2014**, *83*, 241–248.
16. Reyes, A.; Vásquez, J.; Pailahueque, N.; Mahn, A. Effect of Drying Using Solar Energy and Phase Change Material on Kiwi Fruit Properties. *Dry. Technol.* **2018**, 1–13.
17. Shringi, V.; Kothari, S.; Panwar, N. L. Experimental Investigation of Drying Of Garlic Clove in Solar Dryer Using Phase Change Material as Energy Storage. *J. Therm. Anal Calorim.* **2014**.
18. Rabha, D. K.; Muthukumar, P. Performance Studies on a Forced Convection Solar Dryer Integrated with a Paraffin Wax–Based Latent Heat Storage System. *Sol. Energy* **2017**, *149*, 214–226.

19. Jain, D.; Tewari, P. Performance of Indirect through Pass Natural Convective Solar Crop Dryer with Phase Change Thermal Energy Storage. *Renew. Energy* **2015,** *80,* 244–250.

20. Baniasadi, E.; Ranjbar, S.; Boostanipour, O. Experimental Investigation of the Performance of a Mixed-Mode Solar Dryer with Thermal Energy Storage. *Renew. Energy* **2017.**

21. El Khadraoui, A.; Bouadila, S.; Kooli, S; Farhat, A.; Guizani, M. Thermal Behavior of Indirect Solar Dryer: Nocturnal Usage of Solar Air Collector with PCM. *J. Clean Prod.* **2017,** *148,* 37–48.

22. Soda, M. S.; Ram, C. Solar Drying Systems and their Testing Procedures: A Review. *Energy Convers. Manag.* **1994,** *35,* 219–267.

23. Augustus Leon, M.; Kumar, S.; Bhattacharya, S. C. A Comprehensive Procedure for Performance Evaluation of Solar Food Dryers. *Renew. Sustain. Energy Rev.* **2002,** *6,* 367–393.

24. Cengel, Y. A., Boles, M. A. *Thermodynamics an Engineering Approach*; Tata McGraw Hill: New Delhi, 2008.

25. Midilli, A.; Kucuk, H. Energy and Exergy Analyses of Solar Drying Process of Pistachio. *Energy* **2003,** *28,* 539–556.

26. Celma, A. R.; Cuadros, F. Energy and Exergy Analyses of OMW Solar Drying Process. *Renew. Energy* **2009,** *34,* 660–666.

27. Benli, H. Experimentally Derived Efficiency and Exergy Analysis of a New Solar Air Heater having Different Surface Shapes. *Renew. Energy* **2013,** *50,* 58–67.

28. Dincer, I., Rosen, M. A. *Thermal Energy Storage Systems and Applications*; John Wiley and Sons: UK, 2011.

29. Duffie, J. A., Beckman, W. A. *Solar Engineering of Thermal Processes*; John Wiley & Sons: New Jersey, 2006.

30. Fudholi, A.; Sopian, K.; Alghoul, M. A.; Ruslan, M. H.; Othman, O. Y. Performances and Improvement Potential of Solar Drying System for Palm Oil Fronds. *Renew. Energy* **2015,** *78,* 561–565.

31. Nag, P. K. *Basic and Applied Thermodynamics;* Tata McGraw Hill: New Delhi, 2010.

32. Akpinar, E. K.; Koçyigit, F. Energy and Exergy Analysis of a New Flat-Plate Solar Air Heater having Different Obstacles on Absorber Plates. *Appl. Energy* **2010,** *87,* 3438–3450.

33. Park, S. R.; Pandey, A. K.; Tyagi, V. V.; Tyagi, S. K. Energy and Exergy Analysis of Typical Renewable Energy Systems. *Renew. Sustain. Energy Rev.* **2014,** *30,* 105–123.

34. Esen, H. Experimental Energy and Exergy Analysis of a Double-Flow Solar Air Heater having Different Obstacles on Absorber Plates. *Build. Environ.* **2008,** *43,* 1046–1054.

35. Koca, A.; Oztop, H. F.; Koyun, T.; Varol, Y. Energy and Exergy Analysis of a Latent Heat Storage System with Phase Change Material for a Solar Collector. *Renew. Energy* **2008,** *33,* 567–574.

36. Mohanraj, M.; Chandrasekar, P. Performance of a Solar Drier with and without Heat Storage Material for Copra Drying. *Int. J. Global Energy Issues* **2009,** *31,* 112–121.

37. Mohanraj, M.; Chandrasekar, P. Performance of a Forced Convection Solar Drier Integrated with Gravel as Heat Storage Material for Chili Drying. *J. Eng. Sci. Technol.* **2009,** *4,* 305–314.

38. Sreekumar, A. Techno-Economic Analysis of a Roof-Integrated Solar Air Heating System for Drying Fruit and Vegetables. *Energy Convers. Manag.* **2010,** *51*, 2230–2238.
39. Sreekumar, A.; Manikantan, P. E.; Vijayakumar, K. P. Performance of Indirect Solar Cabinet Dryer. *Energy Convers. Manag.* **2008,** *49*, 1388–1395.
40. Imre, L. Technical and Economical Evaluation of Solar Drying. *Dry. Technol.* **2007,** *4*, 503–512.

# CHAPTER 11

# Application of Phase Change Materials in Thermal Stability of Batteries and in the Automobile Sector

LUCIA IANNICIELLO[1*], PASCAL HENRY BIWOLÉ[1,2], and PATRICK ACHARD[1]

[1]*MINES ParisTech, PSL Research University PERSEE–Center for Processes, Renewable Energies and Energy Systems, CS 10207, F-06904 Sophia Antipolis, France*

[2]*Université Clermont Auvergne, CNRS, SIGMA Clermont, Institut Pascal, F-63000 Clermont–Ferrand, France*

*Corresponding author. E-mail: lucia.ianniciello@outlook.fr*

## ABSTRACT

It is necessary to provide a thermal management system for Li-ion batteries employed in electric vehicles. In fact, uncontrolled temperature increase in a stack of battery cells can lead to capacity and power losses, electrical imbalance, reduced lifespan or even thermal runaway of the battery. The use of phase change materials (PCMs) instead of conventional methods, like forced air convection or liquid cooling, is very promising. In fact, PCMs can be used passively to ensure a better thermal management with a lower overall temperature in the pack of cells and a more uniform temperature. It is also possible to combine PCMs with active cooling to be able to regenerate the PCM after each thermal load. Moreover, to achieve a better performing system, it is possible to enhance the PCMs' thermal conductivity by the addition of highly conductive materials like metals or carbons. Many of these options have been widely studied and are still under development.

## 11.1 INTRODUCTION

Thermal energy storage (TES) systems can be divided into three categories: systems based on sensible heat, latent heat, or on chemical reactions, as shown on Figure 11.1 The sensible heat-based TES systems use the specific heat capacity of the materials to store thermal energy. However, the specific heat capacity of most materials is lower than their latent heat. Therefore, the use of sensible heat storage alone is not sufficient in many thermal management systems. The TES systems through chemical reactions can present a better efficiency but are more difficult to operate. In fact, they require one or two reaction chambers for the different reactants. TES through the use of latent heat storage is more commonly employed through the use of phase change materials (PCMs). The constant temperature phase change is one of the advantages of these TES systems. The phase change can be either: solid–solid, solid–liquid, liquid–gas, or solid–gas. The liquid–gas or solid–gas permit to store an important quantity of thermal energy but they are not practical due to the risk of gas leakage and because of their volumetric thermal expansion, which is important. The solid–solid phase change does not induce a real phase change but a reorganization in the material crystalline structure. The solid–solid phase change materials are mainly used in high temperature operating systems. The solid–liquid phase change is the one which is mostly used to store energy or for thermal management applications.

PCMs can be used in several applications as TES systems. PCMs can be found in building thermal systems as well as are integrated into building walls to help regulate the room temperature. Such PCMs can be operated impregnated in the building walls, microencapsulated, or shape stabilized.[2] They are also used in the textile industry for the creation of "smart" textiles.[3] PCMs have also been tested in mattresses to improve the human thermoregulation.[4] In the food industry, PCMs are integrated in freezers for food transportation[5] or preservation[6] as well as in bottles for beverages transportation.[7] PCM applications are also found in the medical field for the transport of blood, organs, and so forth.[7] They are used for the thermal management of solar thermal systems: photovoltaic panels,[8] solar cooker, solar water heater, or solar air heater.[9, 10]

The use of PCMs in battery thermal management is also widely studied. In the recent years, the number of articles about PCMs for battery thermal management has greatly increased. Batteries can overheat and therefore need thermal management systems. In the first section, the present chapter details the battery heat generation processes and the effects of poor thermal stability. The second section explores the different battery thermal management systems (BTMS) used to maintain an optimal operative temperature. The section focuses on the PCMs used in electric vehicles batteries, including nano-enhanced composite PCMs. The different strategies used for BTMS integrated with PCMs are also discussed. Indeed, PCMs can be used alone but are often coupled with other active cooling systems. The section details the different hybrid systems currently tested. Finally, the use of PCMs for thermal comfort or air conditioning of vehicles is investigated. In fact, when the car stops, the start and stop technology turns the motor off, but air heating or cooling must be maintained. In such cases, PCMs can be employed passively, which is very interesting.

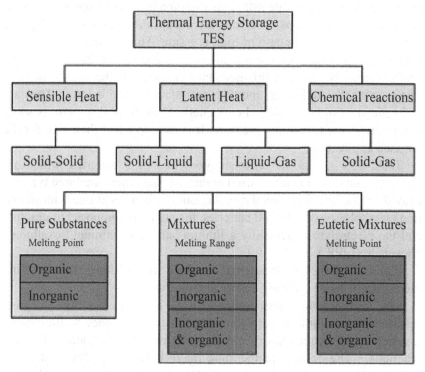

**FIGURE 11.1** Thermal energy storage systems. (Reprinted with permission from Ref. 1. © 2018 Elsevier.)

## 11.2 BATTERY THERMAL STABILITY

There are four different battery categories: lead batteries, alkaline batteries, lithium batteries and sodium batteries. Lithium-ion batteries are the most employed in portable devices as they show great life span and autonomy. However, the first lithium battery was subjected to thermal runaways. In fact, dendrites formed on the lithium and those dendrites could go through the separator to the other electrode and created a short circuit. Then, the temperature increased and the lithium melted, which led to the thermal runaway. Second-generation lithium-ion batteries were developed and successfully commercialized by Sony, with a lithium-cobalt-oxide cathode.

The batteries can be assembled according to three different geometries: cylindrical, prismatic, and pouch. The cylindrical cells permit obtaining better electrical performances, a better mechanical stability, and good cycle repeatability. The cells are usually small; the system is composed of bands of anode/separator/cathode rolled up around a pivot. The size varies from ten to hundreds of loops. The bans are encapsulated in a steel wrapping with a security valve in case of high pressure. The prismatic cells have a shorter life span and their thermal management is more difficult to ensure. The envelope is rigid and this geometry ensures a better energy density. The pouch cells are lighter, and their manufacture cost is lower. They are cells put in flexible and sealed containers. Their flexibility permits some deformation due to internal pressure. It is easy to adapt their size for the targeted application.

Table 11.1 shows the different types of Li-ion batteries, which are characterized by the material used at the cathode. There are three types of cathodes: one based on metal dioxide, one "spinel", and one with transition metal phosphates. For all batteries, the anode is composed of graphite and the electrolyte is generally a lithium salt called lithium hexafluoro-phosphate ($LiPF_6$).

Figure 11.2 shows the internal structure of a Li-ion battery and the mechanisms inside it during charge and discharge.

The two principal heat sources in Li-ion batteries are: the heat released by the Joule effect (coming from the resistance to the charge transfer in the accumulator) and the heat released by electrochemical reactions.[11] The optimal temperature range is between 15°C and 35°C. If the temperature goes below 15°C, there is a diminution of the battery capacity and of its performances. If the temperature goes above 35°C,

**TABLE 11.1** Different Types of Li-ion Batteries.

| | LCO (lithium cobalt oxide) | LMO (lithium manganese oxide) | NMC (lithium nickel manganese cobalt oxide) | LFP (lithium iron phosphate) | NCA (lithium nickel cobalt aluminum oxide) | LTO (lithium titanate) |
|---|---|---|---|---|---|---|
| Cathode | $LiCoO_2$ | $LiMn_2O_4$ | $LiNiMnCoO_2$ | $LiFePO_4$ | $LiNiCoAlO_2$ | $Li_4Ti_5O_{12}$ |
| Nominal tension (V) | 3.60 | 3.70 | 3.60–3.70 | 3.20–3.30 | 3.60 | 2.40 |
| Specific energy (capacity) (Wh/kg) | 150–200 | 100–150 | 150–220 | 90–120 | 200–260 | 70–80 |
| Life span (cycles) | 500–1000 | 300–700 | 1000–2000 | 1000–2000 | 500 | 3000–7000 |
| Charge (C-rate) | 0.7–1C | 0.7–1C | 0.7–1C | 1C | 0.7C | 1C |
| Discharge (C-rate) | 1C | 1C | 1C | 1C | 1C | 10C |
| Applications | Portable phones, computers, and cameras | Medical device, electric drive unit | Electric bikes, medical device, electric vehicles | Portable or stationary applications with high charge rate | Medical device, electric drive unit | Electric drive unit |
| Thermal runaway (°C) | 150 | 250 | 210 | 270 | 150 | - |
| Characteristics | High specific energy, limited specific power, Co expensive | High power and higher security than LCO but low capacity | High capacity and high power | Flat discharge curve, secure but low capacity and high auto discharge | Similar to LCO. | High life span, fast charge, secure but low specific energy and expensive |

Reprinted with permission from Ref. 12. © 2018 Elsevier.

irreversible reactions occur that eventually lead to a diminution of the battery life span. Lithium-ion batteries can overheat depending on several parameters:

-   the charge and discharge current: for high solicitations the Joule effect is predominant and the temperature increases;
-   the state of charge (SOC), which is the energy quantity that remains in the battery;
-   the depth of discharge (DOD) corresponding to the energy quantity discharged from the battery, given in percentage of capacity;
-   the temperature inside the battery, because the electrochemical reactions are stimulated and the internal resistance decreases when the temperature inside the battery is high;
-   the accumulator chemistry because the materials nature has a great influence on the heat sources; and
-   the electrolyte solution that can undergo self-heating processes; different additives can be used to improve its performances.

The elevation of temperature affects the electrochemical system, the round trip efficiency, the charge acceptance, the power and the energy capability, the reliability, and the life cycle. In fact, the temperature increase can create a film in the solid electrolyte interface, can cause the electrolyte and the anode decomposition, and can trigger a reaction between the cathode and the electrolyte, or between the cathode and the adhesive. The major thermal issues are: capacity or power loss, thermal runaway, electrical imbalance between several cells, and decreased performances under low temperatures.

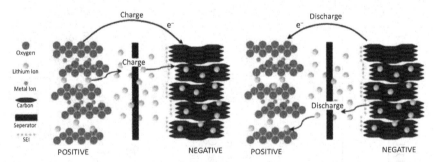

**FIGURE 11.2**   Diagram of a Li-ion battery and the reaction mechanisms.[13]

## 11.2.1 CAPACITY OR POWER LOSS

The capacity and power loss can lead to autonomy loss, a shorter life span and even self-discharge. These performances losses can occur with time with the aging of the battery, which impacts the thermal properties and in turn accelerates the loss of capacity and power. There are two types of battery aging: the calendar aging and the active aging. Calendar aging occurs even when the battery is not used and is accentuated when the battery is stored at high state of charge and high temperatures.[14] The temperature is very important in the first four weeks of storage as it is the main cause for self-discharge. Moreover, the reduced performance of the graphite electrodes is accentuated when the battery is fully charged. The solid electrolyte interface formed on their surfaces does not passivate them completely.[15] Therefore, when the temperature increases, the surface species dissolve. This increases the electronic conductivity and the rate of self-discharge. Active battery aging corresponds to the aging effect due to the utilization of the battery. The two aging phenomena affect the battery in different ways and both modify the battery response to the same solicitations at different aging states.

As there are several types of Li-ion batteries and therefore different types of electrode materials and cell chemistries, it is still difficult to predict and understand all the mechanisms behind the capacity and power fade. In general, it can be induced by the combination of high current discharge and high ambient temperature as it leads to the overheating of the battery. Usually, the capacity loss is attributed to the loss of ions $Li^+$ and active material. The availability of free lithium ions and active material can be decreased by the solid electrolyte interface development or dissolution on carbonaceous anode, along with the crystal structure instability, side reactions, the active material dissolution, and the solid, the temperature increases which accelerate the loss of available energy.[16] The power fade is usually linked to the increase of the cell internal resistance. In fact, there are three resistive phenomena in electrochemical batteries: the electric resistance inherent to the components, the resistance related to the diffusion phenomena of $Li^+$ through the different components of the cell, and the resistance to the charge transfer at the interface between the solution and the insertion material. The first resistance is an ohmic resistance while the last two are electrochemical resistances.

For further explanation, the mechanisms of capacity and power loss are more detailed in Bandhauer et al.[17], Vetter et al.[18], and Barré et al.[19]

## 11.2.2   THERMAL RUNAWAY

Thermal runaway is mainly due to exothermic reactions during improper charge and discharge and to short circuits. The temperature of the thermal runaway depends on the state of charge. There can be a series of chemical reactions which produce a great amount of heat and gases. Several phenomena can occur[20–22]:

– the separator can shrink, leading to more heat release and to temperatures up to 500°C;
– when the temperature of the battery reaches 85°C, the solid electrolyte interface on the graphite electrodes can decompose following an exothermic reaction;
– the metastable components included in the solid electrolyte interface may decompose exothermically when the temperature is around 90°C;
– the graphite electrode can be exposed because of an incomplete solid electrolyte interface, and it can react with the solvent at a temperature around 100°C and may reach 200°C. This reaction is exothermic but does not always occur because of the presence of salt, $LiPF_6$, which inhibits the reaction;
– if the temperature increases to 110°C, a secondary film, which would have been created, will also decompose;
– between 130°C and 190°C, the separator can melt, which can create a short circuit;
– at 140°C the electrolyte can evaporate and the vapor can cause combustion;
– the positive electrode can react with the electrolyte or can release oxygen that can react with the electrolyte. This reaction is highly exothermic but happens only when the temperature of the battery reaches 180°C; and
– if the temperature reaches 660°C the aluminum collector will melt.

## 11.2.3   ELECTRICAL IMBALANCE

The electrical imbalance is a difference of capacity between the cells of a battery pack. The capacity depends on the temperature; therefore, a temperature difference inside the battery can lead to electrical imbalance. This nonuniform temperature distribution can result in a localized deterioration and eventually lead to a general deterioration of the entire pack. In fact, during the discharge, the pack will produce less energy, equivalent to the cells with the lowest capacity. When charged, the weakest cell will be overcharged. Therefore, in the case of Li-ion batteries it is important to keep the temperature under a given value and to have a uniform temperature field as shown on Figure 11.3.

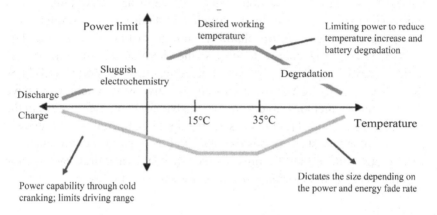

**FIGURE 11.3**   Battery power curves in charge and discharge depending on the temperature.[25]

## 11.2.4   PERFORMANCES UNDER LOW TEMPERATURES

The heat generation and the electrochemical performances of a Li-ion battery depend on the temperature. When the battery is operated under low temperature, the transportation of ions and the charge transfer at the solid electrolyte interface is more difficult. Several studies focused on the development of electrolyte mixtures with low freezing points and high ionic conductivities.[23] The heat generation rate at low temperature can be raised by the increase of internal resistance. Under temperature below -10°C the charge transfer resistance between graphite and electrolyte increases and becomes the dominant factor. Moreover, under cold condition, the charge performances decrease faster as it is more difficult to charge a battery than discharging one.[24]

   Therefore, Li-ion batteries operability strongly depends on the temperature. To overcome the issue of Li-ion batteries thermal stability, several thermal management systems can be employed. There are two types of solutions to decrease the heat released by Li-ion batteries: interior modification to reduce the heat generation or exterior thermal management system to enhance the heat dissipation from the cells. The external thermal management solutions can include air circuits or coolant circuits, for example. The internal ones can consist of a modification of the electrode thickness, of the materials, of the anodes, or the cathodes. The electrolyte can be employed as a coolant through micro channels embedded inside the electrodes.[26] Currently, the most common systems are the external ones. They can use natural or forced air convection, liquid passive or active cooling, or heat pipes. The main issues with these systems are their complex implementation, their cost, their lack of reliability, and most of all, the fact that they constantly consume energy to ensure the thermal management of the battery even when the car is parked. Latent heat thermal storage (LHTES) through the use of PCMs, represents an attractive way to achieve passive thermal management and to ensure that the battery temperature stays in the desired range. Figure 11.4 presents the maximum temperature difference in a mock-up battery stack versus the SOC and the type of thermal management used, while Table 11.2 presents the advantages and drawbacks of the different thermal management solutions for Li-ion batteries.

## 11.3 PCMs IN AUTOMOBILE SECTOR

### 11.3.1 PCMs FOR LI-ION BATTERIES

The employment of PCMs for Li-ion batteries thermal management has been widely studied. In fact the PCMs represent an attractive way to achieve thermal management as it can be a passive thermal management. However, sometimes the PCM alone is not enough to keep the temperature in the desired temperature range. The "desired temperature range" varies according to the application considered.

PCMs can not only be used for the battery employed in vehicles, but also in scooters, laptops, and so forth. There are several studies on the use of PCMs for vehicles and several existing patents. Some industries already employ PCMs for automobiles.

Several strategies can be elaborated: the use of only one PCM, the combination of several PCMs, the use of PCMs-based composites. The use of composites has been investigated in numerous studies as it represented an interesting way to overcome the quite low thermal conductivity of PCMs. For most studies, the phase change selected is the solid–liquid one, but there are also some studies including solid–solid PCMs.[28]

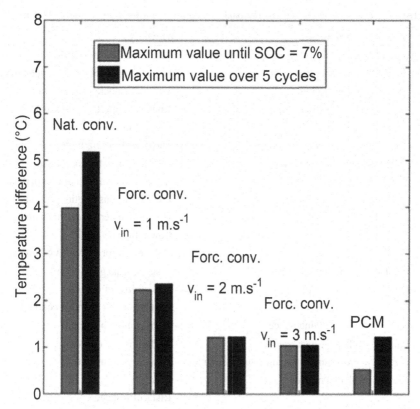

**FIGURE 11.4** Maximum temperature difference according to the thermal management system employed. (Reprinted with permission from Ref. 27. © 2014 Elsevier.)

The thermal resistance between the PCM and the cells is a constraint, and they have to be very close to each other to ensure perfect contact. In the industry, thermal pads are employed to overcome this issue. There is also the huge constraint of space limitation especially in automobiles.

## 11.3.1.1 UTILIZATION OF ONLY ONE PCM

There are many studies on the use of PCMs for battery thermal management. It has been introduced by Al-Hallaj et al.[29] in 2000. They used the PCM alone and the cylindrical battery cells were immersed in the PCM. For cylindrical and prismatic cells, having the cells surrounded by PCM is the easiest way to perform the thermal management, as shown on Figure 11.5.

**TABLE 11.2**   Advantages and Drawbacks of Battery Thermal Management Systems.

| | Advantages | Drawbacks |
|---|---|---|
| Natural convection | - Low initial cost<br>- No operative cost<br>- Easy to integrate<br>- Passive cooling | - Low heat transfer coefficient<br>- Depends on ambient air temperature<br>- Limited temperature reduction<br>- Uneven temperature distribution<br>- Low efficiency<br>- Insufficient for high rates |
| Air forced convection | - Simple operation<br>- Low initial cost<br>- Low maintenance | - Low heat transfer coefficient<br>- Depends on ambient air temperature<br>- Electricity consumption for fans<br>- Uneven temperature distribution<br>- Low efficiency<br>- Insufficient for extreme conditions |
| Liquid passive cooling | - Low initial cost<br>- Low operative cost<br>- Passive cooling<br>- Low maintenance | - Risks of leakage<br>- Risk of loss of the gas<br>- Insufficient for extreme conditions<br>- Regeneration of the gas |
| Liquid active cooling | - Higher heat transfer<br>- More uniform temperature distribution<br>- Higher efficiency | - Highest initial cost<br>- Highest operative cost<br>- Complex<br>- Risks of leakage<br>- Short lifespan<br>- Electricity consumption for pumps<br>- Insufficient for extreme conditions |
| Heat pipe | - Higher conductivity<br>- Higher heat transfer<br>- Higher efficiency | - Highest initial cost<br>- High operative cost<br>- Complex<br>- Risks of leakage<br>- Electricity consumption |
| PCM | - Low cost<br>- Low maintenance<br>- Passive cooling<br>- Higher efficiency<br>- More uniform temperature distribution<br>- Performant in extreme conditions | - Low conductivity<br>- Risk of leakage<br>- Regeneration of the PCM<br>- Risk of supercooling<br>- Volume difference with phase change |

 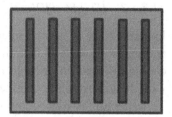

**FIGURE 11.5**   Example of configurations of PCM (in green color) battery cooling.

Al-Hallaj et al. studied the use of a paraffin wax with a phase change occurring at 56 °C. They compared the use of PCM to natural convection and forced air convection. The PCM permitted to obtain a more uniform temperature distribution among the battery cells.

In 2010, Duan et al.[86] used a heater to represent a cylindrical battery cell in their experiments. They compared two configurations, one with the cell dived in the PCM and another one with just a layer of PCM around the cell. The phase change temperature of the PCM was 50°C. Different tests were done, one at constant discharge rate with room temperature at 25°C. The second was done with a discharge rate that changed every 5 min. The third was conducted with temperature cycle between 5°C and 25°C. In every test, it proved that the use of a PCM permitted to have a lower temperature peak and a more uniform temperature distribution among the battery cells.

Javani et al.[30] investigated the influence of the PCM thickness on the thermal management of a Li-ion battery. The PCM tested in this study was the n-octadecane. The thicknesses tested were: 3, 6, 9, and 12 mm. They applied various charge and discharge cycles at different power densities to the battery at a room temperature of 21°C. As expected, the temperature peak decreased with the increase of the PCM mass.

The use of only one PCM has shown some limitations and was sometimes not enough to produce a proper thermal management system for Li-ion batteries, for example, the low thermal conductivity of PCMs or the difficulty to regenerate the PCM once it is in its liquid phase. Another strategy was then to use several PCMs.

## 11.3.1.2   *UTILIZATION OF SEVERAL PCMs*

Implementing a thermal management system with several PCMs can be an interesting idea. In fact, the phase change temperature of the two PCMs can be different in order to have the other PCM take over when the one with the lowest phase change temperature is completely melted.

Ramandi et al.[31] combined two PCMs for battery cooling in an electric vehicle. The use of one PCM only was compared to the use of two PCMs. Four different PCMs were considered with phase change temperature ranging from 30°C to 40°C: capric acid (melting temperature $T_m = 31.5°C$), eicosane ($T_m = 36.8°C$), $Na_2(SO_4)10H_2O$ ($T_m = 32.4°C$) and $Zn(NO_3)26H_2O$ ($T_m = 36.4°C$). The PCMs were tested for constant discharges of the battery during 2.5 h, at a room temperature of 30°C. The results showed that the double layer layout is slightly more efficient than the single layer one.

Moraga et al.[32] compared systems employing one or three layers of different PCMs with different thicknesses (0, 3, and 12 mm) for battery cooling in racing solar cars. The PCMs tested were capric acid, eicosane, decahydrated sodium carbonate and octadecane. The tests were executed at a constant discharge rate at room temperatures of 21°C, then 30°C. Using a three-layer configuration, the best performances were reached when the PCM with the highest thermal conductivity was against the battery and the one with the lowest thermal conductivity was in the third layer.

Nasehi et al.[33] studied the use of three layers of PCMs for battery thermal management. The PCMs tested were capric acid, eicosane, and $Na_2(SO_4)10H_2O$. They investigated different arrangements and layer thicknesses. They compared the use of a single layer to the three-layer configuration at a room temperature of 30°C. The results showed that when the boundaries are insulated, the use of only one PCM was more efficient. However, when there was free convection at the boundaries, the use of the three-layer configuration was more efficient. They also noticed, similarly to Moraga et al. that with the three-layer configuration, placing the PCM with the highest thermal conductivity closest to the battery gave better results. Moreover, the time to regenerate the PCM was higher with a three-layer configuration.

As discussed earlier, the results are interesting when using several PCMs because as one PCM is completely melted, another one can take over and thus maintain the thermal management. However, using several PCMs requires a lot of space, which is not always available in a vehicle.

Plus, in the case of insulated boundaries, the use of only one PCM showed a higher efficiency.

Another key issue in the use of PCMs is their thermal conductivity, which is quite low for most PCMs. The use of high thermal conductivity additives to enhance the PCMs performance in battery applications has been investigated by several researchers.

### 11.3.1.3 PCMs-BASED COMPOSITES

It was noticed that when employing a layer of PCM to absorb the extra heat released by a battery, increasing the thickness of the PCM layer, which means increasing the mass of PCM, results in improved performances of the system until a certain peak point. Beyond a certain thickness, the extra PCM added becomes useless. It has been proven that it is due to the low thermal conductivity of the PCMs, which limits the performance of the system. In order to improve the performances of PCMs, composites can be elaborated to enhance the PCMs thermal conductivity, thus increasing the heat transfer rates through such materials.

PCM thermal conductivity enhancement is interesting for several applications pertaining to thermal energy storage, such as solar water heating and solar thermal power generation systems,[34, 35] thermal management of electronic equipment,[36] and thermal management of Li-ion batteries.[37] In general, PCM thermal conductivity can be enhanced via two different techniques. The first method involves embedding the PCM in a highly conductive material matrix such as a metal foam[37, 38] or in expanded graphite matrices;[39–41] the second consists in embedding in the PCM a highly conductive matrix such as carbon nanostructures.[42] In both cases, the highly conductive material employed is either metallic or carbon based.

For the first method, fins, metal, or graphite foam are usually employed. Many authors worked on this method to enhance the PCM performances. Khateeb et al.[43] replaced a lead battery by a Li-ion battery coupled with a PCM in an electric scooter. They tested two different PCMs with phase change temperatures of 12°C and 42°C for three configurations: the PCM alone, the PCM embedded in aluminum foam, and the PCM embedded in aluminum foam and with aluminum fins. The results showed that the foam and the fins are necessary to achieve a good repeatability of the cycles for three cycles. Javani et al.[44] worked on n-octadecane wax infused in foam layers and tested five samples of foam. Li et al.[45] used copper foam saturated with a commercial paraffin. A comparison was made between a system using natural air convection only, a system with PCM only, and a system the PCM embedded in copper foam. They tested five different foams with different porosities and pore densities. As a result, the battery surface temperature decreased with the decrease of the porosity at a fixed pore density while the battery surface temperature increased when the pore density increased. Rao et al.[46] used paraffin/copper foam for LiFePO$_4$ thermal management. Alipanah et al.[47] compared the performances of octadecane, gallium, and octadecane with aluminum foam with porosities from 0.88 to 0.97. When using the metal foam, the solidification time of octadecane was higher but the temperature was more uniform and the maximum temperature of the battery was lower. Al-Hallaj et al.[48] worked on a graphite matrix filled with a commercial PCM. Kizilel et al.[49] also studied a paraffin wax PCM encapsulated in a graphite matrix. Somasundaram et al.[50] worked on paraffin wax ($T_m$ = 52°C–55°C) impregnated in a graphite matrix. The implementation of this method proved easy and the thermal conductivity could be greatly enhanced, with an increase of up to 4500% of the initial PCM thermal conductivity using copper foam with a porosity of 88.89%.[51] The use of nickel foam was less effective but still resulted in an increase of 658% of the initial PCM thermal conductivity for a foam porosity of 90.61%.[51] The properties of the foam, especially its density, or that to the expanded graphite (mass fraction, density) can be changed to improve the performances of the system.[38] In the case of electrochemical battery thermal management, metal foam,[52] or carbon-based materials such as graphite matrix,[41] expanded graphite or carbon fibers,[53] are usually employed. However, heat transfer promoters such as these considerably increase the weight and volume of LHTES systems and can limit the natural convection within the PCM.

The second method involves using a powder made of nanoparticles or nanotubes. Many different nanostructures can be employed.[51, 42] This method has been less applied to battery thermal management. In 2015, Babapoor et al.[54] worked on the thermal management of a Li-ion battery using carbon fiber PCM composite. They found that the carbon fibers length and their mass fraction influence the efficiency of the heat removal. The carbon fibers have to be long enough to enable a good thermal management, but not too long otherwise there is no optimal dispersion in the PCM.

They proved that the maximum temperature difference in the system is a function of the carbon fibers length and the mass fraction. Samimi et al.[55] also evaluated the performance of carbon fibers–PCM composite for battery thermal management and found similar results. With this method, the resulting thermal conductivity increase ranges from 7% using carbon nanotubes, to 1000% using graphene nanoplatelets[42]. Zeng et al.[56] worked on the thermal conductivity enhancement of organic PCMs using Ag nanowires. Ag nanowires are one-dimensional nanostructured materials. A thermal conductivity of up to 1.46 W/(m.K) with 62.73wt.% of Ag nanowires was obtained with the composite. Carbon nanotubes have attracted a lot of interest for PCM thermal conductivity management[57–59] and have shown promising results. Adding high thermal conductivity nanoparticles can overcome the low thermal conductivity of the PCM. More examples of PCMs thermal conductivity enhancement with carbon nanostructures are shown in Table 11.3. However, the realization of the composites is not easy, one of the key issues being to ensure a homogeneous dispersion of the carbon nanostructures in the PCM.

It is also possible to combine the two methods in order to benefit from the advantages of both methods. Such configuration was tested by Situ et al.,[60] who used both expanded graphite and copper mesh. Adding the copper mesh to the PCM with expanded graphite could increase its thermal conductivity by 36%.

## 11.3.1.4   HYBRID SYSTEMS

As previously mentioned, there are several thermal management systems types for batteries, and those systems can be combined to obtain a better efficiency. PCMs can be combined with forced air convection, liquid cooling, or heat pipes. All combinations have been investigated, as detailed hereafter.

### Combination of PCM and air

The combination of PCM and forced or natural air convection is the easiest combination to operate. Several authors worked on it, there can be different strategies: using forced air convection in continuous operation or using it only when the PCM is completely melted to regenerate it. Fathabadi[74] tested an active–passive thermal management system combining PCM and natural air convection. A composite PCM was used,

**TABLE 11.3**  PCMs Thermal Conductivity Enhancement Using Carbon Nanostructures.

| PCM | Nanostructure | Nanostructure content | Thermal conductivity increase (%) | Reference |
|---|---|---|---|---|
| Paraffin wax | R (random) GNs | 5.0wt.% | 390 | [61] |
| | O (oriented) GNs | | 1210 | |
| Paraffin wax | MWCNTs | 0.6vol.% | 40 – 45 | [62] |
| Paraffin wax | SMWCNTs | 5.0wt.% | 30 | [63] |
| | LMWCNTs | | 15 | |
| | CNFs | | 15 | |
| | GNPs | | 164 | |
| Paraffin wax | S-MWCNTs | 4.0wt.% | 20 | [64] |
| | C-S-MWCNTs | | 20 | |
| | L-MWCNTs | | 7 | |
| | CNFs | | 20 | |
| | GNPs | | 93 | |
| Paraffin wax | CNFs | 10.0wt.% | 40 | [65] |
| Soy wax | CNFs | 10.0wt.% | 45 | |
| | CNTs | 10.0wt.% | 24 | |
| Paraffin wax | MWCNTs | 2.0wt.% | 35 – 45 | [66] |
| Paraffin wax | CNFs | 4.0wt.% | 45 | [67] |
| 1-Hexadecanol | MWCNTs | 3.0wt.% | 31 | [68] |
| N-octadecane | CNTs | 5.0wt.% | 48 – 66 | [69] |

**TABLE 11.3** *(Continued)*

| PCM | Nanostructure | Nanostructure content | Thermal conductivity increase (%) | Reference |
|---|---|---|---|---|
| Eicosane | GNPs | 10.0wt.% | 400 | [70] |
| Palmitic acid | Treated MWCNTs | 1.0wt.% | 24 – 50 | [71] |
| Stearic acid | MWCNTs | 1.0wt.% | 9 | [72] |
| Water | Short SWCNT | 0.48vol.% | 104 | [73] |
| | Long SWCNT | 0.48vol.% | 108 | |
| | MWCNT | 0.48vol.% | 103 | |

C–PCM, composite composed of carbon and PCM; CNF, carbon nanofiber; CNT, carbon nanotubes; GN, graphite nanotubes; GNP, graphene nanoplatelets; LMWCNT, long multiwall carbon nanotubes; MWCNT, multiwall carbon nanotubes; PCM, phase change material; SMWCNT, short multiwall carbon nanotubes; SWCNT, single-wall carbon nanotubes.

composed of paraffin wax and expanded graphite. The air velocity being 0.01 m/s, it was therefore considered as natural convection. The tests were run for a constant discharge rate at room temperatures from 20°C to 57°C. The comparisons with system without PCM, with only natural or forced air convection, showed that this hybrid system reduces the battery maximal temperature and the temperature difference inside the battery. Ling et al.[75] compared two systems, one composed of an air circuit only and a second one combining air and PCM. The PCM surrounded the cells and the air was circulated around the PCM. The PCM was a composite made of paraffin and expanded graphite. For low discharge rates, both systems were equally efficient, even if the hybrid one permitted obtaining lower temperatures. For high discharges rates, the system with the forced air convection only was not efficient enough while a minimum air speed of 3 m/s was necessary for the hybrid system.

### Combination of PCM with liquid cooling

The use of liquid cooling has proven to be more efficient that air cooling, due to the higher specific heat of water compared to air, but it is also more difficult to operate because of the risk of leakage. Different strategies can be employed with liquid cooling, as there can be direct or indirect cooling. Therefore, the configurations and geometry can vary a lot from one study to another. Hémery et al.[27] experimentally compared air cooling, to a system combining PCM with liquid cooling to regenerate the PCM. The semi-passive system was composed of two cooling plates in contact with aluminum cans with fins. The fins permitted to facilitate the PCM solidification. The liquid cooling was performed with water regulated at 22°C during the charge of the battery. They tested air velocities ranged from 1 m/s to 3 m/s. The results showed that the hybrid system greatly improves the temperature uniformity inside the battery. Rao et al.[76] numerically studied a system composed of a PCM and mini channels of fluid. They tested different PCM and liquid physical properties. They noticed that the hybrid system is more performant as it permits obtaining a lower temperature inside the battery. Coleman et al.[77] compared two heat sink materials combined with indirect liquid cooling. The materials were a solid aluminum block and a copper/wax composite PCM. They tested the configuration for three different cell spacing (1 mm, 5 mm, and 10 mm). The results showed that in compact design, the quantity of PCM is reduced and is not enough to dissipate heat during single cell failure. The aluminum block showed good performances for all loads and during a failure, the heat was quickly dissipated into the coolant.

### Combination of PCM and heat pipe

Wang et al.[78] worked on the combination of PCM and an oscillating heat pipe. The PCM employed was paraffin and the working fluid of the heat pipe was acetone. The room temperature was 25°C. They compared the performances of the oscillating pipe with and without PCM. The results showed that the system with PCM is more efficient as it permits having a lower average temperature and a more uniform temperature inside the battery surrogate.

Thus, it proves that the utilization of PCM is interesting as battery thermal management. Moreover, it can be improved by several ways: employing several PCMs, creating a PCM-based composite, or adding an active system to obtain a semi-passive system. Some systems based on PCMs are already commercialized, for example, by AllCell®. PCMs are still widely studied to regulate the temperature of the cars battery, but they can also be used to regulate the temperature of the passenger compartment. In fact, the VEGETO project in France aimed to employ PCMs for the thermal regulation of the battery and for the regulation of the air temperature inside the car.

## 11.4 PCMs FOR THERMAL COMFORT IN VEHICLES

In electric vehicles as well as in thermal motor vehicles, there is a necessity for reducing the energy consumption. In the first case, the aim is to preserve the autonomy of the battery whereas in the second case, the aim is to reduce the emissions of noxious gases. Moreover, with the widespread of the Start and Stop function in most vehicles, it is necessary to have air conditioning operation that does not depend on a working engine, otherwise the air conditioning would not be maintained when the vehicles stops at traffic signals. When the car is parked outdoor in summer, the indoor temperature increases a lot, and it can lead to the thermal discomfort of the passengers getting inside the car. Moreover, thermal regulation in vehicles is at most times operated by forced air systems that can be unsatisfactory and ineffective, especially when the car is already heated up. Therefore, the use of PCMs as a passive system is interesting to keep the temperature of the passenger compartment in a given range. The PCM storage capacity is recharged while driving or while charging the battery in electrical vehicles. Some studies worked on the integration of PCM in the car structure to improve vehicles thermal inertia while other studies worked on the development of thermal systems (batteries) filled with PCM for vehicles indoor air conditioning.

### 11.4.1  *PCMs INTEGRATED IN THE VEHICLE*

Purusothaman et al.[79] added a layer of PCM in the roof of a vehicle. The addition of the PCM led to up to 30% of energy savings. Saleel et al.[80] used coconut oil as PCM for the vehicle thermal comfort. The space between the rooftop and ceiling of the car was filled with foam and coconut oil pouches. The use of coconut oil as PCM could reduce by about 15°C the cabin temperature, compared to the case without PCM. Oró et al.[81] tested the use of PCM encapsulated in aluminum plates placed under the roof of the car, as well as PCM encapsulated in a plastic bag was placed in the steering wheel. The results of their experiments showed a significant decrease in air temperature and in the steering wheel surface temperature after a car park of 2 h. Some patents have been developed on the integration of PCMs in the car passenger compartment[82]. PCMs can be easily integrated in the vehicles to regulate passively the temperature of the vehicle.

### 11.4.2  *PCMs INTEGRATED IN THERMAL SYSTEMS*

Several types of thermal batteries are developed for the vehicle's thermal comfort. Usually, they are based on heat exchangers with liquids or air. However, those systems consume energy and can be unsatisfactory. The integration of the PCM in a heat battery can permit saving energy. Taylor et al.[83] developed a thermal battery based on PCM with a fluid in a closed loop as heat exchanger. A recent study developed a heat battery with PCMs particles in a fixed bed to regulate the temperature in the vehicle.[84] Lee[85] developed a cold storage heat exchanger using PCM to provide cold air in the vehicle when it stops. This system could delay the temperature rise in the passenger compartment when the vehicle stopped. Some companies such as Sunamp or HM Heizkörper, which developed and now commercialize heat battery technologies based on PCM.

The different studies mentioned earlier have shown that there are different ways to use PCMs to improve the thermal comfort in vehicles. They can be used for auxiliary heating or cooling during short stops or they can be integrated in the vehicle to entirely replace the air conditioning systems. The easiest systems are the one which just include PCM pouches in the roof, the wheel or the seats of a car as such systems are completely passive; however, in that case the regeneration of the PCM can be difficult to achieve.

## 11.5 CONCLUSION

PCMs can be widely used in the automobile sector. In fact, they are of great interest to perform the battery thermal management or the thermal comfort of the vehicle cabin.

Li-ion batteries cannot be used without a thermal management system. Keeping the battery temperature in a given range can prevent it from thermal runaways and from early degradations. Current systems are not completely satisfactory, as there are complex, energy demanding, and cannot keep the battery temperature under 35°C. Therefore, the use of PCMs in passive or semi-passive thermal management systems can be relevant. Many studies investigated the integration of PCMs in the battery thermal management systems. The key issues of their application remain the regeneration of the PCM and their low thermal conductivity. Several improvements can then be made: multiple PCMs can be used simultaneously, the PCM thermal conductivity can be enhanced by elaborating composites, or PCM can be combined with an active system such as air cooling, liquid cooling, or heat pipes to regenerate the PCM. The systems employing PCMs give good results, especially in terms of battery maximal temperature and in terms of temperature uniformity inside the battery stack.

PCMs can also be used in the car's passenger compartment for thermal comfort. There are more studies on the use of PCMs in battery thermal management systems than for the thermal comfort in vehicles. However, the use of PCMs in the passenger compartment as an additional system to the air conditioning that can keep functioning even when the engine is off, is an interesting solution for the car thermal comfort and to reduce the vehicle's energy consummation.

PCMs are interesting materials for cooling applications because of their high latent heat of phase change and their ease of use. They have proved a promising technology to be employed in various domains, especially in the automobile sector.

## KEYWORDS

- **phase change materials**
- **battery thermal management**
- **electric vehicles**
- **Li-ion batteries**
- **battery cells**

## REFERENCES

1. Jaguemont, J.; Omar, N.; Van den Bossche P.; Mierlo, J. Phase Change Materials (PCM) for Automotive Applications: A Review. *Appl. Therm. Eng.* **2018,** *132*, 308–320.

2. Kuznik, F.; David, D.; Johannes, K.; Roux, J. -J. A Review on Phase Change Materials Integrated in Building Walls. *Renew. Sustain. Energy Rev.* **2011,** *15* (1), 379–391.

3. Mondal, S. Phase Change Materials for Smart Textiles–An Overview. *Appl. Therm. Eng.* **2008,** *28* (11–12), 1536–1550.

4. Priego Quesada, J. I.; Gil-Calvo, M.; Lucas-Cuevas, A. G.; Aparicio, I.; Pérez-Soriano, P. Assessment of a Mattress with Phase Change Materials Using a Thermal and Perception Test. *Exp. Therm. Fluid Sci.* **2017,** *81*, 358–363.

5. Oró, E.; Miró, L. M.; Farid, M.; Cabeza, L. F. Improving Thermal Performance of Freezers Using Phase Change Materials. *Int. J. Refrig.* **2012,** *35* (4), 984–991.

6. Gin, B.; Farid, M. M. The Use of PCM Panels to Improve Storage Condition of Frozen Food. *J. Food Eng.* **2010,** *100* (2), 372–376.

7. Oró, E.; de Gracia, A.; Castell, A.; Farid, M. M.; Cabeza, L. F. Review on Phase Change Materials (PCMs) for Cold Thermal Energy Storage Applications. *Appl. Energy* **2012,** *99*, 513–533.

8. Kant, K.; Shukla, A.; Sharma, A.; Biwole, P. H. Heat Transfer Studies of Photovoltaic Panel Coupled with Phase Change Materia. *Sol. Energy* **2016,** *140* (Suppl. C), 151–161.

9. Mazman, M.; Cabeza, L. F.; Mehling, H.; Nogues, M.; Evliya, H.; Paksoy, H. Ö. Utilization of Phase Change Materials in Solar Domestic Hot Water Systems. *Renew. Energy* **2009,** *34* (6), 1639–1643.

10. Vadhera, J.; Sura, A.; Nandan, G.; Dwivedi, G. Study of Phase Change Materials and Its Domestic Application. *Mater. Today Proc.* **2018,** *5* (2), Part 1, 3411–3417.

11. Hémery, C.-V. Etudes des phénomènes thermiques dans les batteries Li-ion. , phdthesis, Université de Grenoble, 2013.

12. Ianniciello, L.; Biwolé, P. H.; Achard, P. Electric Vehicles Batteries Thermal Management Systems Employing Phase Change Materials. *J. Power Sources* **2018,** *378,* 383–403.

13. Badey, Q. Étude des mécanismes et modélisation du vieillissement des batteries lithium-ion dans le cadre d'un usage automobile , PhD thesis, Université Paris Sud-Paris XI, 2012.

14. Thomas, E. V.; Case, H. L.; Doughty, D. H.; Jungst, R. G.; Nagasubramanian, G.; Roth, E. P. Accelerated Power Degradation of Li-ion Cells. *J. Power Sources* **2003,** *24* (1), 254–260.

15. Aurbach, D. A Review on New Solutions, New Measurements Procedures and New Materials for Rechargeable Li Batteries. *J. Power Sources* **2005,** *146* (1–2), 71–78.

16. M. Broussely et al. Main Aging Mechanisms in Li-ion Batteries. *J. Power Sources* **2005,** *146* (1–2), 90–96.

17. Bandhauer, T. M.; Garimella, S.; Fuller, T. F. A Critical Review of Thermal Issues in Lithium-Ion Batteries. *J. Electrochem. Soc.* **2011,** *158* (3), R1–R25.

18. Vetter, J. *et al.* Ageing Mechanisms in Lithium-ion Batteries. *J. Power Sources* **2005,** *147* (1–2), 269–281.

19. Barré, A.; Deguilhem, B.; Grolleau, S.; Gérard, M.; Suard, F.; Riu, D. A Review on Lithium-ion Battery Ageing Mechanisms and Estimations for Automotive Applications. *J. Power Sources* **2013,** *241,* 680–689.

20. Lamb, J.; Orendorff, C. J.; Steele, L. A. M.; Spangler, S. W. Failure Propagation in Multi-cell Lithium-ion Batteries. *J. Power Sources* **2015,** *283,* 517–523.

21. Doughty, D.; Roth, E. P. A General Discussion of Li-Ion Battery Safety. *Electrochem. Soc. Interface* **2012,** *21* (2), 37–44.

22. Abraham, D. P.; Roth, E. P.; Kostecki, R.; McCarthy, K.; MacLaren, S.; Doughty, D. H. Diagnostic Examination of Thermally Abused High-Power Lithium-ion Cells. *J. Power Sources* **2006,** *161* (1), 648–657.

23. Senyshyn, A.; Mühlbauer, M. J.; Dolotko, O.; Ehrenberg, H. Low-temperature Performance of Li-ion Batteries: The Behavior of Lithiated Graphite. *J. Power Sources* **2015,** *282,* 235–240.

24. Zhang, S. S.; Xu, K.; Jow, T. R. Electrochemical Impedance Study on the Low Temperature of Li-ion Batteries. *Electrochimica Acta* **2004,** *49* (7), 1057–1061.

25. Ahmad, Pesaran, Tools for Designing Thermal Management of Batteries In Electric Drive Vehicles, présenté à Large Lithium Ion Battery Technology & Application Symposia Advanced Automotive Battery Conference, Pasadena, CA, 04-févr-2013.

26. Mohammadian, S. K.; He, Y. -L.; Zhang, Y. Internal Cooling of a Lithium-ion Battery Using Electrolyte as Coolant Through Microchannels Embedded Inside the Electrodes. *J. Power Sources* **2015,** *293,* 458–466.

27. Hémery, C. -V.; Pra, F.; Robin, J. -F.; Marty, P. Experimental Performances of a Battery Thermal Management System Using a Phase Change Material. *J. Power Sources* 2014, *270,* 349–358.

28. Fallahi, A.; Guldentops, G.; Tao, M.; Granados-Focil, S.; Van Dessel, S. Review on Solid-Solid Phase Change Materials for Thermal Energy Storage: Molecular Structure and Thermal Properties. *Appl. Therm. Eng.* **2017,** *127,* 1427–1441.

29. Hallaj, S. A.; Selman, J. R. A Novel Thermal Management System for Electric Vehicle Batteries Using Phase-Change Material. *J. Electrochem. Soc.* **2000,** *147* (9), 3231–3236.

30. Javani, N.; Dincer, I.; Naterer, G. F.; Yilbas, B. S. Heat Transfer and Thermal Management with PCMs in a Li-ion Battery Cell for Electric Vehicles. *Int. J. Heat Mass Transf.* **2014,** *72,* 690–703.

31. Ramandi, M. Y.; Dincer, I.; Naterer, G. F. Heat Transfer and Thermal Management of Electric Vehicle Batteries with Phase Change Materials. *Heat Mass Transf.* **2011,** *47* (7), 777–788.

32. Moraga, N. O.; Xamán, J. P.; Araya, R. H. Cooling Li-ion Batteries of Racing Solar Car by Using Multiple Phase Change Materials. *Appl. Therm. Eng.* **2016,** *108,* 1041–1054.

33. Nasehi, R.; Alamatsaz, A.; Salimpour, M. Using Multi-Shell Phase Change Materials Layers for Cooling a Lithium-ion Battery. *Research Gate* **2016,** *20* (2), 391–403.

34. Lin, Y.; Jia, Y.; Alva, G.; Fang, G. Review on Thermal Conductivity Enhancement, Thermal Properties and Applications of Phase Change Materials in Thermal Energy Storage. *Renew. Sustain. Energy Rev.* **2017.** doi: 10.1016/j.rser.2017.10.002.

35. Huang, X.; Lin, Y.; Alva, G.; Fang, G. Thermal Properties and Thermal Conductivity Enhancement of Composite Phase Change Materials Using Myristyl Alcohol/Metal Foam for Solar Thermal Storage. *Sol. Energy Mater. Sol. Cells* **2017,** *170* (Suppl. C), 68–76.

36. Krishna, J.; Kishore, P. S.; Solomon, A. B. Heat Pipe with Nano Enhanced-PCM for Electronic Cooling Application. *Exp. Therm. Fluid Sci.* **2017,** *81* (Suppl. C), 84–92.

37. Wang, Z.; Zhang, Z.; Jia, L.; Yang, L. Paraffin and Paraffin/Aluminum Foam Composite Phase Change Material Heat Storage Experimental Study Based on Thermal Management of Li-ion Battery. *Appl. Therm. Eng.* **2015,** *78* (Suppl. C), 428–436.

38. Hong S. -T.; Herling, D. R. Open-Cell Aluminum Foams Filled with Phase Change Materials as Compact Heat Sinks. *Scr. Mater.* **2006,** *55* (10), 887–890.

39. Karaipekli, A.; Sarı, A.; Kaygusuz, K. Thermal Conductivity Improvement of Stearic Acid Using Expanded Graphite and Carbon Fiber for Energy Storage Applications.

40. Sarı, A.; Karaipekli, A. Thermal Conductivity and Latent Heat Thermal Energy Storage Characteristics of Paraffin/Expanded Graphite Composite as Phase Change Material. *Appl. Therm. Eng.* **2007,** *27* (8–9), 1271–1277.

41. Mills, A.; Farid, M.; Selman, J. R.; Al-Hallaj, S. Thermal Conductivity Enhancement of Phase Change Materials Using a Graphite Matrix. *Appl. Therm. Eng.* **2006,** *26* (14–15), 1652–1661.

42. Amaral, C.; Vicente, R.; Marques, P. A. A. P.; Barros-Timmons, A. Phase Change Materials and Carbon Nanostructures for Thermal Energy Storage: A Literature Review. *Renew. Sustain. Energy Rev.* **2017,** *79* (Suppl. C), 1212–1228.

43. Siddique, S. A.; Khateeb, A. Thermal Management of Li-ion Battery with Phase Change Material for Electric Scooters: Experimental Validation. *J. Power Sources* **2005,** *142* (1), 345–353.

44. Javani, N.; Dincer, I.; Naterer, G. F.; Rohrauer, G. L. Modeling of Passive Thermal Management for Electric Vehicle Battery Packs with PCM Between Cells. *Appl. Therm. Eng.* **2014,** *73* (1), 307–316.

45. Li, W. Q.; Qu, Z. G.; He, Y. L.; Tao, Y. B. Experimental Study of a Passive Thermal Management System for High-Powered Lithium-ion Batteries Using Porous Metal Foam Saturated with Phase Change Materials. *J. Power Sources* **2014**, *255*, 9–15.

46. Rao, Z.; Huo, Y.; Liu, X.; Zhang, G. Experimental Investigation of Battery Thermal Management System for Electric Vehicle Based on Paraffin/Copper Foam. *J. Energy Inst.* **2015**, *88* (3), 241–246.

47. Alipanah, M.; Li, X. Numerical Studies of Lithium-ion Battery Thermal Management Systems Using Phase Change Materials and Metal Foams. *Int. J. Heat Mass Transf.* **2016**, *102*, 1159–1168.

48. Al-Hallaj, S.; Kizilel, R.; Lateef, A.; Sabbah, R.; Farid, M.; Selman, J. R. Passive Thermal Management Using Phase Change Material (PCM) for EV and HEV Li- ion Batteries, *2005 IEEE Vehicle Power and Propulsion Conference*, 2005, p. 5.

49. Kizilel, R.; Sabbah, R.; Selman, J. R.; Al-Hallaj, S. An Alternative Cooling System to Enhance the Safety of Li-ion Battery Packs. *J. Power Sources* **2009**, *194* (2), 1105–1112.

50. Somasundaram, K.; Birgersson, E.; Mujumdar, A. S. Thermal–Electrochemical Model for Passive Thermal Management of a Spiral-Wound Lithium-ion battery. *J. Power Sources* **2012**, *203*, 84–96.

51. Liu, L.; Su, D.; Tang, Y.; Fang, G. Thermal Conductivity Enhancement of Phase Change Materials for Thermal Energy Storage: A Review. *Renew. Sustain. Energy Rev.* **2016**, *2* (Suppl. C), 305–317.

52. Alipanah, M.; Li, X. Numerical Studies of Lithium-ion Battery Thermal Management Systems Using Phase Change Materials and Metal Foams. *Int. J. Heat Mass Transf.* **2016**, *102*, 1159–1168.

53. Goli, P.; Legedza, S.; Dhar, A.; Salgado, R.; Renteria, J.; Balandin, A. A. Graphene-Enhanced Hybrid Phase Change Materials for Thermal Management of Li-ion Batteries. *J. Power Sources* **2014**, *248*, 37–43.

54. Babapoor, A.; Azizi, M.; Karimi, G. Thermal Management of a Li-ion Battery Using Carbon Fiber-PCM Composites. *Appl. Therm. Eng.* **2015**, *82*, 281–290.

55. Samimi, F.; Babapoor, A.; Azizi, M.; Karimi, G. Thermal Management Analysis of a Li-ion Battery Cell Using Phase Change Material Loaded with Carbon Fibers. *Energy* **2016**, *96*, 355–371.

56. Zeng, J. L.; Cao, Z.; Yang, D. W.; Sun, L. X.; Zhang, L. Thermal Conductivity Enhancement of Ag Nanowires on An Organic Phase Change Material. *J. Therm. Anal. Calorim.* **2009**, *101* (1), 385–389.

57. Frusteri, F.; Leonardi, V.; Vasta, S.; Restuccia, G. Thermal Conductivity Measurement of a PCM Based Storage System Containing Carbon Fibers. *Appl. Therm. Eng.* **2005**, *25* (11–12), 1623–1633.

58. Fukai, J.; Kanou, M.; Kodama, Y.; Miyatake, O. Thermal Conductivity Enhancement of Energy Storage Media Using Carbon Fibers. *Energy Convers. Manag.* **2000**, *41* (14), 1543–1556.

59. Fukai, J.; Hamada, Y.; Morozumi, Y.; Miyatake, O. Improvement of Thermal Characteristics of Latent Heat Thermal Energy Storage Units Using Carbon-Fiber Brushes: Experiments and Modeling. *Int. J. Heat Mass Transf.* **2003**, *46* (23), 4513–4525.

60. Situ, W. *et al.*; A Thermal Management System for Rectangular LiFePO4 Battery Module Using Novel Double Copper Mesh-Enhanced Phase Change Material Plates. *Energy* **2017**, *141*, 613–623.

61. Chen, Y. -J.; Nguyen, D. -D.; Shen, M. -Y.; Yip, M. -C.; Tai, N. -H. Thermal Charac-
    terizations of the Graphite Nanosheets Reinforced Paraffin Phase-Change Composites.
    *Compos. Part Appl. Sci. Manuf.* **2013,** *44* (Suppl. C), 40–46.
62. Kumaresan, V.; Velraj, R.; Das, S. K. The Effect of Carbon Nanotubes in Enhancing
    the Thermal Transport Properties of PCM During Solidification. *Heat Mass Transf.*
    **2012,** *48* (8), 1345–1355.
63. Raam Dheep, G.; Sreekumar, A. Influence of Nanomaterials on Properties of Latent
    Heat Solar Thermal Energy Storage Materials – A Review. *Energy Convers. Manag.*
    **2014,** *83* (Suppl. C), 133–148.
64. Yu, Z.-T. et al. Increased Thermal Conductivity of Liquid Paraffin-Based Suspensions
    in the Presence of Carbon Nano-Additives of Various Sizes and Shapes. *Carbon*
    **2013,** *53* (Suppl. C), 277–285.
65. Cui, Y.; Liu, C.; Hu, S.; Yu, X. The Experimental Exploration of Carbon Nanofiber
    and Carbon Nanotube Additives on Thermal Behavior of Phase Change Materials.
    *Sol. Energy Mater. Sol. Cells* **2011,** *95* (4), 1208–1212.
66. Wang, J.; Xie, H.; Xin, Z. Thermal Properties of Paraffin Based Composites
    Containing Multi-Walled Carbon Nanotubes. *Thermochim. Acta* **2009,** *488* (1), p.
    39-42, mai.
67. Elgafy, A.; Lafdi, K. Effect of Carbon Nanofiber Additives on Thermal Behavior of
    Phase Change Materials. *Carbon* **2005,** *43* (15), 3067–3074.
68. Fan, L. -W. et al. Transient Performance of a PCM-Based Heat Sink with High
    Aspect-Ratio Carbon Nanofillers. *Appl. Therm. Eng.* **2015,** *75* (Suppl. C), 532–540.
69. Babaei, H.; Keblinski, P.; Khodadadi, J. M. Thermal Conductivity Enhancement of
    Paraffins by Increasing the Alignment of Molecules through Adding CNT/Graphene.
    *Int. J. Heat Mass Transf.* **2013,** *58* (1), 209–216.
70. Fang, X. et al.; Increased Thermal Conductivity of Eicosane-Based Composite Phase
    Change Materials in the Presence of Graphene Nanoplatelets. *Energy Fuels* **2013,** *27*
    (7), 4041–4047.
71. Wang, J.; Xie, H.; Xin, Z.; Li, Y. Increasing the Thermal Conductivity of Palmitic
    Acid by the Addition of Carbon Nanotubes. *Carbon* **2010,** *48* (14), 3979–3986.
72. Li, T.; Lee, J. -H.; Wang, R.; Kang, Y. T. Enhancement of Heat Transfer for Thermal
    Energy Storage Application Using Stearic Acid Nanocomposite with Multi-Walled
    Carbon Nanotubes. *Energy* **2013,** *55* (Suppl. C), 752–761.
73. Xing, M.; Yu, J.; Wang, R. Experimental Study on the Thermal Conductivity Enhance-
    ment of Water Based Nanofluids using Different Types of Carbon Nanotubes. *Int. J.
    Heat Mass Transf.* **2015,** *88* (Suppl. C), 609–616.
74. Fathabadi, H. High Thermal Performance Lithium-ion Battery Pack Including Hybrid
    Active–Passive Thermal Management System for Using in Hybrid/Electric Vehicles.
    *Energy* **2014,** *70*, 529–538.
75. Ling, Z.; Wang, F.; Fang, X.; Gao, X.; Zhang, Z. A Hybrid Thermal Management
    System for Lithium-ion Batteries Combining Phase Change Materials with Forced-Air
    Cooling. *Appl. Energy* **2015,** *148*, 403–409.
76. Rao, Z.; Wang, Q.; Huang, C. Investigation of the Thermal Performance of Phase
    Change Material/Mini-Channel Coupled Battery Thermal Management System. *Appl.
    Energy* **2016,** *164*, 659–669.

77. Coleman, B.; Ostanek, J.; Heinzel, J. Reducing Cell-to-Cell Spacing for Large-Format Lithium-ion Battery Modules with Aluminum or PCM Heat Sinks under Failure Conditions. *Appl. Energy* **2016**, *180*, 14–26.

78. Wang, Q.; Rao, Z.; Huo, Y.; Wang, S. Thermal Performance of Phase Change Material/ Oscillating Heat Pipe-Based Battery Thermal Management System. *Int. J. Therm. Sci.* **2016**, *102*, 9–16.

79. Purusothaman, M.; kota, S.; Cornilius, C. S.; Siva, R. Experimental Investigation of Thermal Performance in a Vehicle Cabin Test Setup With Pcm in the Roof. *IOP Conf. Ser. Mater. Sci. Eng.* **2017**, *197*, 012073.

80. Saleel, C. A.; Mujeebu, M. A.; Algarni, S. Coconut Oil as Phase Change Material to Maintain Thermal Comfort in Passenger Vehicles. *J. Therm. Anal. Calorim.* **2018**, *136* (2), 629–636.

81. Oró, E.; de Jong, E., Cabeza, L. F. Experimental Analysis of a Car Incorporating Phase Change Material. *J. Energy Storage* **2016**, *7*, 131–135.

82. Pause, B. H. Thermal Control of Automotive Interiors with Phase Change Material, US7320357B2, 22-janv-2008.

83. Taylor, R. A.; Chung, C.-Y.; Morrison, K.; Hawkes, E. R. Analysis and Testing of a Portable Thermal Battery. *J. Therm. Sci. Eng. Appl.* **2014**, *6* (3), 031004–031004–8.

84. Osipian, R. Etude dynamique d'un système de stockage par chaleur latente liquide-solide: application au véhicule électrique, 2018.

85. Lee, D. W. Experimental Study on Performance Characteristics of Cold Storage Heat Exchanger for ISG Vehicle. *Int. J. Automot. Technol.* **2017**, *18* (1), 41–48.

86. Duan, X.; Naterer, G. F. Heat Transfer in Phase Change Materials for Thermal Management of Electric Vehicle Battery Modules. *Int. J. Heat Mass Transf.* **2010**, *53* (23–24), 5176–5182. doi: 10.1016/j.ijheatmasstransfer.2010.07.044.

# CHAPTER 12

# Herbal Materials as Thermal Energy Storage Materials

ABHAY KUMAR CHOUBEY* and VARTIKA SRIVASTAVA

*Division of Sciences & Humanities, Rajiv Gandhi Institute of Petroleum Technology, Jais, Amethi 229304, India*

*Corresponding author. E-mail: achoubey@rgipt.ac.in*

## ABSTRACT

The most productive way of storing thermal energy is through phase change materials. The ability of these materials to store and release high amount of energy makes these materials suitable to be used as energy storage devices. There is a range of materials which can be classified as phase change materials. This chapter reviews the work on herbal materials being reported as phase change materials and are used as energy storage devices. Natural materials such as fatty alcohols, fatty acids and its derivative, polyethylene glycols, waxes, etc. have been reported to be used as energy storage device. These phytochemicals have the ability to absorb and release large amount of energy due to continuous change in the phase of the materials. This paper summarizes the phytochemicals along with their natural sources which can be considered as an alternative to the currently known phase change materials.

## 12.1 INTRODUCTION

Phase change materials (PCMs) are matters having high heat of fusion. They have a tendency to melt and congeal at a certain temperature and have the potential of storing and releasing heavy amount of energy. Latent heat thermal energy storage (TES) and release by means of PCM is an efficient way to lower energy consumption and is a measure toward more sustainable society.[1] Material having high heat of fusion and phase change behavioral property softens and hardens at certain temperature. This material has ability of storing and releasing enormous amount of energy. When PCM transforms its state from solid to liquid and liquid to solid, heat is absorbed or released. Naturally occurring and commonly available PCM is water. It transits from one phase to another at a temperature of around 0°C and during these transitions too, much energy is stored and released.

Previously lot of single materials and mixture of two or more have been examined and evaluated for their potential as PCM. In recent time, they are being used because of their valuable applications in the storage of thermal energy. As far as TES density of PCMs is concerned, PCMs provide comparatively higher TES density than sensible thermal materials. Thus, they have been found to be widely used in a variety of fields such as solar energy utilization,[2] waste heat recovery,[3] building air conditioning,[4] electric energy storage,[5] temperature control of greenhouses,[6-8] telecommunications and microprocessor equipment,[9] kitchen utensils,[10] insulating clothing and textiles for thermal comfort applications,[11] biomedical and biological-carrying systems,[12] and food transport and storage containers.[13] Latent heat storage is dependent on the absorption or release of heat, when a storage material undergoes a phase change from solid to liquid, liquid to gas, solid to gas, or liquid to gas, and vice versa. The most commonly used latent heat storage systems undergo solid–liquid phase transitions due to large heat storage density and small volume change between phases. These types of materials are often categorized as inorganics, organics, and eutectic PCMs, which have been studied well and reported in a few review papers.[3,9,13–16] TES systems correct the disparity between the supply and demand of energy. Sensible heat storage materials are cluster of materials that undergo no change of phase in the temperature range of the storage process. The ability to store sensible heat for a given material strongly depends on the value of its energy density that is the heat capacity per unit volume. Material, which has to be used as TES material, must be reasonably priced and should have good thermal conductivity. Storage systems based on PCMs with solid–liquid transition are considered to be an efficient alternative to sensible thermal storage systems. From an energy-efficiency point of view, PCM storage systems have the advantage that they operate with small temperature differences between charging and discharging. Furthermore, these storages have high energy densities as compared to sensible heat storages.

## 12.2 EXISTING PCMs

Selection of a PCM is generally done with the physical requirements like phase change temperature, enthalpy, cycling stability, and subcooling. An appropriate phase change temperature and a large melting enthalpy are two evident requirements for a PCM. If we think of temperature ranges for different materials, inorganic materials cover a wide range of temperature. In comparison to organic materials, inorganic materials have similar melting enthalpies per mass, but higher ones per volume due to their high density. Commonly used phase change inorganic materials include sodium sulphate, sodium chloride, potassium nitrate, sodium silicate pentahydrate, etc. Organic PCMs are paraffins, fatty acids, sugar alcohols, polyehthylene glycols, etc. Resins also come under this category. These material classes cover the temperature range between 0 to 200°C approx.

There are several other classification schemes that may be applied such as organic/inorganic, trigger of the phase change, etc. However, these schemes are of little significance for a new application. The stimuli suitable for inducing a phase change within the defined boundary conditions are changes in temperature, light, concentration of solvents, pH, electric field strength, mechanical strain, and the concentration of certain ions (not necessarily influencing the pH value). Some PCMs are activated by multiple stimuli. The number of existing materials is more than 150,000 with more every year. Materials are classified in four major classes, which are metals and alloys, ceramics and glasses, polymers and elastomers, and hybrids that include composites and natural materials too.

## 12.3 HERBAL PCMs

Because of presence of covalent bond in organic materials, most of the PCMs are not stable at higher temperature. Among all reported organic materials, only paraffins show good storage density with respect to mass and melt and solidify congruently with little or no subcooling. Their thermal conductivity is comparatively low. In most cases, the density of organic PCM is less than 103 kg/m³, and thus smaller than the density of most inorganic materials like water and salt hydrates. The result is organic materials that usually have smaller melting enthalpies per volume than inorganic materials with an exception of sugar alcohols. Their main disadvantage is material compatibility with metals since severe corrosion has been found to be developed in some PCM-metal combinations. Some natural materials, which behave as herbal PCM, are discussed in the following section.

## 12.4   NATURAL MATERIALS THAT UNDERGO PHASE CHANGE

### 12.4.1   WAX

In general, waxes are specific, natural, or synthetic mixtures of alkenes, fatty alcohols, sterols, fatty acids, and esters.[17] The chemical composition of waxes varies widely. Thus, the definition of wax refers to the physical properties of the substance. Basically it states that waxes are substances, which melt between 43.3°C and 65.5°C. They are more or less translucent having crystalline structure and are greasy in nature.[18] However, because of the wide range of animals and plants that are able to produce waxes, this list is seems incomplete. A more comprehensive list of natural waxes can be found in the work carried out by Warth.[19] In general, waxes can be considered as renewable resource. When using natural waxes, one needs to consider that individual samples differ in quality, composition, and purity. Thus, big variations of the physical properties, such as the melting temperature, are usually observed. These variations may negatively influence controllability.

### 12.4.2   FATTY ALCOHOLS, FATTY ACIDS, AND FATTY ACID DERIVATIVES

The presence of fatty alcohols, fatty acids, and fatty acid derivatives in herbal PCMs in leaves of the plants has been extensively investigated and reported (Tables 12.1–12.3). Fatty alcohols are basically aliphatic alcohols, which occur naturally in free form (component of cuticular lipids) but more usually in esterified or etherified form. Besides paraffins, fatty alcohols, fatty acids, and fatty acid esters belong to the principal constituents of natural waxes. As they are produced by animals or plants, they belong to

**TABLE 12.1** Fatty Acids in Oils and Fats.

| Materials | Molecular Formula | Melting Point in °C | Origin |
|---|---|---|---|
| Capric acid | $CH_3(CH_2)_8COOH$ | 31.6 | Coconut oil (5–10%), palm kernel oil (3–6%), and mammal's milk fat |
| Lauric acid | $CH_3(CH_2)_{10}COOH$ | 43.2 | Coconut oil (45–53%), palm kernel oil (40–52%), and human breast milk (6.2% of total fat) |
| Myristic acid | $CH_3(CH_2)_{12}COOH$ | 54.4 | Coconut oil (15–21%), palm kernel oil (14–18%), and animal fats |
| Plamitic acid | $CH_3(CH_2)_{14}COOH$ | 61–62.5 | Palm oil (38–48%), cotton seed oil (17–29%), butter, cheese, milk, and meat |
| Stearic acid | $CH_3(CH_2)_{16}COOH$ | 67–69 | More abundant in animal fat (up to 30%) than in vegetable fat (typically <5%) Cocoa butter and shea butter (28–45%) (as a triglyceride) |
| Caprylic acid | $CH_3(CH_2)_6COOH$ | 15–17 | Milk of various mammals, and as a minor constituent in coconut oil and palm kernel oil |

**TABLE 12.2** Fatty Alcohols in Herbal Materials.

| Fatty Alcohol | Molecular Formula | Melting Point in °C | Origin |
|---|---|---|---|
| 1-Tetradecanol [24] | $CH_3(CH_2)_{12}CH_2OH$ | 35–39 | Red cabbage (*Brassica oleracea var. capitata f*),[25] coriander |
| 1-Pentadecanol | $CH_3(CH_2)_{13}CH_2OH$ | 41–44[32] | Fruit of *Elaeocarpus serratus* L[26] |
| 1-Hexadecanol | $CH_3(CH_2)_{14}CH_2OH$ | 48–50[33] | Leaves and seeds of *Moringa Oleifera*[27] |
| 1-Heptadecanol | $CH_3(CH_2)_{15}CH_2OH$ | 51–55[34] | *Calotropis procera* Ait, an Asclepiadeae, a drought resistant, salt-tolerant weed[28] |
| 1-Octadecanol | $CH_3(CH_2)_{16}CH_2OH$ | 56–59[35] | *Thesium humile* Vahl herb[29] |
| 1-Nonadecanol | $CH_3(CH_2)_{17}CH_2OH$ | 60–61[36] | Flowers of *Tecomastans* Linn popularly known as Yellow bell flowers[30] |
| 1-Eicosanol | $CH_3(CH_2)_{18}CH_2OH$ | 62–65[37] | Leaves of *Melia azedarach*[31] |
| 1-Docosanol | $CH_3(CH_2)_{20}CH_2OH$ | 65–72[38] | Evening primrose oil |
| 1-Tetracosanol | $CH_3(CH_2)_{22}CH_2OH$ | 75[39] | Evening primrose oil |
| 1-Hexacosanol | $CH_3(CH_2)_{24}CH_2OH$ | 79–81[40] | Evening primrose oil |
| 1-Octcosanol | $CH_3(CH_2)_{26}CH_2OH$ | 83[41] | Evening primrose oil |
| 1-Triacontanol | $CH_3(CH_2)_{28}CH_2OH$ | 86–87[42] | Grape berry |

**TABLE 12.3**   Fatty Acid Esters in Leaves and Algae.

| Fatty Acid Esters | Molecular Formula | Melting Point in °C | Origin |
|---|---|---|---|
| Glycerol tripalmitate | $[CH_3(CH_2)_{14}COOCH_2]_2CHOCO(CH_2)_{14}CH_3$ | 66.5 | Onobrychis viciifolia (Sainfoin) |
| Glycerol tritetradecanoate | $[CH_3(CH_2)_{12}COOCH_2]_2CHOCO(CH_2)_{12}CH_3$ | 58.5 | Betulaceae plant of The Birch family |
| Ethyl palmitate | $CH_3(CH_2)_{13}CH_2COOCH_2CH_3$ | 24 | Leaf of *P. foetida* L. |
| Ethyl stearate | $CH_3(CH_2)_{15}CH_2COOCH_2CH_3$ | 33 | Alga *Spirogyra Longata* |
| Isopropyl stearate | $CH_3(CH_2)_{16}COOCH(CH_3)_2$ | 28 | *Indoneesiella echioides* (L) Nees |

the small group of PCMs that can be regarded as renewable.[20] The usage of these materials in micro-electro mechanical systems has so far been limited to drug delivery systems. Especially the fatty alcohol 1-Tetra-decanol has often been investigated in this context. Fatty acids having the general formula, $CH_3(CH_2)_nCOOH$ have the carboxylate functional group in the structure. All fatty acids have a -COOH group at one end of their molecules and a long hydrocarbon chain at the other end. Similar to PCE, negatively charged carboxylate functional group of fatty acids can be attracted to positively charge colloidal particles in concrete. Depending on whether they contain a carbon–carbon double bond or not, they are referred to as being unsaturated or saturated fatty acids. Compared to other PCMs discussed in this chapter, PCMs have outstanding performance in chemical and thermal stability, nontoxicity, and biodegradability. Their melting enthalpy is similar to that of paraffins and their melting temperature increases with the length of molecule. Fatty acids are stable upon cycling. Like paraffins, fatty acids also exhibit little or no subcooling and have a low thermal conductivity. The physical and especially the thermal proper-ties of these materials are well known from the research in energy storing materials and have been reviewed by Sarier and Onder in the year 2012.[20] An interesting feature of fatty acids is that similar to deep eutectic solvents, they are able for eutectic mixtures that have a lower melting point than the pure components. In contrast to paraffin, short-chain fatty acid comes with good solubility in organic solvents such as ethanol or acetone.[21] Special attention needs to be paid to the corrosive nature and a high sublimation rate of these materials. During experimentation, it has been found that siloxane elastomers such as polydimethylsiloxane do not form a sufficient barrier against the sublimation of fatty acids. In contrast to fatty acids, fatty acid derivatives such as fatty acid esters are less likely to sublimate.[18] Furthermore, these materials are insoluble in water. The biocompatibility of fatty acids and their derivatives is very good. These materials have been proposed as matrix material for self-degrading implants and drug delivery implants. This makes fatty acid esters a promising material for biological lab-on-a-chip systems. A more comprehensive list of the thermal behavior as well as the solubility of fatty acid esters and glycerides has been given by Bennett.[18]

As far as usage of fatty acids is concerned, fatty acids can be utilized as stabilizers for nanomagnetite particles because of availability of a long chain carbon skeleton having carboxylic acid as functional group.[22] Fatty acids, in their liquid phase, have a surface tension in the order of $2–3 \times 10{-4}$ N/m. This is high enough to retain them in the structure of a composite host material.[19] This may be beneficial in preventing the leakage of fatty acids from the porous structure of concrete. Thus, the use of fatty acid mixture as PCM can improve the workability of fresh concrete mixtures as well as improving the thermal performance of hardened concretes. Fatty acids derived from edible vegetable and animal oils are bio-based sustainable materials.[23] Opposite to paraffins, fatty acids are not fossil fuel derivatives and are not affected by the price volatilities of the petroleum market. Advantages of fatty acids as TES material are their smaller volume change during phase transitions, high latent heats per unit mass, and suitable melting temperature ranges. Low flammability is also an important property of fatty acids especially in building applications. Bio-based PCMs are reported as significantly less flammable than paraffinic PCMs.[24]

### 12.4.3 POLYETHYLENE GLYCOLS

Polyethylene glycols (PEGs) also known as polyoxyethylenes are basically polymers having general chemical formula $H-(O-CH_2-CH_2)_n-OH$. The two terminal hydroxyl groups make polyethylene glycol a polar substance. As opposed to paraffin, PEGs are highly hydrophilic in nature. They dissolve readily in water. On the other hand, the hydrophilic PEG does not diffuse into hydrophobic polymers such as poly dimethylsiloxane (PDMS), which is often used to set up micro-electromechanical system membrane actuators. Thus, PDMS devices using PEG will survive more duty cycles than comparable devices using paraffin.[43] In experiment, it has been observed that PEG inhibits the curing process of some commercially available PDMS precursors, which causes problems during assembling. Another drawback of PEGs is that they are subject to thermal degradation, if reheated several times, which is due to a thermally induced reduction in molecular weight as a consequence of oxidation during heating cycles in specific environments. This results in a decreasing melting point.[44]

## 12.5  CONCLUSIONS

An approach to improve the mechanical stability by forming a composite material is the impregnation of mechanically stable, porous materials with the PCM. All nonmetallic liquids, including PCM, have a low thermal conductivity. Since PCMs store large amounts of heat or cold, it is necessary to transfer this heat to the outside of the storage to use it. The low thermal conductivity can be a problem. In the liquid phase, convection can significantly enhance heat transfer; however, often this is not sufficient. In the solid phase, there is no convection. When heat transfer is necessary, one possibility to increase the thermal conductivity of the PCM is to add materials with larger thermal conductivity. This can be done at a macroscopic scale, for example by adding metallic pieces, or on a submillimeter scale with composite materials.

Keeping all the existing materials in mind, there is a need to look for an alternative of existing material which should be cheap, light weight, abundant, nontoxic, environmental friendly, and nonflammable in nature and easy to manufacture. Recently, phytochemicals (naturally occurring natural products) have attracted wide attention of researchers toward PCMs. Unlike other existing PCMs, phytochemicals may absorb excess heat during the day and release this energy back in the evening as and when buildings start cooling. These materials may absorb and release large quantities of energy very quickly and for extensive periods of time due to phase changing. It may elegantly work day and night to stabilize indoor temperatures and save time and money. It may achieve sustainability goals.

A chain of novel photocrosslinked bio-based shape stabilized PCMs based on octadecanol, eicosanol, and docosanol have been prepared using UV technique for TES applications. Acrylic acid was added with the epoxidized soyabean oil to produce acrylated soyabean oil (ASO). The heating and freezing process phase change enthalpy were found to be in between 30 and 68J/g and 18 and 70J/g, respectively. The decomposition of UV-cured PCMs started at 260°C and reached a maximum of 430°C. All the bio-based UV-cured PCMs were found suitable in improving latent heat storage capacity in comparison to the pristine ASO sample. By this way, these materials exhibited potential in TES applications.[45]

Bio-based PCMs play a vital role as a TES device by employing their high storage density and latent heat property. One of the prospective applications of this type of PCM is for upcoming buildings by incorporating them in the envelope for energy conservation purposes.[46]

PCMs in building envelopes operate by changing phase from solid to liquid while absorbing heat from the outside and thus reducing the heat flow into the building, and releasing the absorbed heat when it gets cold outside reducing the heat loss through the building envelope.

Bio-based materials such as oleochemical carbonates have not been examined for their applicability as PCMs, which are prepared through a carbonate interchange reaction between renewable C10-C18 fatty alcohols and dimethyl or diethyl carbonate in the presence of a catalyst. The latent heat of melting and freezing for a series of symmetrical oleochemical carbonates ranging from 21 to 37 carbon atoms have been evaluated to develop a fundamental understanding of the solid–liquid transitions for utilization in TES applications. These oleochemical carbonates have shown sharp phase transitions and good latent heat properties. The latent heats of melting and freezing for different types of oleochemical carbonates, that is, decyl, dodecyl, tetradecyl, hexadecyl, and octadecyl carbonates have been found to be 144, 200, 227, 219, and 223 and 146, 199, 229, and 215 J/g, respectively. These carbonates signify novel renewable-based PCM chemicals that compliment fatty acids, fatty alcohols, and their fatty acid esters while substituting a valuable bio-based alternative to currently dominating paraffin wax and salt hydrate PCM.[47]

As we know, the most commonly used PCMs are salt hydrates, fatty acids and esters, and various paraffins. In chemical world, fatty acid is basically a carboxylic acid with a long aliphatic chain, which is either saturated or unsaturated in nature. Most naturally occurring fatty acids have an unbranched chain of an even number of carbon atoms, from 4 to 28. Among all those materials as PCMs, only fatty acids are the only materials, which exist in plants. Plants have the capability to create double bonds after the third and sixth carbon on the chain. Only plants can synthesize omega-3 (alpha-linolenic acid) and omega-6 fats (linoleic acid), which are called "essential" fatty acids.

**Sources of Omega-3 fatty acids:** Materials include flax seeds, hemp seeds, canola, walnuts, pumpkin seeds, soybeans, etc.

**Sources of Omega-6 fatty acids:** Materials consisting of Omega-6 fatty acids are sunflower, safflower, soybeans, hemp seeds, walnuts, pumpkin seeds, sesame seeds, flax seeds, etc.

The long-chain fatty acids (>C20) in bryophyte plant samples have been obtained based on distillation extraction with 1:1 (v/v) chloroform/methanol as combination of solvent. 1-[2-($p$-toluenesulfonate)-ethyl]-2-phenylimidazole-[4,5-f]-9,10-phenanthrene as tagging reagent could easily and quickly label long-chain fatty acids at 90 °C in the presence of $K_2CO_3$ catalyst in dimethyl formamide. Eleven free long-chain fatty acids from the extracts of bryophyte plants have been sensitively determined.[48]

## KEYWORDS

- energy storage
- phase change materials
- herbal material
- enthalapy
- latent heat

## REFERENCES

1. Kang, Y.; Jeong, S. G.; Wi, S.; Kim, S. Energy Efficient Bio-Based PCM with Silica Fume Composites to Apply in Concrete for Energy Saving in Buildings. *Sol. Energy Mat. Sol. Cells* **2015,** *143,* 430–434.
2. El-Dessouky, H.; Al-Juwayhel, F. Effectiveness of a Thermal Energy Storage System Using Phase-Change Materials. *Energy Convers. Manage.* **1997,** *38* (6), 601–617.
3. Farid, M. M.; Khudhair, A. M.; Razack, S. A. K.; Al-Hallaz, S. A Review on Phase Change Energy Storage: Materials and Applications. *Energy Convers. Manage.* **2004,** *45* (9–10), 1597–1615.
4. Lorsch, H. G.; Kauffman, K. W.; Denton, J. C. Thermal Energy Storage for Heating and Air Conditioning, Future Energy Production System. *Heat Mass Transf. Process.* **1976,** *1,* 69–85.

5. Kandasamy, R.; Wang, X.-Q.; Mujumdar, A. S. Application of Phase Change Materials in Thermal Management of Electronics. *Appl. Therm. Eng.* **2007,** *27* (17–18), 2822–2832.

6. Bascetincelik, A.; Paksoy, H. O.; Ozturk, H. H.; Demirel, Y. *Seasonal Latent Heat Storage System for Greenhouse Heating.* In Phase Change Materials and Chemical Reactions for Thermal Energy Storage First Workshop, Adana Turkey, 1998.

7. Benli, H.; Durmus, A. Performance Analysis of a Latent Heat Storage System with Phase Change Material for New Designed Solar Collectors in Greenhouse Heating. *Sol. Energy* **2009,** *83* (12), 2109–2119.

8. Mazman, M.; Cabeza, L. F.; Mehling, H.; Paksoy, H. Ö.; Evliya, H. Heat Transfer Enhancement of Fatty Acids When Used as PCMs in Thermal Energy Storage. *Int. J. Energy Res.* **2008,** *32,* 135–143.

9. Sharma, A.; Tyagi, V. V.; Chen, C. R.; Buddhi, D. Review on Thermal Energy Storage with Phase Change Materials and Applications. *Renew. Sustain. Energy Rev.* **2009,** *13* (2), 318–345.

10. Alkan, C.; Sari, A.; Biçer, A. Thermal Energy Storage by Poly (Styrene-Co-p-Stearoyl-styrene) Copolymers Produced by the Modification of Polystyrene. *J. Appl. Polym. Sci.* **2012,** *125* (5), 3447–3455.

11. Mondal, S. Phase Change Materials for Smart Textiles – An Overview. *Appl. Therm. Eng.* **2008,** *28* (11–12), 1536–1550.

12. Schossig, P.; Henning, H. M.; Gschwander, S.; Haussmann, T. Micro-Encapsulated Phase-Change Materials Integrated into Construction Materials. *Sol. Energy Mater. Sol. Cells.* **2005,** *89* (2–3), 297–306.

13. Zalba, B.; Marin, J. M.; Cabeza, L. F.; Harald, M. Review on Thermal Energy Storage with Phase Change: Materials, Heat Transfer Analysis and Applications. *Appl. Therm. Eng.* **2003,** *23* (3), 251–283.

14. Zhou, D.; Zhao, C. Y.; Tian, Y. Review on Thermal Energy Storage with Phase Change Materials (PCMs) in Building Applications. *Appl. Energy.* **2012,** *92,* 593–605.

15. Li, G.; Hwang, Y.; Radermacher, R.; Chun, H.–H. Review of Cold Storage Materials for Subzero Applications. *Energy* **2013,** *51,* 1–17.

16. Kenisarin, M. M.; Kenisarina, K. M. Form-Stable Phase Change Materials for Thermal Energy Storage. *Renew. Sustain. Energy Rev.* **2012,** *16* (4), 1999–2040.

17. Garti, N.; Amar-Yuli, I. *Nanotechnologies for Solubilization and Delivery in Foods, Cosmetics and Pharmaceuticals.* DEStech Publications, Inc.: Jerusalem, 2012.

18. Bennett, H. *Industrial Waxes: Natural and Synthetic Waxes.* Chemical Publ. Co.: New York, 1963.

19. Warth, A. H. The Chemistry and Technology of Waxes. *Soil Sci.* **1956,** *82* (4), 344.

20. Sarier, N.; Onder, E. Organic Phase Change Materials and Their Textile Applications: An Over View. *Thermochimica Acta.* **2012,** *540,* 7–60.

21. Ralston, A. W.; Hoerr, C. W. The Solubilities of the Normal Saturated Fatty Acids. *J. Org. Chem.* **1942,** *07* (6), 546–555.

22. Kohl, M. *Shape Memory Microactuators.* Springer: Berlin, Heidelberg, New York, 2004.

23. Wei, Z. G.; Sandstroröm, R.; Miyazaki, S. Shape-Memory Materials and Hybrid Composites for Smart Systems: Part I Shape-Memory Materials. *J. Mater. Sci.* **1998,** *33,* 3743–3762.

24. Jani, J. M.; Leary, M.; Subic, A.; Gibson, M. A. A Review of Shape Memory Alloy Research. Applications and Opportunities. *Mater. Des.* **2014,** *56,* 1078–1113.
25. Rajamani, L. Phytochemical and GC-MS Analysis of Brassica Oleracea Var. Capitata f. Rubra. *World J. Pharmaceut. Res.* **2018,** *7* (3), 1392–1400.
26. Geetha, D. H.; Jayashree, I.; Rajeshwari, M. GC–MS Analysis of Ethanolic Extract of Elaeocarpus Serratus L.. *Eur. J. Pharmaceut. Med. Res.* **2015,** *2* (2), 296–302.
27. Aja, P. M.; Nwachukwu, N.; Ibiam, U. A.; Igwenyi, I. O.; Offor, C. E.; Orji, U. O. Chemical Constituents of Moringa Oleifera Leaves and Seeds from Abakaliki, Nigeria. *Am. J. Phytomed. Clin. Therapeut.* **2014,** *2* (3), 310–321.
28. Moronkola, D. O.; Ogukwe, C.; Awokoya K. N. Chemical Compositions of Leaf and Stem Essential Oils of *Calotropis Procera* Ait R. Br [Asclepiadeae], *Der Chemica Sinika.* **2011,** *2* (2), 255–260.
29. Belakhdar, G.; Benjouad, A.; Abdennebi, E. H. Determination of Some Bioactive Chemical Constituents from *Thesium humile* Vahl. *J. Mater. Environ. Sci.* **2015,** *6* (10), 2778–2783.
30. Anburaj, G.; Marimuthu, M.; Rajasudha, V.; Manikandan, D. R. R. Phytochemical Screening and GC-MS Analysis of Ethanolic Extract of *TecomaStans (Family: Bignoniaceae)* "Yellow Bell Flowers." *Int. J. of Res. Pharmacol. Pharmacotherapeu.* **2016,** *5* (3), 256–261.
31. Sen, A.; Batra, A. Chemical Composition of Methanol Extract of the Leaves of Melia Azedarach L. *Asian J. Pharmaceut. Clin. Res.* **2012,** *5* (3), 42–45.
32. Sigma-Aldrich (Ed.). *1-Pentadecanol,* 5.1 ed. Sigma-Aldrich Chemie GmbH, 2015, www.sigmaaldrich.com.
33. Sigma-Aldrich (Ed.). *1-Hexadecanol,* 5.3 ed. Sigma-Aldrich Chemie GmbH, 2015, www.sigmaaldrich.com.
34. Sigma-Aldrich (Ed.). *1-Heptadecanol,* 5.1 ed. Sigma-Aldrich Chemie GmbH, 2015, www.sigmaaldrich.com.
35. Sigma-Aldrich (Ed.). *1-Octadecanol,* 5.2 ed. Sigma-Aldrich Chemie GmbH, 2014, www.sigmaaldrich.com.
36. Sigma-Aldrich (Ed.). *1-Nonadecanol,* 5.1 ed. Sigma-Aldrich Chemie GmbH, 2013, www.sigmaaldrich.com.
37. Sigma-Aldrich (Ed.). *1-Eicosanol,* 5.2 ed. Sigma-Aldrich Chemie GmbH, 2013, www.sigmaaldrich.com.
38. Sigma-Aldrich (Ed.). *1-Docosanol,* 5.0 ed. Sigma-Aldrich Chemie GmbH, 2012, www.sigmaaldrich.com.
39. Sigma-Aldrich (Ed.). *1-Tetracosanol,* 5.1 ed. Sigma-Aldrich Chemie GmbH, 2013, www.sigmaaldrich.com.
40. Sigma-Aldrich (Ed.). *1-Hexacosanol,* 5.0 ed. Sigma-Aldrich Chemie GmbH, 2012, www.sigmaaldrich.com.
41. Sigma-Aldrich (Ed.). *1-Octacosanol,* 5.0 ed. Sigma-Aldrich Chemie GmbH, 2012, www.sigmaaldrich.com.
42. Sigma-Aldrich (Ed.). 1-*Triacontanol,* 5.0 ed. Sigma-Aldrich Chemie GmbH, 2012, www.sigmaaldrich.com.
43. Song, W. H.; Kwan, J.; Kaigala, G. V; Hoang, V. N.; Backhouse, C. J. Readily Integrated, Electrically Controlled Microvalves. *J. Micromech. Microeng.* **2008,** *18* (4), 1–9.

44. Velraj, R.; Seeniraj, R. V.; Hafner, B.; Faber, C.; Schwarzer, K. Heat Transfer Enhancement in a Latent Heat Storage System. *Sol. Energy* **1999,** *65* (3), 171–180.

45. Baştruk, E.; Kahraman, M. V. Photocrosslinked Biobased Phase Change Material for Thermal Energy Storage. *J. Appl. Polym. Sci.* **2016,** *133* (32), 1–8.

46. Muruganantham, K.; Phelan, P.; Horwath, P.; Ludlam, D.; McDonald, T. *Experimental Investigation of a Bio-Based Phase Change Material to Improve Building Energy Performance.* ASME2010 4th International Conference on Energy Sustainability, Volume1, Phoenix, Arizona, USA, May 17–22, 2010.

47. Kenar, J. A. Latent Heat Characteristics of Bio-Based Oleochemical Carbonates as Potential Phase Change Materials. *Sol. Energy Mat. Sol. Cells* **2010,** *94* (10), 1697–1703.

48. You, J.; Zhao, X.; Suo, Y.; Li, Y.; Wang, H.; Chen, G. Determination of Long-Chain Fatty Acids in Bryophyte Plants Extracts by HPLC with Fluorescence Detection and Identification with MS. *J. Chromatograp. B Analyt. Technol. Biomed. Life Sci.* **2007,** *848* (2), 283–291.

# CHAPTER 13

# Today's Renewable Energy Market: Innovations, Commercialization, and Impact on Market

ROHIT BANSAL*, VIKAS, and RACHIT JAISWAL

*Department of Management Studies, Rajiv Gandhi Institute of Petroleum Technology, Raebareli, Uttar Pradesh, India*

*Corresponding author. E-mail: rbansal@rgipt.ac.in*

## ABSTRACT

The multi-fold increase in the energy requirements of the global economy has made it a necessity to search for innovative sources of energy. There has been introduction of various new renewable sources of energy but the technological and financial viability are still in the infant stage. Market failures combined with unpropitious institutional, financial, and regulatory environments mandate government intervention to establish renewable sources of energy. Commercialization is a planned way by which technologies and innovations developed through research make their way to market. The commercialization of renewable energy consists the positioning of three generations of renewable energy technologies. The first-generation technologies involve hydroelectricity, biomass, geothermal power, and heat which have reached their maturity and are economically competitive. Second-generation technologies include photovoltaics, solar heating, wind power, modern forms of bioenergy and solar thermal power stations. These technologies have entered into the market almost a decade ago. Third-generation technologies are at infant stage and need continuous R&D efforts to mark its presence on global scale. These technologies mainly consist of ocean energy, advanced biomass gasification and hot-dry-rock-geothermal power energy. This chapter includes the recent developments in various sources of renewable energy like solar photovoltaic cells, solar heating system, solar distillation, biomass, biomedical waste, tidal energy, geothermal energy, wind energy, hydroelectricity, their commercialization and their impacts on customers, costing, environment, etc.

## 13.1   OCEAN ENERGY

It is the form of energy generated in the ocean from ocean waves, tidal streams, ocean currents, temperate, and salinity inclination. By the end of 2017, only handful facilities for production of commercial energy have been initiated. Eventually in 2016, ocean energy power amounted to nearly 536 MW, of which 494 MW was contributed by two tidal barrage amenities, Sihwa Plant in the Republic of Korea which was 254 MW and built in 2011 being the first one and 240 MW La Rance (France) completed in 1966 being the second.[1] These two tidal range facilities use inline stream turbine technology. However, other ocean energy methods are still in nascent stages. The advancement in these technologies relies on government support for infrastructural and funding support. Some projects were undertaken to utilize ocean thermal energy and technologies of salinity gradient. Also, converters of wave energy have advanced to precommercial demonstration stage. Various programs for research and development activities focusing on tidal stream, wave energy, thermal energy, and salinity gradients are going on especially in countries like Canada, China, Chile, Korea, the United States, and some European countries. Companies operating in ocean energy industry are increasing all over the world. These companies are constantly striving for advancement in the deployed technology and are using modern and refined devices. However, the victory for these technologies is hindered by a number of challenges like towering risks, elevated upfront costs, and a call for better planning, agreeable licensing procedures.

Tidal industry made some remarkable achievements in the year 2016 with numerous deployments in countries like Scotland, France, and Canada. Some of the achievements are listed below:

1.   In Scotland, Nova Innovation (UK) with the European Law Students' Association (ELSA), a Belgian partner, claimed to operate first of its kind grid that was connected to tidal array with two direct-drive (100 kW M100) turbines installed in Shetland's Bluemull Sound. In early 2017, the third turbine, other than above two, was also installed.[2]

2.  In Orkney, Scotland, Scotrenewables Tidal Power (UK) claimed to install the world's largest tidal turbine SR2000 that was 2 MW and had two turbines that were horizontal-axis being attached on a platform that is floating hull.[3]
3.  In Scotland, the world's largest tidal energy plant: MeyGen installed first 1.5-MW tidal turbine in later half of 2016.
4.  In France: OpenHydro (Direction des Construction Navales Services—DCNS subsidiary, France) initiated and installed two tidal turbines at tidal array of Electricité de France (EDF)'s (France), located at Paimpol-Brehat.[4] They both were open center.

In 2016, wave energy progressed with several pilot projects around the world. We will look into the major developments that took in countries like Spain, Sweden, the United States, Korea, and China.

1.  Spain—Oceantec developed 30-kW prototype converters for wave energy which was first of its kind in Spain and was connected at the Biscay Marine Energy Platform (BiMEP).[5] Spain hosts Mutriku multiturbine wave-power plant which is operating since 2011. This plant uses wave-driven compressed air to generate electricity.
2.  Sweden—In Norwegian waters, Waves4Power installed WaveEL and Seabased linked wave-power array of 1 MW Sotenas with the grid.[6] The plant unites linear generators in sea floor with surface floats. This array is considered to be first of its kind operating numerous converters for wave energy.
3.  United States—In the Pacific region, Fred Olsen (Norway) had deployed Bolt Lifesaver device for pilot testing over one year at US Navy's Wave Energy Test Center (WETS) located in Hawaii. It yielded positive results and was concluded by March 2017. This unit generated power for six months continuously.[7] At WETS, Northwest Energy Innovations (the United States) continued the half-scale grid-connected testing of its energy device for Azura wave of 20 kW.
4.  Korea—Korea launched remarkable projects in 2016 focusing on ocean-energy technologies. Studies were based on integrating converters of wave energy like oscillating water column (OWC) devices having energy storage.[8] The setup of 500-kW OWC pilot plant near Jeju Island was efficaciously completed.
5.  China—Based on Ocean Thermal Energy Conversion (OTEC) which considers the variation in temperature of cooler surface and

warmer facade to generate energy, China installed a 10-kW OTEC device. Out of seven turbines of 3.4-MW demonstration project related to wave energy located in Zhejiang, electricity generation started from the first two turbines. As per the released 13th Five-Year Plan on ocean energy, the country aims to achieve 50 MW installed capacity by the year 2020.[9] The five-year plan also intends a research spotlight for conversion of tidal wave as well as thermal energy. Through systematic collaborations between government and industry, the industry has seen advancement in the plans and roadmaps in rest of the world. The European Commission's Ocean Energy Forum finalized the core agenda in 2016 and released Ocean Energy Strategy Roadmap. The roadmap framed action tactics to set up an advanced route to reduce project risk and fritter away. A fund in European investment was set up to support ocean energy farms, European fund for insurance and guarantee was built to endorse hazard of the plan, and a unified planning and accepting program.

There are few major challenges affecting the ocean energy development. One such challenge is the adverse impacts that ocean energy devices can have on marine animals, for instance—risk of animals hitting with the moving substances, possible effects of sound transmission from energy devices also all biological outcomes of electromagnetic areas that are created by underwater cables. In addition, the other challenges are the lack of clarity in the ocean energy development process. There is a need to properly plan the marine spatial to clarify the role and authority of diverse authorities and to synchronize and restructure licensing and compliant processes.

## 13.2 GEOTHERMAL ENERGY

One of the important sources of renewable energy is geothermal which supplies electricity and thermal energy services. Worldwide projected geothermal electricity and thermal production were 567 PJ (157 TW h) with equal share in the year 2016. Some of the geothermal plants provide thermal fluid to run various heat applications like fish farming, space heating and cooling, and so on.

The global geothermal power-generating capacity increased by approximately 3.05% and reached to 13.5 GW. The leading countries in geothermal plant installations were Indonesia and Turkey. During 2015, Japan, Mexico, and Kenya have completed various projects and many other projects are in progress in other parts of the world (Table 13.1).

**TABLE 13.1** Geothermal Power Plant World Total Installed Capacity (in MW Electric) from 2000 to 2015.[10]

| Country | 2000 | 2005 | 2010 | 2013 | 2015 |
|---|---|---|---|---|---|
| Australia | 0.2 | 0.2 | 1.1 | 1 | 1.1 |
| Austria | 0 | 1 | 1.4 | 1.4 | 1.2 |
| China | 29.2 | 28 | 24 | 27 | 27 |
| Costa Rica | 142.5 | 163 | 166 | 207.1 | 207 |
| El Salvador | 161 | 151 | 204 | 204.4 | 204 |
| Ethiopia | 8.5 | 7 | 7.3 | 8 | 7.3 |
| France | 4.2 | 15 | 16 | 17 | 16 |
| Germany | 0 | 0.2 | 6.6 | 11.9 | 27 |
| Guatemala | 33.4 | 33 | 52 | 48 | 52 |
| Iceland | 170 | 322 | 575 | 664.4 | 665 |
| Indonesia | 589.5 | 797 | 1197 | 1341 | 1340 |
| Italy | 785 | 790 | 843 | 875.5 | 916 |
| Japan | 546.9 | 535 | 536 | 537 | 519 |
| Kenya | 45 | 127 | 167 | 248.5 | 594 |
| Mexico | 755 | 953 | 958 | 1017.4 | 1017 |
| New Zealand | 437 | 435 | 628 | 842.6 | 1005 |
| Nicaragua | 70 | 77 | 88 | 149.5 | 159 |
| Papua New Guinea | 0 | 39 | 56 | 56 | 50 |
| Philippines | 1909 | 1931 | 1904 | 1848 | 1870 |
| Portugal (Azores) | 16 | 16 | 29 | 28.5 | 29 |
| Russia | 23 | 79 | 82 | 81.9 | 82 |
| Turkey | 20.4 | 20.4 | 82 | 405 | 394 |
| USA | 2228 | 2544 | 3093 | 3389 | 3450 |
| Total | 7973.8 | 9063.8 | 10,716.4 | 12,010.1 | 12,632.6 |

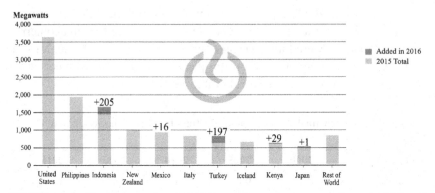

**FIGURE 13.1** Top 10 countries of 2016 for geothermal power capacity and additions.[11] (Reprinted with permission from Renewables 2017 Global Status Report; REN21: Paris, 2017.)

| Levelised Cost of Energy → USD/kWh 0 | 0.05 | 0.10 | 0.15 | 0.20 | 0.25 | 0.30 |
|---|---|---|---|---|---|---|
| **GEOTHERMAL POWER** | | | | | | |
| Africa | | | | | | |
| Asia | | | | | | |
| Central America and the Caribbean | | | | | | |
| Eurasia | | | | | | |
| Europe | | | | | | |
| Middle East | | | | | | |
| North America | | | | | | |
| Oceania | | | | | | |
| South America | | | | | | |
| China | | | | | | |
| India | | | | | | |
| United States | | | | | | |

**FIGURE 13.2** Levelized cost of geothermal energy in different countries, 2016.[12] (Reprinted with permission from Renewables 2017 Global Status Report; REN21: Paris, 2017.)

| Investment Cost → USD | min | max | wa | Capacity Factor → | min | max | wa |
|---|---|---|---|---|---|---|---|
| Africa | 1719 | 7689 | ● 3818 | | 0.8 | 0.92 | ● 0.84 |
| Asia | 2047 | 5045 | ● 3116 | | 0.58 | 0.9 | ● 0.85 |
| Central America and the Caribbean | 3260 | 3537 | ● 3413 | | 0.57 | 0.6 | ● 0.58 |
| Eurasia* | 2613 | 3278 | ● 3113 | | 0.8 | 0.8 | ● 0.8 |
| Europe | 3613 | 8919 | ● 5209 | | 0.6 | 0.8 | ● 0.66 |
| Middle East | | | | | | | |
| North America | 2029 | 8353 | ● 5017 | | 0.74 | 0.92 | ● 0.83 |
| Oceania* | 3303 | 4676 | ● 3796 | | 0.8 | 0.8 | ● 0.8 |
| South America | 3027 | 4348 | ● 3587 | | 0.8 | 0.95 | ● 0.82 |
| China | 1501 | 9722 | ● 1943 | | | | |
| India | 1501 | 7475 | ● 2169 | | | | |
| United States | 2941 | 8353 | ● 5961 | | 0.74 | 0.9 | ● 0.79 |

**FIGURE 13.3** Geothermal investment cost and capacity factor in different countries, 2016.[13] (Reprinted with permission from Renewables 2017 Global Status Report; REN21: Paris, 2017.)

Table 13.2 shows the major developments in geothermal energy for the year 2016.

**TABLE 13.2** The Major Developments in Geothermal Energy for the Year 2016.

| Country | Major developments in recent years |
|---------|-----------------------------------|
| Indonesia | • Net addition of capacity was around 200 MW in 2016[14]<br>• Commercial production started at Sarulla Plant having capacity of 110 MW, which was one of the world's largest geothermal plant |
| Turkey | • Total installed capacity was about 621 MW in 2015 which increased to 821 MW after a net addition of 200 MW in 2016 as country has added 10 new geothermal power plants<br>• All the newly added plants were binary organic cycle (ORC) units having capacity up to 25 MW<br>• The country applies traditional flash turbine technology to develop apt projects for the leftover resources of high temperature of the country |
| Kenya | • Total installed capacity was around 630 MW at the end of 2016 after a capacity addition of 29 MW at the Olkaria III complex plant[15] |
| Mexico | • Added a condensing flash unit of 25 MW under the Domo San Pedro plant, after adjusting two 5-MW temporary wellhead units<br>• Total capacity reaches to 950 MW with net addition of 15 MW<br>• During 2016, the Mexican government has amended the geothermal energy law to permit private companies to explore and use geothermal resources. The country had granted permission to three private Mexican companies to explore and use country's geothermal resources |
| Japan | • The geothermal development of the country has become mixed, with the requirements of competition for the choice of fossil and nuclear energy on one side, and apprehensions regarding safety and uncertain environmental and economic consequences on others<br>• Country had exempted the small scale project having capacity of less than 7.5 MW from environmental impact assessments which result in tremendous growth in small-scale geothermal power projects[16]<br>• A small geothermal facility in Tsuchiyu and at least one ORC generator became operational in 2015 and 2016, respectively[17]<br>• No large-scale project was under development in early 2017 |
| Ethiopia | • The country enjoys geothermal wealth of Great Rift Valley with Kenya, but till now there has been insufficient growth having capacity of 7 MW<br>• To boost the geothermal energy development<br>• In 2015, the country had signed power purchase agreement for Corbetti project's first phase having capacity of 500 MW and which is expected to be constructed in two phases within the period of 8–10 years[18]<br>• In 2016, Ethiopia has made amendment in mining law to exempt geothermal energy form royalty payment |

**TABLE 13.2**    *(Continued)*

| Country | Major developments in recent years |
|---|---|
| Croatia | • In 2016, Croatia initiated 16 MW binary plant construction to exploit Pannonian basin's 170°C geothermal brine and steam, as its first-ever geothermal power project |
| China | • Country had a capacity of less than 30 MW till 2015 and these were mostly in Tibet[19] |
| | • But, to cut local air pollution and greenhouse gas emissions, the Chinese government calls for a 500-MW geothermal energy addition by 2020 in country's 13th Five-Year Plan[20] |
| Philippines | • Country has the second largest geothermal power capacity in operation after USA |
| | • No addition has been made in capacity in the year 2016 |
| | • In an attempt to encourage the progress of more-challenging low-temperature resources, the geothermal industry association of Philippines invited for FITs (feed in tariffs) for geothermal power, alike to those permitted to other renewables |
| | • Resources with low temperature may necessitate the use of deep drilling and binary-cycle technology, thereby increasing the risk of development and the final cost of the energy produced[21] |

## 13.3  HYDROPOWER

The year 2016 saw an addition of more than 25 GW in global capacity of hydropower causing the total capacity to reach approximately 1096 GW. The leading countries for hydropower capacity are China, Brazil, the United States, Canada, the Russian Federation, India, and Norway. At the end of the year 2016, these countries contributed to about 62% of the installed capacity. In 2016, the global hydropower generation and global pumped storage capacity were estimated at 4102 TW h (with 3.2% increase) and 150 GW (with 6.4 GW increase), respectively.[22] China contributed to more than one-third of new hydropower capacity in 2016. Apart from China, Brazil, Ecuador, Ethiopia, Vietnam, Peru, Turkey, Lao PDR, Malaysia, and India were the leading contributors to capacity addition. For pumped storage capability addition, the top countries were China, South Africa, Switzerland, Portugal, and the Russian Federation.

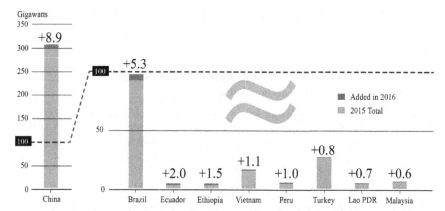

**FIGURE 13.4** Top nine countries of 2016 for hydropower capacity and additions.[23] (Reprinted with permission from Renewables 2017 Global Status Report; REN21: Paris, 2017.)

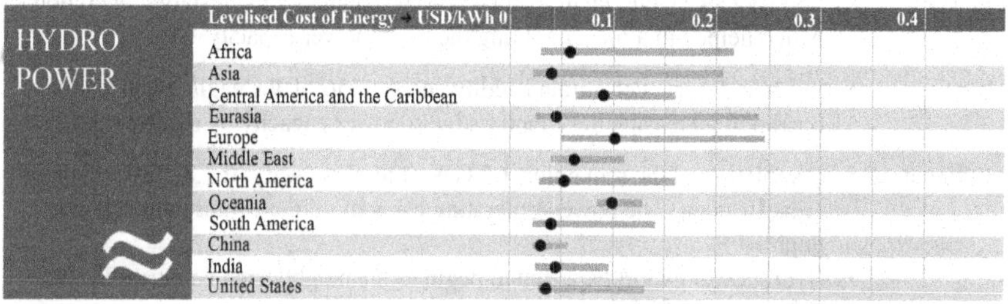

**FIGURE 13.5** Levelized cost of hydropower in different countries, 2016.[24] (Reprinted with permission from Renewables 2017 Global Status Report; REN21: Paris, 2017.)

| Investment Cost → USD | min | max | wa | Capacity Factor → | min | max | wa |
|---|---|---|---|---|---|---|---|
| Africa | 920 | 6730 | ●1593 | | 0.3 | 0.65 | ●0.43 |
| Asia | 483 | 7553 | ●1446 | | 0.16 | 0.81 | ●0.47 |
| Central America and the Caribbean | 1650 | 4474 | ●3230 | | 0.32 | 0.57 | ●0.53 |
| Eurasia | 1111 | 5934 | ●1530 | | 0.3 | 0.72 | ●0.54 |
| Europe | 570 | 5388 | ●1847 | | 0.16 | 0.7 | ●0.38 |
| Middle East | 1238 | 1656 | ●1526 | | 0.2 | 0.76 | ●0.36 |
| North America | 1051 | 5195 | ●2309 | | 0.38 | 0.78 | ●0.49 |
| Oceania | 3470 | 4119 | ●3689 | | 0.39 | 0.48 | ●0.45 |
| South America | 799 | 5743 | ●1755 | | 0.49 | 0.91 | ●0.61 |
| China | 971 | 2581 | ●1273 | | 0.32 | 0.6 | ●0.5 |
| India | 1014 | 2556 | ●1519 | | 0.25 | 0.81 | ●0.44 |
| United States | 723 | 6757 | ●1384 | | 0.38 | 0.78 | ●0.39 |

**FIGURE 13.6** Hydropower investment cost and capacity factor in different countries, 2016.[25] (Reprinted with permission from Renewables 2017 Global Status Report; REN21: Paris, 2017.)

Table 13.3 shows the major developments that took place in various countries.

**TABLE 13.3**  The Major Developments That Took Place in Various Countries.

| Countries | Major developments in recent years |
|---|---|
| China | • Total capacity of hydropower reached 305 GW with an addition of 8.9 GW and total capacity of pumped storage reached 27 GW with an addition of 3.7 GW<br>• By the year 2020, China's 13th Five-Year Plan for hydropower development aims for the capacity of hydropower and pumped storage to reach 340 and 40 GW, respectively[26]<br>• Hydropower generation in 2016 increased by about 6% to 1193 TW h |
| Brazil | • Hydropower capacity increased by 5.8% to reach the total of 96.9 GW<br>• Hydropower generation increased by 7.4% to 410 TW h because of improved hydrological conditions<br>• The output from thermal power plants reduced relatively by 30% because of increase in wind power and hydropower generation |
| Ecuador | • 1.5 GW Coca Codo facility and 487 MW Sopladora plant started operation which helped in almost doubling the hydropower capacity[27] |
| Peru | • 525 MW Cherro del Aguila Facility and the 456 MW Chaglla Plant started operation increasing the country's hydropower capacity by almost one-quarter |
| Ethiopia | • Completion of the remaining 1.5 GW of total capacity of 1.87 GW Ethiopia's Gibe III plant. The power-generating capacity of the country almost doubled[28]<br>• Hydropower export expected to begin to the neighboring countries Kenya, Djibouti, and Sudan |
| Vietnam | • 260 MW Huoi Quang Plant and 30 MW Coc San Facility began operation<br>• Addition of 1.1 GW capacity during the year helping the total capacity to reach 16.3 GW[29] |
| Malaysia | • Completion of 372 MW Ulu Jelai Project<br>• Construction of two new dams at Tasik Kenyir added 265 MW[30] |
| Turkey | • Total hydropower capacity reached 26.7 GW with an addition of 0.8 GW during the year[31] |
| India | • Total hydropower capacity reached 47.5 GW with an addition of 0.6 GW during the year<br>• Hydropower generation remained almost unchanged at 129 TW h during the year 2016 |
| United States | • With an addition of 380 MW, hydropower capacity reached 80 GW. The United States ranks third globally in terms of hydropower capacity[32]<br>• Power generation increased by 6.7% to 266 TW h in the year 2016[33] |

**TABLE 13.3**   *(Continued)*

| Countries | Major developments in recent years |
|---|---|
| Russian Federation | • Construction of 30 MW Zaragizhskaya plant in Kabardino-Balkaria got completed in 2016. Another 140 MW (160 MW in pump mode) which was built in Zelenchukskaya pools two reversible turbines to integrate hydropower generation with pumped storage capability[34]<br>• Hydropower capacity increased by 230 MW to reach the total capacity of 48.1 GW[35]<br>• There had been a significant increase in the hydropower generation by 11.3% in the year |

There are some challenges imposed by hydropower projects like submerging of large areas of land, substantial alterations to river ecosystems, resettlement of communities, and so on. These challenges must be judiciously considered and alleviated. The World Bank has continued its support to hydropower projects for both climate mitigation and local development. In Paris Climate Conference 2015, the World Bank emphasized on the significance of using hydropower along with other renewable power technologies in its efforts to escalate climate-resilient and low-carbon development in sub-Saharan Africa. There has been an increase in the demand for additional storage capacity caused by growing shares of renewable energy. Many new projects for pumped storage, which is one of major sources of large-scale energy storage, are under development. The global pumped storage capacity rose by approximately 6 GW in the year 2016 with new additional capacity installation in South Africa, China, and Europe. The installation of three turbines of 333 MW each of the 1.3 GW Ingula Plant for pumped storage in South Africa was completed in 2016, while the remaining two turbines became operational in January 2017.[36]

In Portugal, the pumped storage plant of 189 MW Baixo Sabor started operating and the 780 MW Frades II station's construction was completed. Both these open-loop storage plants use reversible pumps to supplement generation from natural flow with pumping capability.[37]

## 13.4   SOLAR PHOTOVOLTAICS

The year 2016 saw an addition of 75 GWdc in solar PV capacity worldwide aiding the global capacity to reach more than 303 GW.[38] Asia contributed for about two-third of the global additions. Considering the country-wise contribution in additions in 2016, five countries, namely, China, the United States, Japan, India, and the United Kingdom, accounted for approximately 85% of additions.[39] The other top countries were Germany, Korea, Australia, the Philippines, and Chile. In terms of cumulative capacity, the leading countries were China, Japan, the United States, and Italy. The factors responsible for market expansion of solar PV were increasing competitiveness of solar PV, growing demand for electricity, solar PV's potential to extenuate pollution, and minimize $CO_2$ emissions.

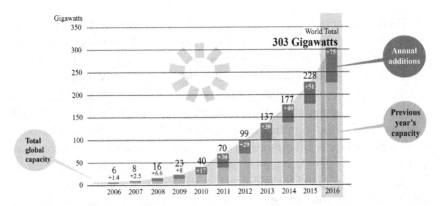

**FIGURE 13.7**  Global solar PV capacity and year-wise additions, 2006–2016.[40] (Reprinted with permission from Renewables 2017 Global Status Report; REN21: Paris, 2017.)

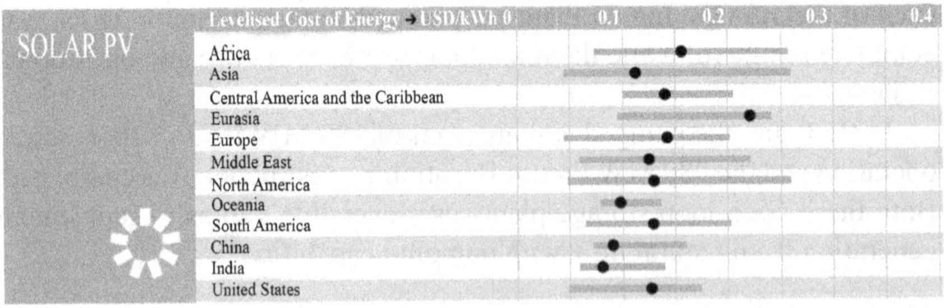

**FIGURE 13.8**  Solar PV levelized cost in different countries, 2016.[41] (Reprinted with permission from Renewables 2017 Global Status Report; REN21: Paris, 2017.)

| Investment Cost → USD | min | max | wa | Capacity Factor → | min | max | wa |
|---|---|---|---|---|---|---|---|
| Africa | 818 | 6848 | ● 2344 | | 0.14 | 0.28 | ● 0.2 |
| Asia | 832 | 6124 | ● 1414 | | 0.1 | 0.25 | ● 0.16 |
| Central America and the Caribbean | 1337 | 4000 | ● 2001 | | 0.16 | 0.23 | ● 0.19 |
| Eurasia | 1484 | 3697 | ● 2537 | | 0.1 | 0.18 | ● 0.14 |
| Europe | 944 | 2827 | ● 1370 | | 0.1 | 0.3 | ● 0.12 |
| Middle East | 1311 | 4000 | ● 2554 | | 0.17 | 0.35 | ● 0.26 |
| North America | 965 | 5900 | ● 2203 | | 0.12 | 0.34 | ● 0.2 |
| Oceania | 1600 | 2785 | ● 2477 | | 0.2 | 0.25 | ● 0.23 |
| South America | 1407 | 4951 | ● 2477 | | 0.12 | 0.34 | ● 0.24 |
| China | 1022 | 1953 | ● 1083 | | 0.17 | 0.19 | ● 0.17 |
| India | 916 | 1832 | ● 1064 | | 0.16 | 0.22 | ● 0.19 |
| United States | 1241 | 2971 | ● 1998 | | 0.16 | 0.32 | ● 0.19 |

**FIGURE 13.9** Solar PV investment cost and capacity factor in different countries, 2016.[42] (Reprinted with permission from Renewables 2017 Global Status Report; REN21: Paris, 2017.)

Table 13.4 displays the major developments in solar PV in various countries in the year 2016.

**TABLE 13.4** The Major Developments in Solar PV in Various Countries in the Year 2016.

| Countries | Major developments in recent years |
|---|---|
| China | • Solar PV capacity of 34.5 GW was added during the year (up 126% over 2015). The total capacity reached 77.4 GW (highest among all the countries)[43] |
| | • Xinjang (3.3 GW), Shandong (3.2 GW), and Henan (2.4 GW) provinces were the top markets in China.[44] Fifteen provinces in China added more than 1 GW each (out of which 9 provinces are in eastern regions) |
| | • The added capacity during the year was represented mostly by the large-scale plants. These large-scale solar PV plants constituted approximately 86% of the total capacity at the year-end even though central government continuous efforts to boost smaller scale distributed systems[45] |
| | • The rapid surge in solar PV capacity led to interconnection delays and grid congestion problems.[46] The challenge of curtailment increased in the year 2015 and because of inadequate transmission, the problems increased further.[47] To overcome these challenges, China fixed minimum assured utilization hours (purchase requirements) for solar (and wind) power plants in affected areas. Further to connect northwestern areas to coastal areas, China continued to construct numerous ultrahigh-voltage transmission lines |

**TABLE 13.4**    *(Continued)*

| Countries | Major developments in recent years |
|---|---|
| The United States | • Solar PV capacity of more than 14.8 GW was added during the year making the total capacity to reach 40.9 GW. California (1.2 GW), Utah (1.2 GW) and Georgia (1 GW) led the capacity addition[48] |
| | • The nonresidential market of the United States, including both industrial and commercial markets, increased 49% to 1.6 GW. However, the residential sector experienced a decline in the expansion rate. The sector grew at a rate of 19%, after record growth in previous years. The success of distributed solar PV and decreasing costs have impelled some US utilities to set up their own solar programs |
| Japan | • The world's third solar PV market, Japan had an installation of 8.6 GW during the year. The total capacity reached 42.8 GW. The growth rate experienced a contraction of 20% after the 2015 boom because of factors like declining FIT payments led by falling prices, land shortages, and problems getting grid connections[49] |
| | • The country's solar PV expansion in recent years is driven mostly by large-scale projects.[50] However, residential sector also witnessed a growth in demand and contributed to 11.8% of the new installations[51] |
| | • The residential solar-plus-storage options also noticed an increased demand. More than 50,000 residential systems contained storage |
| India | • India ranked fourth and seventh for solar PV capacity addition and total additions, respectively. The added capacity during the year was 4.1 GW and the total capacity reached 9.1 GW. The top states as per the total capacity were Tamil Nadu (approx. 1.6 GW), Rajasthan (1.3 GW), Gujarat (1.1 GW), and Andhra Pradesh (1 GW)[52] |
| | • The sharp fall in prices and policy support at national as well as state level have led to increase in demand for large-scale solar projects |
| | • In spite of significant expansion in solar rooftop market, its contribution in country's solar PV capacity remained at about 10%.[53] By 2022, India aims rooftop target of 40 GW. Various challenges such as curtailment and grid congestion are affecting the rooftop solar market growth. To overcome these challenges, construction of eight "green energy corridors" transmission lines to transmit power from high-production areas to high-demand regions had begun[54] |
| Other Asian countries | • In Korea, total capacity reached 4.4 GW with an addition of 0.9 GW[55] |
| | • Philippines added 0.8 GW capacity (total of 0.9 GW) |
| | • Thailand added 0.7 GW capacity (total of 2.15 GW) |

**TABLE 13.4**   *(Continued)*

| Countries | Major developments in recent years |
|---|---|
| European Union | • The total capacity of the European Union reached an estimate of 106 GW with an addition of 5.7 GW during the year. The region's new grid-connected capacity was about 70% contributed by the United Kingdom, Germany, and France.[56] Other top countries were Belgium, Italy, and the Netherlands<br><br>• There are several reasons pertaining to which Europe has become a competitive market. Some of these reasons are transition from feed-in-tariff incentives to tenders and feed-in-premiums for large-scale systems, stagnating electricity demand, etc. |
| The United Kingdom | • Addition of 2 GW during the year made the assisted the total capacity to reach 11.7 GW.[57] From April to September, solar PV produced more electricity than coal depicting changing face of UK electricity supply |
| Germany | • Total capacity of Germany reached about 41.3 GW with an addition of 1.5 GW. Germany and Denmark joined hands and allowed companies to bid in cross-border auctions related to solar PV[58]<br><br>• The shift of consumers from FITs to self-consumption has led to rapid growth in solar-plus-storage market. The share of newly installed residential storage paired systems substantially increased from 14% in 2014 to 41% in 2015 to 50% in 2016. In Europe, Germany alone contributed to approximately 80% of home energy storage market |
| Australia | • With an addition of 0.9 GW in 2016, the country's total solar PV capacity reached 5.8 GW.[59] Solar PV market of Australia has been primarily dominated by residential sector; however, the large-scale and commercial sectors have begun to increase their proportions in 2015 and 2016.[60] At the year end, nearly 1.6 million solar PV installations were functioning.[61] In South Australia and Queensland, about 30% of the houses had solar PV installations. Various other states and territories also had high shares of PV installations[62]<br><br>• The consumers feel encouraged to move to solar PV because of higher retail prices and low wholesale electricity prices. Further, there is a little incentive for the consumers to sell their generation to the grid |
| Chile | • The flourishing mining industry has led to swift development in the northern regions of the country.[63] With an addition of 0.7 GW in the year 2016, the total capacity reached 1.6 GW |

**TABLE 13.4**    *(Continued)*

| Countries | Major developments in recent years |
|---|---|
| Mexico | • Mexico added about 150 MW and the total capacity reached 0.3 GW[64] |
| | • The country held its first tender, although one-third of the addition was accounted by distributed systems |
| Middle East Countries | • Although Middle East was operating at low capacity, the interest in solar PV picked up at the end of 2016 |
| | • Israel retained the spot of the leading market in the region. There was an addition of 0.1 GW with the total capacity of 0.9 GW[65] |
| | • Dubai commenced a 200-MW plant during the year |
| | • During the year, both Kuwait and Jordan brought large projects online |
| Africa | • To meet the increasing electricity demand, diversify the energy mix and to impart energy access, the African countries have started focusing on solar energy |
| | • South Africa (0.5 GW) and Algeria were the leading countries for new capacity during the year[66] |

## 13.5   CONCENTRATING SOLAR THERMAL POWER

Concentrating solar thermal power (CSP), which is also termed as solar thermal electricity, positions the mirrors in an arranged way in an attempt to focus the sun energy to drive conventional steam turbines that produce electricity. The global capacity of CSP increased by 110 MW and reached to total of more than 4.8 GW at the year 2016 end. The total capacity annual growth rate was just over 2% which is the lowest in the past 10 years. The countries leading in terms of new additions were South Africa, China, Spain, the United States, and so on. The newly developed facilities have incorporated thermal energy storage (TES). TES adds to the CSP competitiveness by providing flexibility of dispatchability. In terms of global capacity of CSP, Spain retained its position at the top with 2.3 GW capacity, followed by the United States (1.7 GW). These two countries contributed for over 80% of global installed capacity. However, there has not been any increase in Spain's capacity since 2013.[67]

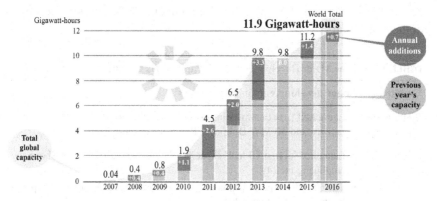

**FIGURE 13.10** CSP global capacity and annual additions, 2007–2016.[68] (Reprinted with permission from Renewables 2017 Global Status Report; REN21: Paris, 2017.)

| CONCENTRA- TING SOLAR THERMAL POWER (CSP) | Levelised Cost of Energy → USD/kWh 0 | 0.1 | 0.2 | 0.3 | 0.4 | 0.5 | 0.6 |
|---|---|---|---|---|---|---|---|
| | Africa | | | | | | |
| | Asia | | | | | | |
| | Central America and the Caribbean | | | | | | |
| | Eurasia | | | | | | |
| | Europe | | | | | | |
| | Middle East | | | | | | |
| | North America | | | | | | |
| | Oceania | | | | | | |
| | South America | | | | | | |
| | ADD* | | | | | | |
| | India | | | | | | |
| | United States | | | | | | |

**FIGURE 13.11** CSP levelized cost in different countries, 2016.[69] (Reprinted with permission from Renewables 2017 Global Status Report; REN21: Paris, 2017.)

| Investment Cost → USD | min | max | wa | Capacity Factor → | min | max | wa |
|---|---|---|---|---|---|---|---|
| Africa | 7164 | 11300 | ● 8392 | | 0.36 | 0.53 | ● 0.4 |
| Asia | 3501 | 13693 | ● 4423 | | 0.17 | 0.54 | ● 0.28 |
| Central America and the Caribbean | | | | | | | |
| Eurasia | | | | | | | |
| Europe | 4811 | 17341 | ● 8839 | | 0.15 | 0.63 | ● 0.31 |
| Middle East | 3491 | 4097 | ● 3705 | | 0.19 | 0.26 | ● 0.22 |
| North America | 4714 | 9009 | ● 6794 | | 0.18 | 0.41 | ● 0.3 |
| Oceania | 9735 | 10767 | ● 9829 | | 0.11 | 0.23 | ● 0.12 |
| South America | | | | | | | |
| China* | 2550 | 7800 | ● 3004 | | 0.17 | 0.28 | ● 0.26 |
| India | 3539 | 7475 | ● 4328 | | 0.21 | 0.54 | ● 0.28 |
| United States | 4714 | 9009 | ● 6794 | | 0.18 | 0.41 | ● 0.3 |

**FIGURE 13.12** CSP investment cost and capacity factor in different countries, 2016.[70] (Reprinted with permission from Renewables 2017 Global Status Report; REN21: Paris, 2017.)

Table 13.5 depicts the major developments in CSP in various countries in the year 2016.

**TABLE 13.5**    The Major Developments in CSP in Various Countries in the Year 2016.

| Countries | Major developments in recent years |
|-----------|-----------------------------------|
| South Africa | • Launched first commercial tower plant with 50 MW Khi Solar One facility and 50 MW Bokpoort Parabolic Trough Plant[71] |
| China | • Launched its first 10 MW capacity and also added 10 MW in Shouhang Dunhuang facility<br>• The CSP program of China aims to install total 1.4 GW of CSP capacity by the end of the year 2018<br>• Projects constituting approx. 650 MW of tower, trough, and Fresnel capacity were under construction |
| India | • 25 MW Gujarat solar one plant under construction<br>• 14 MW Integrated Solar Combined-Cycle plant constructed by National Thermal Power Corporation in Dadri in process |
| Israel | • Construction of 121 MW Ashalim Plot B tower in progress<br>• Also, the construction of 110 MW Ashalim Plot A parabolic trough facility was under progress[72] |
| Kuwait | • Construction continued on Kuwait's 50 MW Shagaya plant |
| France | • Construction continued on 9 MW Pyrenees-Orientales district[73] |
| Denmark | • A 17-MW combined heat and power plant capable to generate both low-temperature heat and electricity for district heating was in progress[74] |

The top CSP companies in the operation, construction, and manufacturing are Saudi Arabia's ACWA Power, Rioglass Solar (Belgium), ACS, Acciona, Sener and TSK (all Spain), Supcon (China), and Brightsource, Solar Reserve, and GE (all the United States). The CSP activity witnessed a substantial shift from the developing countries like Spain and the United States to developing countries especially South Africa, Middle East and North Africa, and China. Till date, trough and tower plants have been preferred by commercial developers with many facilities exceeding the size of 100 MW. However, for nontraditional or smaller facilities, Fresnel facilities are now being built. At the end of the year 2016, construction of four Fresnel plants totaling 90 MW China and a 9-MW facility in France were in progress.[75]

Research and development in CSP sector continued to emphasize on the ways to minimize cost of vital CSP components, on improvements and cost saving in TES, on efficiency of heat transfer process, and so on. Some of the researches have yielded significant positive results, for instance, researchers were able to achieve 97% efficiency in converting sunlight into steam. Further, advances in hybridized CSP systems and thermochemical energy storage were developed. The larger TES systems facilities showed their ability to even produce electricity $24 \times 7$ in South Africa and in absence of sunlight in many other parts of the world. For instance, in South Africa, Khi Solar One facility completed a 24-h cycle of continuous solar power generation.[76]

## 13.6  WIND ENERGY

The global wind energy capacity reached to approximately 487 GW with a net addition of 55 GW in the year 2016. The total increase in wind energy capacity was about 12% in 2016 which was below the record high in 2015. The main reason for such a noteworthy decline in global wind energy capacity addition was the Chinese market which had experienced a strong addition in the year 2015.[77] However, China retained its leading position in new installations.[78] Other countries like United States, Germany, India, Brazil, France, Turkey, the Netherlands, the United Kingdom, and Canada were among the top 10 countries leading in addition of wind energy capacity after China. Bolivia and Georgia were the countries who installed their first wind plants in 2016.

Asia accounts for nearly half of the world's wind energy capacity addition for successively eight years. Uncertainty about upcoming changes in policy, and cyclical or policy-related slowdowns were the main factors that affected some largest wind energy market. Cost-competitiveness and eco-friendly features of wind energy are the main drivers for development of wind energy market.

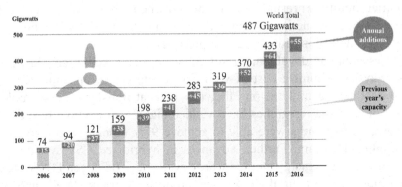

**FIGURE 13.13**   Global wind power capacity and additions, 2006–2016.[79] (Reprinted with permission from Renewables 2017 Global Status Report; REN21: Paris, 2017.)

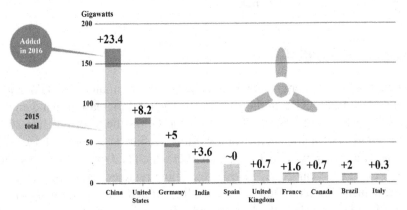

**FIGURE 13.14**   Top 10 countries of 2016 for wind power capacity and additions.[80] (Reprinted with permission from Renewables 2017 Global Status Report; REN21: Paris, 2017.)

| Levelised Cost of Energy → USD/kWh 0 | 0.05 | 0.10 | 0.15 | 0.20 | 0.25 |
|---|---|---|---|---|---|
| **ONSHORE WIND POWER** Africa | | | | | |
| Asia | | | | | |
| Central America and the Caribbean | | | | | |
| Eurasia | | | | | |
| Europe | | | | | |
| Middle East | | | | | |
| North America | | | | | |
| Oceania | | | | | |
| South America | | | | | |
| China | | | | | |
| India | | | | | |
| United States | | | | | |

**FIGURE 13.15**   Onshore wind power cost in different countries, 2016.[81] (Reprinted with permission from Renewables 2017 Global Status Report; REN21: Paris, 2017.)

| Investment Cost → USD | min | max | wa | Capacity Factor → | min | max | wa |
|---|---|---|---|---|---|---|---|
| Africa | 1345 | 2506 | ●1924 | | 0.19 | 0.48 | ●0.37 |
| Asia | 909 | 2784 | ●1263 | | 0.13 | 0.46 | ●0.25 |
| Central America and the Caribbean | 1680 | 3265 | ●2144 | | 0.24 | 0.54 | ●0.35 |
| Eurasia | 1315 | 2651 | ●1891 | | 0.24 | 0.49 | ●0.35 |
| Europe | 1054 | 3702 | ●1866 | | 0.14 | 0.51 | ●0.28 |
| Middle East | 1857 | 3148 | ●2531 | | 0.29 | 0.4 | ●0.34 |
| North America | 1270 | 3148 | ●1805 | | 0.17 | 0.52 | ●0.39 |
| Oceania | 1600 | 3581 | ●2150 | | 0.3 | 0.44 | ●0.35 |
| South America | 1108 | 2903 | ●1912 | | 0.27 | 0.54 | ●0.43 |
| China | 1166 | 1414 | ●1244 | | 0.23 | 0.29 | ●0.25 |
| India | 1044 | 1420 | ●1120 | | 0.19 | 0.33 | ●0.24 |
| United States | 1481 | 2445 | ●1715 | | 0.23 | 0.5 | ●0.4 |

**FIGURE 13.16** Onshore wind power investment cost and capacity factor in different countries, 2016.[82] (Reprinted with permission from Renewables 2017 Global Status Report; REN21: Paris, 2017.)

**FIGURE 13.17** Offshore wind power cost in different countries, 2016.[83] (Reprinted with permission from Renewables 2017 Global Status Report; REN21: Paris, 2017.)

| Investment Cost → USD | min | max | wa | Capacity Factor → | min | max | wa |
|---|---|---|---|---|---|---|---|
| Africa | | | | | | | |
| Asia | 2787 | 4258 | 3286 | | 0.20 | 0.31 | ●0.26 |
| Central America and the Caribbean | | | | | | | |
| Eurasia | | | | | | | |
| Europe | 2053 | 6480 | 4207 | | 0.27 | 0.55 | ●0.36 |
| Middle East | | | | | | | |
| North America | | | | | | | |
| Oceania | | | | | | | |
| South America | | | | | | | |
| China | 1890 | 4258 | 3083 | | 0.23 | 0.29 | ●0.26 |
| India | | | | | | | |
| United States | | | | | | | |

**FIGURE 13.18** Offshore wind power investment cost and capacity factor in different countries, 2016.[84] (Reprinted with permission from Renewables 2017 Global Status Report; REN21: Paris, 2017.)

Table 13.6 depicts the recent development in the wind energy development in major countries.

**TABLE 13.6**    The Recent Development in the Wind Energy Development in Major Countries.

| Countries | Major developments in recent years |
|---|---|
| China | • Total installed capacity was about 145.6 GW in 2015 which increased to 169 GW after a net addition of 23.4 GW in 2016[85] |
| | • Net addition in 2016 was approximately 24% less as compared to 2015 |
| | • Reasons for such decline in new installations were weak electricity demand growth and problem of grid integration[86] |
| | • Leading areas for additions in capacity were in the Yunnan (3.3 GW), Hebei (1.7 GW), and Jiangsu (1.5 GW), with the latter two moderately close to demand centers[87] |
| | • For the first time, the country increased new installations in eastern and southern region also in accordance with the new regulations to promote investment in these regions |
| Asia | • India |
| |    1. Total installed capacity was about 25.1 GW in 2015 which increased to 28.7 GW after a net addition of 3.6 GW in 2016 |
| |    2. Holds fourth position in total capacity addition |
| |    3. The record installations were mainly due to benefits that were ending by 2017[88] |
| | • Turkey with a net addition of about 1.4 GW in 2016 had a total capacity of 6.1 GW and gained position among top 10 countries[89] |
| | • Other countries of Asia like Pakistan has added 0.3 GW, Japan and Republic of Korea both added around 0.2 GW capacity, assisting to drive Asia's total of more than 203 GW[90] |
| | • By the end of 2016, major additional capacity was under construction in Indonesia's first utility-scale wind farm, and Vietnam had just contracted another 940 MW |
| United States | • Ranked second after China, with total capacity of 82.1 GW (including net addition of 8.2 GW in 2016) |
| | • Wind power generation of the country was 226.5 TW h which is only 6% less than China |
| | • Various states of the United States like Texas, Oklahoma, Iowa, Kansas, and North Dakota added capacity of 2.6, 1.5, and 0.7 GW, respectively, led to one-fourth of US capacity[91] |
| | • Nebraska turned out to be the 18th US state to have a capacity of exceeding 1 GW of cumulative wind power capacity[92] |
| | • Total additional wind power capacity under construction was 10.4 GW at the end of 2016 |

**TABLE 13.6** *(Continued)*

| Countries | Major developments in recent years |
|---|---|
| Canada | • Total installed capacity was about 11.9 GW in 2016 with net addition of 0.7 GW which is nearly half of the 2015 capacity addition[93] |
| | • Even though the slow growth of 2016 in comparison to 2015 and 2014 level, wind energy was the largest source of new electricity generation for last 11 years[94] |
| | • The Ontario and Quebec region of the country added 0.4 and 0.2 GW, respectively |
| European Union | • Total installed capacity was 153.7 GW (92% onshore and 8% offshore) with a gross addition of 12.5 GW during 2016[95] |
| | • Net addition of 12.5 GW in 2016 was nearly 3% less than 2015 record high |
| | • 16 member states had capacity of more than 1 GW at the end of 2016[96] |
| | • Factors like continuing economic crises, strict measures united with the shift from regulated prices to bids has affected the growth most |
| | • Nearly 75% of the total capacity addition contributed by top five markets, that is, Germany, France, the Netherlands, the United Kingdom, and Poland |
| | • Germany, the largest wind market increased its operating wind power capacity by 5 GW to increased total capacity to 49.5 GW and its growth was mainly driven by looming shift from guaranteed FITs to competitive auctions for most renewables installations |
| | • Other members of EU like France (adding 1.6 GW), the Netherlands (0.9 GW, mostly offshore), Finland (0.6 GW), Ireland (0.4 GW), and Lithuania (0.2 GW) had a record new installations |
| | • Total wind power capacity for Finland and Lithuania saw their increase by over 56%, and the Netherlands joined the global top 10 for annual additions for the first time in decades[97] |

## 13.6.1 WIND ENERGY INDUSTRY

Global acceptance of the Paris Agreement, the United Kingdom's vote to exit the EU, polls in important wind power markets, and entry of large energy companies are the several developments that had substantial implications (positive and negative) for the wind power industry in future years. It was a worthy year for top turbine manufacturers, with numerous seeing their orders and revenue up over 2015 driven largely by competition with low-cost natural gas capacity and increasingly with solar PV, companies continued inventing to reduce prices and improve yields.

Cost of energy varies in different countries depending upon the regulation, fiscal framework, cost of capital, and other local influences. Various advancement like know-how for setting up and upkeep, standardization of turbine production, and increased efficiency and capacity leads to continuous reduction in LCOE of wind energy. Offshore wind energy prices fell quickly in countries like Chile, India, Mexico, and Morocco due to record low bid for tenders. Nevertheless, challenges have to be faced, as well as wind energy is sensitive to policy changes or measures to protect fossil fuels in some countries. Apart from this, due to wind production and increase in the share of total production, so many countries have to face the challenges related to the grid.

### 13.6.1.1  CHALLENGES FOR WIND ENERGY

Both in coastal and offshore consist of lack of transmission infrastructure, delay in grid connections, lack of public acceptance and deduction, where the rules and current management systems have introduced large amounts of variable renewables. Most wind turbine manufacturing occurs in China, EU, India, and the United States, and the majority is relatively concentrated among some players.[98] In 2016, Ventas (Denmark) regained its leading position, leaving behind Goldwind (China), due to its strong year in the US market. GE (USA) has taken one step to reach second place, Goldwind (below two) as well as Goias (Spain; up one) and Enricon (Germany) scored in the top five. In the top 10, there were other Siemens and Nordex Xionia (both Germany), followed by United Power, Environment, and Mingyang (all China). Goldwind and other top Chinese companies lost their ground mainly due to substantial dependence on the domestic market.[99] The most global supplier was Vestas in 2016, with the establishment in 34 countries.[100] The top 10 turbine maker countries occupied 75% of the market in 2016. However, machineries are being supplied from numerous countries: for example, blade construction has been moved from Europe to North America, South and East Asia, and lately Latin America and North Africa should be close to new markets.

Due to the growing demand of wind energy technologies and projects, dealers of turbine and various project developers have expanded their existing business and/or started new plants and offices around the world. At least seven companies of the United States have increased existing manufacturing plants. To maintain the European offshore industry, Siemens had taken several steps like started new blade factory in

England, led foundation of nacelle factory in Germany, and also entered into an agreement to build a rotor blade plant in Morocco.[101] Senvion, a German company, started their own regional subsidiaries in Japan and India; Innogy on the other hand entered into Ireland to build an onshore portfolio, and DONG, a Denmark company, opened an office in Chinese Taipei to develop offshore projects.[102]

Mergers, takeover, and consolidation were the main tools used by companies to expand their scale and markets. Acquisition of Acciona helped Nordex to become a major player in emerging wind market.[103] The European Union had given its approval to merger of Siemens and Gamesa in 2017 which will create world's largest wind power company in terms of capacity.[104] After that Siemens–Gamesa announced acquisition of French nuclear firm Areva's share of Adwen (Germany), an offshore industry. To gain assets upstream, GE acquired LM Wind Power, a Denmark blade manufacturing company; Senvion took over blade manufacturer Euros Group, a German company; Nordex acquired SSP Technology a rotor blade molds developer and manufacturer; and Vestas purchased Availon to enlarge its service business. Various companies of Chinese government purchased assets in various parts of the world, and EDF purchased UPC Asia Wind Management company and became the first European wind operator to enter the Chinese market.

In 2016, Smart Wind Technologies, an American company, installed DONG Energy's advanced BEAC on radar system which provides three-dimensional data of wind for every minute as it moves over a wind farm or stretch of sea. It also provides significant insights to advise the siting, design, and operation of upcoming offshore projects. To avoid the need of crane operation, Siemens, a customized transport vessel that allows for rolling nacelles on and off deck, avoids the need for crane operations. Offshore wind energy has improved greatly faster than economists anticipated. The main drivers were attaining scale of economies by large turbines and projects, gained experience, increased rivalry between firms which helped in cost reduction, technical enhancement with turbines, and reduced cost of capital because of poor awareness of risk in financial markets.

During June 2016, nine European countries bid jointly for various offshore wind power joint tenders with mutual understanding to cooperate on offshore wind power. On the same day, 11 companies gave consent to uniform legal framework by signing an open letter and set a target of producing offshore wind power at more reduced cost, that is, for less than

EUR 80 per MW h (USD 84 per MW h as of end-2016) within a period of 10 years, that is, by 2025.[105] The industry reached close to these goals during the year, and at the end of 2016, the tenders were lowered by bids for projects from Denmark and Dutch coasts: EUR 50 per MW h, EUR 72 per MW h, and GBP 100 per MW h (USD 123 per MWh), excluding grid connection costs. According to an estimate, the industry achieved the goal of the UK government of 2012—up to 2020 to reduce offshore LCOE for GBP 100 per MWh (USD 123 per MHz)—four years before the target date.[106]

## 13.7   BIOMASS ENERGY

Biomass can be transformed into useful renewable energy through several approaches. Wide range of residues, wastes and crops grown for energy purposes can be transformed into liquid or gaseous fuels for transportation, as a substitute for petrochemicals or can be utilized directly as fuels for heating, cooling, and electricity production. Factors like environmental concerns: both local and global and the rising demand for energy increased production and use of bioenergy in 2016. However, during the year, the biopower production faced obstacles like the existence of cheaper renewable sources of electricity and the increased competition. Contribution to global renewable energy supply—bioenergy in both its traditional and modern use is the primal provider to global renewable energy supply. Biomass's traditional use includes use of animal dung, fuelwood, and agricultural remains in simple stoves with almost nil ignition ability. In 2016, energy contributed by biomass was approximately 62.5 EJ[107] and since 2010, there is a growth rate of around 2.5% per year for biomass.

### 13.7.1   BIOPOWER MARKETS

In 2016, the capacity of global biopower and generation increased by 6% to reach 112 GW and 504 TW h, respectively. The United States (68 TW h) led the world in terms of capacity generation through biomass in the year 2016, followed by China (54 TW h). The other leading countries were Germany (52 TW h), Brazil (51 TW h), Japan (38 TW h), India (30 TW h), and the United Kingdom (30 TW h). Table 13.7 shows countries and their contribution in biopower markets for the year 2016.

**TABLE 13.7** Countries and Their Contribution in Biopower Markets for the Year 2016.

| Countries | Contribution |
|---|---|
| USA | It was the leading producer in terms of electricity from biomass sources; however, production fell 2% to 68 TW h, because alternative renewable generation sources posed price competition for the existing capacity. But, after the installation of 51 small-scale generation plants, capacity of US biopower significantly increased by 197 MW (0.5%) to reach 16.8 GW[108] |
| Europe (Germany/ the United Kingdom/ Poland) | As led by Renewable Energy Directive, in 2016, the electricity generation growth from biogas and solid biomass continued in Europe[109] |
| | Europe's largest contributor of electricity from biomass was Germany, which registered the hike in the capacity of biopower by 2%, to 7.6 GW[110] |
| | The capacity of biopower registered a growth of 6% to reach total of 5.6 GW owning to large-scale generation and continuing rise in production of biogas for electricity |
| | In Poland, capacity auction schemes enhanced the positioning of newly installed biopower capacity resulting to which the capacity of biopower rose by 0.07 GW to reach 1.34 GW, and the generation increased by 50% to reach 15 TW h[111] |
| China | In China, objectives for 13th Five-Year Plan were altered which augmented the capacity of biopower by approximately 13% in 2016, to 12 GW. Further, the total generation increased to approximately 54 TW h |
| Asia (Japan/ Korea/India) | In Japan, the generation and capacity increased significantly as bioenergy starred in the FIT, i.e., feed-in tariff scheme. Japan's capacity and generation for dedicated biomass plants marked to a total of 4 GW and 38 TW h, respectively, in 2016, recording a 5% increase from 2015[112] |
| | In Korea, generation of bioenergy increased by 44% and reached 8 TW h, portraying political endeavors to dwindle down use of coal in electricity generation[113] |
| | In context of India also, the biopower capacity increased. While the on-grid capacity exhibited an increase by 164 MW to reach 8.3 GW, the off-grid capacity enhanced by 18.9 MW to reach 330 MW. Thus, embarking the growth rate of 8% in generation relative to 2015 to mark the total of 30 TW h[114] |
| Brazil | When it comes to consumption, Brazil takes the lead in consumption of biopower and electricity in Latin America. Both capacity and generation rose by 5%. Bagasse, the leftover portion of sugarcane after the extraction of juice, contributed to over 80% of biomass-based electricity generation in Brazil[115] |

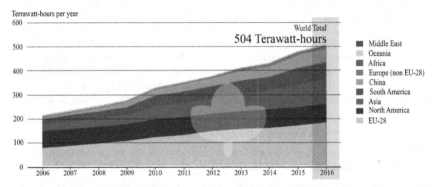

**FIGURE 13.19**    Global biopower generation, by region, 2006–2016.[116] (Reprinted with permission from Renewables 2017 Global Status Report; REN21: Paris, 2017.)

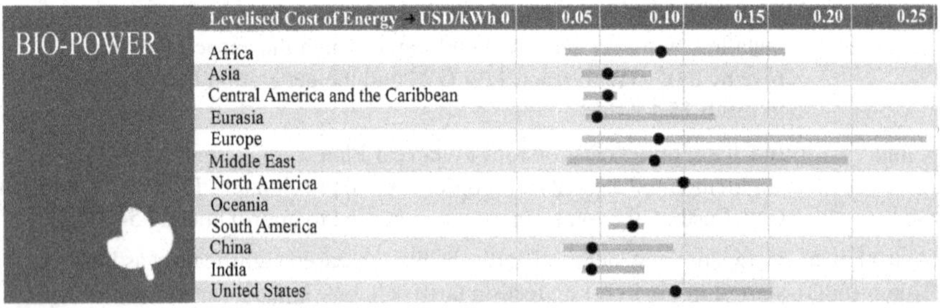

**FIGURE 13.20**    Biopower cost in different countries, 2016.[117] (Reprinted with permission from Renewables 2017 Global Status Report; REN21: Paris, 2017.)

| Investment Cost → USD | min | max | wa | Capacity Factor → | min | max | wa |
|---|---|---|---|---|---|---|---|
| Africa | 625 | 5579 | ●1654 | | 0.45 | 0.91 | ●0.62 |
| Asia | 865 | 3334 | ●1318 | | 0.63 | 0.9 | ●0.67 |
| Central America and the Caribbean | 534 | 7805 | ●1666 | | 0.27 | 0.63 | ●0.6 |
| Eurasia | 1344 | 7106 | ●1756 | | 0.71 | 0.96 | ●0.83 |
| Europe | 956 | 7599 | ●3423 | | 0.45 | 0.93 | ●0.86 |
| Middle East | 885 | 4272 | ●2895 | | 0.29 | 0.93 | ●0.57 |
| North American | 868 | 7375 | ●3666 | | 0.16 | 0.93 | ●0.78 |
| Oceania | | | | | | | |
| South America | 1200 | 1666 | ●1433 | | 0.21 | 0.94 | ●0.53 |
| China | 542 | 6082 | ●1215 | | 0.21 | 0.95 | ●0.62 |
| India | 865 | 2113 | ●1043 | | 0.63 | 0.9 | ●0.77 |
| United States | 1668 | 7375 | ●4135 | | 0.89 | 0.96 | ●0.93 |

**FIGURE 13.21**    Biopower investment cost and capacity factor in different countries, 2016.[118] (Reprinted with permission from Renewables 2017 Global Status Report; REN21: Paris, 2017.)

### 13.7.1.1   TRANSPORT BIOFUEL MARKETS

Table 13.8 depicting countries and their contribution in transport biofuel market.

As compared to ethanol production, the production of biodiesel is more geographically varied since the production is spread across many countries. Top countries in the production of fatty acid methyl ester biodiesel in the year 2016 were the United States, Brazil, Indonesia, Germany, and Argentina with contribution of 18%, 12%, 10%, 10%, and 10%, respectively, in the global production.[123] After a substantial decline in 2015, global production recorded a growth of 7.5% to reach 30.8 billion liters in 2016. The increase was dedicated to reinstated production levels in Argentina and Indonesia and to growth in North America. The biodiesel production in the United States increased by 15% to 5.5 billion liters in 2016 because of improved opportunities for diesel within the Renewable Fuel Standard 2 (RFS2). Growth rate was recorded of 19% in Canada as well. Table 13.9 complies with the following information.

**TABLE 13.8** Countries and Their Contribution in Transport Biofuel Market.

| Countries | Category | Bio Fuel Production |
|---|---|---|
| USA, Brazil, Canada | Ethanol | In the year 2016, the production of ethanol in USA rose by 3.5% to reach a total of 58 billion liters.[119] Brazil's ethanol production showed a minor fall to account for a total of 27 billion liters.[120] Globally, Canada ranked fourth in the year 2015 for ethanol production; the production dwindled by 3% in the year 2016 |
| China | Ethanol | Among the regional producers of ethanol, Asia stands at third position with China being the region's largest and third largest globally. China recorded a hike of 5% over 2015 in 2016. Almost all ethanol production in China is based on conventional starch-based feedstock, the production as well as distribution of which is managed by state-owned oil companies. In 27 cities and 4 provinces of China, E10 mandate is followed but there has been constrain in the production.[121] Earlier, the blending was not permitted to take place outside these areas but this restriction was lessened in the year 2016 |
| Thailand/ India/ France/ Hungary/ the United Kingdom | Ethanol | Driven by robust policy backing in the form of mandates, ethanol production reached 0.9 billion liters in India.[122] In Thailand, ethanol production increased by 3.9% to reach 1.2 billion liters. In Europe, which is the next-largest producing region, ethanol production shrank by 6% in 2016 |
| | | Owing to a poor grain harvest, the production fell sharply by 14% in France |
| | | Production grew at a faster rate in Hungary (38%) and the United Kingdom (23%) |

## 13.7.2  THE BIOENERGY INDUSTRY

The bioenergy industry consists of suppliers and processors of feedstock, firms serving the end-users with biomass, biomass-harvesting distributors and manufacturers, equipment for storage and handling, and appliances manufacturers and hardware components aimed to transform biomass to worthwhile energy service and carriers. Biomass has mainly three forms—solid, liquid, and gaseous.

### 13.7.2.1  SOLID BIOMASS INDUSTRY

Solid biomass industry is concerned with using, processing, and delivering solid biomass for the production of electricity and heat, extending from traditional biomass informal supply to locally centered supply of heating appliances of smaller-scale and to global and regional players engaged in district heating of larger scale and supply and operations of power generation technology.

**TABLE 13.9**  Trends of Countries/Continents for Biodiesel Production.

| Country | Category | Biodiesel production |
| --- | --- | --- |
| Brazil/ Argentina | Biodiesel | Despite an increase in the blending mandate, biodiesel production in Brazil fell by 3% because of a decline in demand for diesel consumption which in turn led to a reduced level of business activity. Led by increased domestic demand and better market prospects in Peru and the United States, the production of biodiesel increased 43% to reach 3 billion liters after recovering from a fall in 2015 in Argentina[124] |
| Europe | Biodiesel | In Europe, the production of biodiesel dropped by 5% to reach a total of 10.7 billion liters.[125] Among the European countries, Germany (3 billion liters) and France (1.5 billion liters) were the top two largest producers. Both of these countries experienced a decline of 11% in production.[126] The biodiesel production in Spain and Poland grew by 1% and 8% to reach 1.1 and 0.9 billion liters, respectively |

**TABLE 13.9** *(Continued)*

| Country | Category | Biodiesel production |
|---------|----------|----------------------|
| Asia | Biodiesel | Post a major decline in 2015, Indonesian production grew by 76% and became region's largest producer again in 2016, encouraged by a number of factors that stimulated the domestic market[127] |
| | | Owing to the absence of prevalent blending mandates and reduced use of diesel fuel which signifies a slowdown in the growth of industrial activities, the production of biodiesel in China fell by 10% to 0.3 billion liters[128] |
| Global perspective | HVO biodiesel (hydrotreated vegetable oil) | The global production of HVO grew by 22% to reach 5.9 billion liters in the year 2016.[129] The production was geographically concentrated in the countries, such as the Netherlands, Singapore, the United States, and Finland |
| | | In the United States, the consumption of biomethane as a transport fuel grew by almost six times between two years, 2014 and 2016 to reach 712 million liters and it overtook Germany and Sweden.[130] EPA ruling on RFS2 in 2014 stimulated the conversion to biomethane from biomass. The EPA ruling incentivized biomethane by promoting it to the category of advanced cellulosic biofuels category |

*Europe*—Europe continued the trend of converting the power station capacity of mostly larger scale to run on wood pellets instead of coal. For instance, a power station of 360 MW was converted from coal to operate on wood pellets in Denmark which aimed to supply heat and electricity based on biomass to over 100,000 homes and about 230,000 homes, respectively.[131] Further, in the United Kingdom, the European Commission approved Drax to transfigure one-third unit of its coal-based plant to operate on wood pellets.

*The Republic of Korea and Japan*—In both the Republic of Korea and Japan, the imports of wood pellet increased during 2016. Japan imports consist of 300,000 t industrial pellets per year and 600,000 t of palm kernel shells.[132]

The United States is the leading wood pellets exporter, producing nearly 6.9 million tons and exporting 70% of it in 2016.[133] The exports of wood pellets also rose in Canada by 47% to reach 2.5 million tons.[134] The pellet industry consists of self-regulating producers and is centered around sawmill operations. There are 142 operating pellet plants in the United States and 58 in Canada.[135] There is a widespread expansion in industrial and heating use of the global market for wood pellets, growing progressively at a rate of closely 1 million tons/year over a period of 10 years.

There is a controversy regarding the pellets large-scale use and bioenergy sustainability. The European Commission, on November 16, in its proposals for a new Renewable Energy Directive specified its aim to reinforce compulsory criteria for bioenergy sustainability by outspreading the scope to include solid biomass and electricity generation, cooling, and heating through biomass. Even though the enforcement of such mandatory criteria is applicable to biofuels only, the member states can bring criteria for electricity and heat sectors, as done by Denmark and the United Kingdom.[136]

The process of torrefaction of wood facilitates the pellets production having greater energy density and leads to a product which is compatible with coal-based designed systems. Major developments in 2016 included the commencement of production of terrified pellets by Airex Energy (Canada) at its Bécancour plant having an annual capacity of 15,000 t.[137] Further, an investment of USD 74–84 million was made by Biopower Oy, a Finnish company to construct a biocoal-based plant in Mikkeli, Finland, with an expectation to produce 200,000 t pellets yearly.[138]

### 13.7.2.2 LIQUID BIOFUELS INDUSTRY

Liquid biofuel production is like an oligopolistic market characterized by few large industrial players having significant market shares.

With an increase in the demand and production in China, there is a development of new patterns of trade for ethanol. Till 2016, China was exporting ethanol to some Asian markets. In 2016, US exports to China increased by 2.4 times, thus making China a major ethanol importer of the United States.[139] To curb down the effects, China lately announced an import tax on ethanol to assist domestic production. Between 2015 and 2016, the net imports of biodiesel to the United States increased two-fold (reached 2.3 billion liters from 1 billion liters).[140] Argentina was a leading supplier of this growth, having a significant capacity of biodiesel production. Since the year 2010, it has been providing biodiesel to markets in the European Union, United States, Peru, and other countries. But after an imposition of heavy tax on import of Argentinian biodiesel by the

European Union in the year 2013, the biodiesel production capacity of Argentina has remained underutilized (at 40–55%).[141]

In Africa, despite significant potential and attempts, production has been slow because of lack of appropriate technology. Some promising developments took place like launch of national biofuels strategy by Nigeria in 2016. Setup of a biorefinery to produce ethanol and other products using agricultural products was announced by the Nigeria National Petroleum Corporation. Aiming to a target of producing 22,000 liters of ethanol daily, Union Dicon Salt and Delta State of Nigeria jointly agreed for a project to plant cassava in 100,000 ha and to build a plant for ethanol processing. Biofuels Nigeria is aiming to construct a biodiesel plant based on Jatropha as its feedstock.[142]

The development and commercialization of advanced biofuels aim at three objectives. The first one is to produce those fuels which can serve more savings in terms of life-cycle carbon than the existing biofuels generated from oils, sugar, and starch. The second objective is the production of fuels which have a lesser impact on the use of land, for instance, from residues and wastes and thus minimizing indirect land-use change impacts. The last objective is the production of biofuels with attributes facilitating them to substitute fossil fuels in advanced transport systems like aviation engines or to be combined with conventional sources of fuels in high proportions. HVO led the market for new biofuels followed by ethanol from cellulosic materials and by fuels from thermochemical processes. Plans were proposed for building numerous additional manufacturing plants of cellulosic ethanol, extending the production geographical coverage outside Europe and the United States.

In the context of Asia, New Tianlong Industry Company Ltd. of China and DuPont of the United States signed an agreement to begin the construction of cellulosic ethanol manufacturing plant in Siping City which would be the largest plant in China.[143] The first cellulosic in India was inaugurated by India Glycols in Kashipur which operates on cane bagasse, cotton stalk, maize stover, wood chips, and bamboo. Further, in India, MoU concerning five extracellulosic ethanol plants were concluded. In Thailand, Mitsui and Toray (both belonging to Japan) announced plans to construct a large-scale plant to produce ethanol from sugar bagasse by August 2018.[144]

Commercialization of thermal processes like pyrolysis and gasification progressed in 2016 with strong interest in the progress of aviation biofuels. By the year end, American Society for Testing and Materials (ASTM) certified two additional technology pathways to produce biojet fuels, bringing the total to five. There are various manufacturers in the aircraft industry like Airbus and Boeing that are pivotal when it comes to production of bio-jet fuels. Also, various air carriers like Aeromexico, British Midland, Southwest Airlines, Qatar Airways, United Airlines, Alaska Airlines, Finnair, Gol, Scandinavian Airlines, Lufthansa, FedEx, and KLM initiated and persistently used biofuels in the year 2016.[145] In Netherlands, an initiative was introduced to produce sustainable biofuels for marine applications.

### 13.7.2.3   GASEOUS BIOMASS INDUSTRY

*The United States and Europe*: The major proportion of production of biogas takes place in the United States (by accumulating landfill gas) and in Europe (by stimulating anaerobic digestion of animal manures and wastes from agriculture).

*Asia and Africa*: Though these two areas started from a very low level, the rate of growth is spiking tremendously. Mounting markets for biomethane and biogas are fostering commercial activities at a global level. In several other parts, it's a trend to generate and utilize biogas produced by handling liquid outflow and wastes.

*USA*: Biomethane is accepted as a fuel for transportation sector recently. It was announced by BP to purchase the business dealing with bio-methane gas for a sum of US $155 million.[146]

*India*: Capacity of biogas is approximately 300 MW which is produced by several industrial processes having powerful standards for quality of water thus restricting the discharge of wastes into waterways.

*Africa*: Growth trend in production of biogas has been registered from wastes of agriculture and municipality. If we talk about South Africa, "energy-from-waste" plant near Cape Town was set up, having a total expenditure of US $29 million and first grid powered by biogas was introduced in Kenya to bring in power to approximately 6000 homes in rural areas.[147]

## 13.8   CONCLUSION

The present study has made an attempt to discuss about the innovations in renewable energy market, ongoing commercialization in terms of new projects establishments, additions in the existing capacity, major developments in recent years, and so on and their impact on market depicting the change in respective energy investment cost and their levelized cost. Six forms of renewable energy, namely, ocean energy, geothermal energy, hydropower, solar photovoltaics (PV), concentrating solar thermal power, wind energy, and biomass energy have been focused in the study. The findings show the top countries in terms of total capacity, countries leading for capacity addition, challenges, and so on for each energy form.

The total capacity of ocean energy reached 536 MW for the year ending 2016. Countries like Scotland, France, and Canada made a mark with numerous deployments. Geothermal energy served as one of the important sources of renewable energy with global capacity reaching to 13.5 GW. Indonesia and Turkey were the top countries for geothermal plant installations. In case of hydropower energy, the leading countries for total capacity were China, Brazil, the United States, Canada, the Russian Federation, India, and Norway. Out of total new installations in the year 2016, China contributed to one-third of it. The global capacity of solar PV reached to more than 303 GW in the year 2016 in which Asia had approximately two-third share. Solar PV experienced market expansion because of reasons like growing demand for electricity, its increasing competitiveness, potential to control pollution, and minimize carbon dioxide emissions. China, the United States, Japan, India, and the United Kingdom were the leading five countries in the year 2016 for capacity additions. CSP global capacity increased by 110 MW and reached to a total of approx. 4.8 GW at the year 2016 end. However, the growth rate came down to 2% which is the lowest in the last 10 years. Wind energy, again one of the critical sources of renewable energy, saw an addition of 55 GW in the year 2016 aiding the total capacity to reach 487 GW. China retained its top position in new installations. The United States, Germany, India, and Brazil were the other leading countries in capacity addition. Bioenergy, in both its traditional and modern use is the primal provider to global renewable energy supply. The bioenergy generation capacity increased by 6% in the year 2016 and reached to 504 TW h. In terms of capacity generation through biomass, the United States, China, Germany, Brazil, and Japan were the leading countries. All three forms of biomass industry, that is, solid, liquid, and gaseous, have been discussed in the present study. An analysis of the rate of development of various sources of renewable energy along their rising demand helps in sustaining the positivity in the renewal energy market and makes the future promising.

## KEYWORDS

- ocean energy
- geothermal energy
- hydropower
- photovoltaics
- wind energy

## REFERENCES

1. International Renewable Energy Agency (IRENA). Renewable Capacity Statistics 2017 (Abu Dhabi: April 2017). http://www.irena.org/DocumentDownloads/Publications/IRENA_RE_Capacity_Statistics_2017.pdf
2. Nova Innovation. "Nova & ELSA third turbine deployed in Shetland tidal array—a world showcase for Scottish renewable innovation", press release (Edinburgh: February 23, 2017). https://media.wix.com/ugd/36770a_87c2b2545db246c7b2fd7707b239077c.docx?dn=Nova%20T3%20Press%20Release%2023rd%20Feb%202017.docx; Nova Innovation. "Nova innovation deploys world's first fully operational offshore tidal array in Scotland", press release (Edinburgh). https://media.wix.com/ugd/efc58c_651189c778284074bc41d080564bcdf8.docx?dn=NOVA%20INNOVATION%20DEPLOYS%20WORLDS%20FIRST%20TIDAL%20ARRAY.docx
3. Scotrenewables Tidal Power. "Scotrenewables installs world's largest tidal turbine at EMEC for first time", press release (Edinburgh: October 13, 2016). http://www.emec.org.uk/pressrelease-scotrenewables-installs-worlds-largest-tidal-turbine-atemec-for-first-time/
4. OpenHydro. "OpenHydro deploys second Paimpol-Brehatturbine", press release (Dublin: May 30, 2016). http://www.openhydro.com/OpenHydro/media/Documents/News%20PDFs/30-May-2016.pdf; OpenHydro. "The first of two OpenHydro tidal turbines on EDF's Paimpol-Brehat site successfully deployed", press release (Dublin: January 20, 2016). http://www.openhydro.com/OpenHydro/media/Documents/News%20PDFs/20-Jan-2016.pdf
5. Technalia. "Oceantec deployed at BiMEP its first wave energy converter". http://www.tecnalia.com/en/energy-and-environment/news/oceantec-deployed-at-bimepits-first-wave-energy-converter.htm (accessed October 13, 2016).
6. Waves4Power. "6 months at Runde". http://www.waves4power.com/w4p-news-updates/1706 (Accessed August 9, 2016); Seabased. "Wave power generated to Nordic electricity grid!". http://www.seabased.com/en/newsroom/218-wave-power-generated-to-nordic-electricity-grid (accessed January 21, 2016).

7. Fred. Olsen & Co. http://www.boltwavepower.com (accessed April 21, 2017); Fred. Olsen & Co. "6 month of continuous power export". http://www.boltwavepower. com/6-month-of-continuous-power-export (accessed January 20, 2017).

8. Korea Research Institute of Ships and Ocean Engineering. "Development of wave energy converters applicable to breakwater and connected to micro-grid with energy storage system". http://www.kriso.re.kr/eng/user/3400ms (accessed February 9, 2017).

9. China's State Oceanic Administration. "Announcement of the 13th Five-Year Plan on ocean energy". http://www.soa.gov.cn/zwgk/zcgh/kxcg/201701/t20170112_54473.html (accessed January 1, 2017).

10. Singh, H. K.; Chandrasekharam, D.; Trupti, G.; Mohite, P.; Singh, B.; Varun, C.; Sinha, S. K. Potential Geothermal Energy Resources of India: A Review. *Curr. Sustain. Renew. Energy Rep.* **2000**, 3 (3–4), 80–91.

11. Page 53; Renewables 2017 Global Status Report; REN21: Paris, 2017.

12. Page 92; Renewables 2017 Global Status Report; REN21: Paris, 2017.

13. Page 93; Renewables 2017 Global Status Report; REN21: Paris, 2017.

14. Capacity of 1.44 GW at end of 2015 from Indonesian Ministry of Energy and Mineral Resources, "Pemerintah Targetkan Kapasitas Terpasang PLTP 1.751 MW Selama 5 Tahun", press release. https://www.esdm.go.id/en/media-center/news-archives/pemer-intah-targetkankapasitas-terpasang-pltp-1751-mw-selama-5-tahun (Jakarta: January 8, 2016); "Capacity of 1.64 GW at end-2016" from Indonesian Ministry of Energy and Mineral Resources, "Sistem Satu Data Tekan Biaya Eksplorasi Panas Bumi", press release. https://www.esdm.go.id/en/media-center/news-archives/sistem-satu-data-tekan-biaya-eksplorasi-panas-bumi (Jakarta: February 15, 2017).

15. Ormat Technologies Inc. "Ormat announces commercial operation of Plant 4 in Olkaria III in Kenya, expanding complex capacity to nearly 140 MW", press release (Reno, NV: February 4, 2016) http://www.ormat.com/news/latest-items/ormatan-nounces-commercial-operation-plant-4-olkaria-iii-kenyaexpanding-complex-c

16. Nikkei Asian Review. "Small geothermal plants gaining steam in Japan". http://asia. nikkei.com/Business/Companies/Small-geothermal-plants-gaining-steam-in-Japan. (accessed February 27, 2017).

17. Negishi, M. *Wall Street Journal.* "Japan's shift to renewable energy loses power". https:// www.wsj.com/articles/japans-shift-to-renewable-energy-losespower-1473818581 (accessed September 14, 2016); Movellan, J. *Renewable Energy World.* "Popular hot springs in Japan co-exist with binary geothermal power plants". http://www. renewableenergyworld.com/articles/2015/12/popular-hotsprings-in-japan-co-exist-with-binary-geothermal-power-plants.html (accessed December 14, 2015).

18. Reykjavik Geothermal. "Corbetti Geothermal Power". http://www.rg.is/static/files/ about-us/rg-corbettigeothermalpower.pdf (accessed March 31, 2017); Richter, A. ThinkGeoEnergy. "Corbetti projects signs 500 MW PPA with Ethiopian state utility". http://www.thinkgeoenergy.com/corbetti-project-signs-500-mw-ppa-with-ethiopian-state-utility/ (accessed July 27, 2016).

19. China National Energy Administration (CNEA). 13th Five-Year-Plan for Geothermal Power. http://www.nea.gov.cn/136035635_14863708180701n.pdf (Beijing: February 6, 2017).

20. Xin, Z. "Sinopec to harvest more heat from earth". *China Daily*. http://europe. chinadaily.com.cn/business/2017-02/15/content_28202201.htm (accessed February 15, 2017).

21. Remo, A. R. "Group seeks perks for new geothermal technology". *Philippine Daily Inquirer*. http://business.inquirer.net/213590/group-seeks-perks-for-geothermal-technology (accessed August 17, 2016).

22. Estimate based on IHA, op. cit. note 1, and on IHA, *Hydropower Status Report 2016*. http://www.hydropower.org (London: May 2016).

23. Page 58; Renewables 2017 Global Status Report; REN21: Paris, 2017.

24. Page 92; Renewables 2017 Global Status Report; REN21: Paris, 2017.

25. Page 93; Renewables 2017 Global Status Report; REN21: Paris, 2017.

26. CNEA, 13th Five-Year-Plan, op. cit. note 2.

27. Renewable Energy Statistics 2017. http://www.irena.org/DocumentDownloads/Publications/IRENA_RE_Capacity_Statistics_2017.pdf (Abu Dhabi: 2017).

28. Impregilo, S. "Ethiopia inaugurates tallest RCC dam in world built by Salini Impregilo", press release. https://www.salini-impregilo.com/en/press/press-releases/ethiopia-inaugurates-tallest-rcc-dam-in-world-built-by-saliniimpregilo.html (Milan: December 17, 2016).

29. IHA, op. cit. note 19; IHA, *2017 Key Trends in Hydropower*, op. cit. note 1.

30. Tenaga Nasional. "Ulu Jelai hydroelectric project to boost TNB's hydro installed capacity to 2,533 MW", press release. https://www.tnb.com.my/assets/press_releases/ PRESS_RELEASE_ULU_JELAI_HYDROELECTRIC_PROJECT_TO_BOOST_ TNBS_HYDRO_INSTALLED_CAPACITY_TO_2%2C533MW_09082016.pdf. (Kuala Lumpur: August 9, 2016); The Star. "New source of green power", http://www. thestar.com.my/metro/community/2016/09/24/new-sources-of-powertwo-dams-in-tasik-kenyir-to-add-265mw-to-national-grid/ (accessed September 24, 2016).

31. Capacity at end-2016 of 26.7 GW from TEİAŞ website. http://www.teias.gov.tr

32. EIA, op. cit. note 2, Tables 6.2.B and 6.3; US Federal Energy Regulatory Commission. "Energy infrastructure update for December 2016". https://www.ferc.gov/legal/staff-reports/2016/dec-energy-infrastructure.pdf (Washington, DC: December 2016).

33. Annual generation data from EIA, op. cit. note 2, Table 1.1.

34. RusHydro. "RusHydro launches Zaragizhskaya small hydropower plant in the South of Russia", press release. http://www.eng.rushydro.ru/press/news/102279.html (Moscow: December 29, 2016); RusHydro. "RusHydro inaugurates Zelenchukskaya hybrid hydropower plant in the South of Russia", press release. http://www.eng.rushydro.ru/ press/news/102176.html (Moscow: December 23, 2016).

35. Based on 2016 year-end capacity of 48,086 MW from system operator of the Unified Energy System of Russia, op. cit. note 2, and 47,855 MW at the end of 2015, from system operator of the Unified Energy System of Russia, Report on the Unified Energy System in 2016. http://www.so-ups.ru/fileadmin/files/company/reports/disclosure/2016/ups_rep2015.pdf (Moscow: February 1, 2016).

36. Eskom. "Ingula: powering South Africa's economy", press release. http://www.eskom. co.za/news/Pages/Marr8.aspx (Johannesburg: March 8, 2017).

37. da Silva, A. F.; et al., "Developing the Baixo Sabor pumped storage cascade in Portugal", HydroWorld. http://www.hydroworld.com/articles/print/volume-25/issue-2/features/ developing-the-baixosabor-pumped-storage-cascade-in-portugal.html (accessed April 1, 2017).

38. At least 303 GW, from IEA PVPS, op. cit. note 1, p. 7.
39. IEA PVPS, op. cit. note 1, p. 15.
40. Page 66; Renewables 2017 Global Status Report; REN21: Paris, 2017.
41. Page 92; Renewables 2017 Global Status Report; REN21: Paris, 2017.
42. Page 93; Renewables 2017 Global Status Report; REN21: Paris, 2017.
43. Xiao, D. China added 34.54 GW for a year-end total of 77.42 GW, "2016 photovoltaic power generation statistics", China National Energy Board. http://www.nea.gov.cn/2017-02/04/c_136030860.htm (accessed February 4, 2017).
44. Top market and installations based on data from Xiao, op. cit. note 13; "no-go" area from AECEA, briefing paper—China solar PV development, January 2017, p. 1; Xinjiang's 2016 installations exceeded government guiding targets by 300%, from idem.
45. Figure of 86% based on data from Ibid.; of the total in operation at end-2016, 67.1 GW was classified as large-scale plants and 10.33 GW as distributed. Pushing for distributed and grid inadequacy from IEA-PVPS, op. cit. note 3, p. 43. http://www.ieapvps.org/fileadmin/dam/public/report/national/Trends_2016_-_mr.pdf
46. *PV Magazine*. Up 11-fold based on cumulative capacity of 6750 MW at end-2012, from IEA PVPS, op. cit. note 3, p. 68; grid congestion and delays from, for example, "China's NDRC order grid operators to purchase curtailed solar power in congested regions". https://www.pv-magazine.com/2016/06/01/chinas-ndrc-order-grid-operators-to-purchasecurtailed-solar-power-in-congested-regions_100024816/ (accessed June 1, 2016); Shen, F. *Renewable Energy World*. "China's grid operator blames bad planning for idled renewable energy". http://www.renewableenergyworld.com/articles/2016/04/china-s-gridoperator-blames-bad-planning-for-idled-renewable-energy.html (accessed April 1, 2016).
47. Challenge in 2015 from China National Energy Board, cited in China Electricity Council, "2015 PV-related statistics". http://www.cec.org.cn/yaowenkuaidi/2016-02-05/148942.html (accessed February 6, 2016); Renewable Energy World, inadequate transmission from Paula Mints, "Notes from the Solar Underground: The solar roller coaster and those along for the ride—First Solar, SunPower, Q-Cells". http://www.renewableenergyworld.com/articles/2016/08/notes-fromthe-solar-underground-the-solar-roller-coaster-and-those-alongfor-the-ride-first-solar-sunpower-q-cells.html (accessed September 1, 2016).
48. Yale e360. Georgia third largest installer without additional subsidies from Cheryl Katz, "Northern lights: large-scale solar power is spreading across the U.S.". http://e360.yale.edu/features/northern-lights-utility-scale-solarpower-spreading-across-the-us (accessed March 23, 2017).
49. *Japan Times*. BMI Research, cited in Anne Beade, "Sun setting on Japan's solar energy boom". http://www.japantimes.co.jp/news/2016/11/30/business/sun-settingjapans-solar-energy-boom/ (accessed November 30, 2016); Watanabe, C.; Stapczynski, S. *Bloomberg*. "Japan's solar boom showing signs of deflating as subsidies wane". https://www.bloomberg.com/news/articles/2016-07-06/japan-s-solar-boom-showingsigns-of-deflating-as-subsidies-wane (accessed July 5, 2016).
50. Beade, op. cit. note 38.
51. Brian Publicover. "Distributed-generation takes the lead in Japan's new power capacity development", Solar Asset Management. http://solarassetmanagement.asia/news/2016/5/18/distributed-generation-takes-the-lead-in-japans-new-powercapacity-

development (accessed May 18, 2016); residential sector's share (based on projects <10 kW), from METI, provided by Matsubara, op. cit. note 37.

52. Top states and capacities of Tamil Nadu (1577 MW), Rajasthan (1324 MW), Gujarat (1101 MW) and Andhra Pradesh (1009 MW) from Mercom Capital Group, op. cit. note 46; Tamil Nadu leads all other states for capacity due to lack of reliable electricity from the grid and high consumer awareness, from Jyoti Gulia, "Rooftop solar market in India witnessing rapid growth but 2022 target seems elusive", Bridge to India. http://www. bridgetoindia.com/rooftop-solar-market-in-india-witnessingrapid-growth-but-2022-target-seems-elusive/ (accessed May 31, 2016).

53. Bridge to India, cited in Ibid.; "OPEX model takes hold in India but faces a key challenge", Bridge to India, http://www.bridgetoindia.com/opex-model-takes-hold-india-faces-keychallenge/ (accessed October 17, 2016); "2016 was a great year...", op. cit. note 46; the rooftop market passed 1 GW in September 2016, from idem.

54. Kenning, op. cit. note 49.

55. Added 850 MW for a total of 4.35 GW, from IEA PVPS, op. cit. note 1, p. 15; and added 0.9 GW for a total of 4.5 GW, from Jaehong Seo, KOPIA, presentation for International Green Energy Conference 2017, Daegu, Republic of Korea, 5–6 April 2017, provided by Haugwitz, op. cit. note 14.

56. Based on data from SolarPower Europe, op. cit. note 5, and on country-specific data and sources provided in this section. The EU installed 5683.3 MW in 2016, and the United Kingdom, Germany and France added a combined 3951 MW, from idem.

57. The United Kingdom added 1.97 GW from SolarPower Europe, op. cit. note 5, and added 1.97 GW for a total of 11.63 GW, from IEA PVPS, op. cit. note 1, p. 15; added 2039 MW in 2016 for total of 11,727 MW, based on data for end-2015 and end-2016 from UK Department for Business, Energy & Industrial Strategy, "Solar photovoltaics deployment in the UK, February 2017". https://www.gov.uk/government/uploads/system/uploads/attachment_data/file/585828/Solar_photovoltaics_deployment_March_2017.xlsx (accessed March 30, 2017); and added 2.4 GW for a total of 11,562 MW, from UK Department for Business, Energy & Industrial Strategy, National Statistics, Energy Trends Section 6: Renewables, Table 6.1, pp. 63, 69. https://www.gov.uk/government/statistics/energy-trends-section-6-renewables (accessed March 30, 2017).

58. BMWI and German Federal Network Agency (Bundesnetzagentur), "Federal Network Agency launches Germany's first cross-border PV auction with Denmark", press release. http://www.bmwi.de/Redaktion/EN/Pressemitteilungen/2016/20161012-federalnetwork-agency-launches-germany-s-first-cross-border-PVauction-with-denmark.html (Berlin: October 12, 2016).

59. Data based on year-end 2015 total of 4939.134 MW and on year-end 2016 totals of 5783.963 MW reported installed and 5794.371 MW estimated installed, for estimated additions of 855 MW, all from Australian PV Institute (APVI), "Australian PV market since April 2001". http://pv-map.apvi.org.au/analyses.

60. *Straits Times*. Masson, op. cit. note 1; Jonathan Pearlman, "Australia taking solar power to the next level". http://www.straitstimes.com/asia/australianz/australia-taking-solarpower-to-the-next-level (accessed January 31, 2016); SolarPower Europe, op. cit. note 62, p. 21. About 1.5 million households had rooftop solar PV, with the highest share (nearly 30%) in Queensland, from Pearlman, op. cit. this note.

61. APVI. op. cit. note 75 (accessed March 9, 2017).

62. APVI. "Percentage of dwellings with a PV system by state/territory", funded by the Australian Renewable Energy Agency. pv-map.apvi.org.au (accessed March 10, 2017). As of late 2016, 30.4% of dwellings in Queensland had solar PV installations, followed by South Australia (29.5%), West Australia (23.8%), Victoria (14.7%), New South Wales (14.6%), Australian Capital Territory (13.5), Tasmania (12.7%), and New Territories (10%), from idem.

63. Forbes. William Pentland, "Solar power thrives in Chile, no subsidies needed". https://www.forbes.com/sites/williampentland/2015/11/07/solar-powerthrives-in-chile-no-subsidies-needed/#41e375853987 (accessed November 7, 2015).

64. Mexico added 150 MW for a total of 320 MW, from IEA PVPS, op. cit. note 1, p. 15; and added about 300 MW, from SolarPower Europe, op. cit. note 5.

65. Israel added 130 MW for a year-end total of 910 MW, from IEA PVPS, op. cit. note 1, p. 15.

66. South Africa added 536 MW for a total of 1450 MW, and Algeria installed some 50 MW, from IEA PVPS, op. cit. note 1, p. 5; South Africa added 505 MW, Algeria added 171 MW, and Senegal added 43 MW, all based on data for end-2016 and end-2015, from IRENA, Renewable Capacity Statistics 2017 (Abu Dhabi: 2017), p. 24. http://www.irena.org/DocumentDownloads/Publications/IRENA_RE_Capacity_Statistics_2017.pdf

67. Op. cit. note 1, all sources. United States from Solar Energy Industries Association (SEIA), "Solar industry data: solar industry growing at a record pace", http://www.seia.org/researchresources/solar-industry-data (accessed April 17, 2017); Spain from op. cit. note 1, all sources.

68. Page 73; Renewables 2017 Global Status Report; REN21: Paris, 2017.

69. Page 94; Renewables 2017 Global Status Report; REN21: Paris, 2017.

70. Page 95; Renewables 2017 Global Status Report; REN21: Paris, 2017.

71. Op. cit. note 1, all sources. Bernardo, C. "Khi Solar One kicks into commercial operation", ESI Africa. https://www.esi-africa.com/news/khi-solar-one-kicks-intocommercial-operation (February 8, 2016); Bernardo, C. "What you need to know about the Bokpoort solar plant", IOL. http://www.iol.co.za/news/south-africa/northern-cape/what-you-need-to-know-about-the-bokpoort-solar-plant-1997714 (accessed March 14, 2016).

72. NREL. "Ashalim Plot B". https://www.nrel.gov/csp/solarpaces/project_detail.cfm/projectID=277 (accessed March 22, 2016).

73. Shumkov I. "French JV raises EUR 60m for solar thermodynamic project". https://renewablesnow.com/news/french-jv-raises-eur-60m-for-solarthermodynamic-project-543113/ (accessed October 13, 2016).

74. Aalborg, C. S. P. "Aalborg CSP supplies concentrated solar power system for combined heat and power generation in Denmark". http://www.aalborgcsp.com/news-events/newstitle/news/aalborg-csp-supplies-concentrated-solar-powersystem-for-combined-heat-and-power-generation-in-denma/ (accessed February 29, 2016).

75. GreentechMedia. Jason Deign. "Concentrating solar power isn't viable without storage, say experts". https://www.greentechmedia.com/articles/read/is-csp-viable-without-storage (accessed November 1, 2016).

76. Energyblog. Abengoa. "Khi Solar One near Upington achieves a technological milestone",Energyblog.http://www.energy.org.za/news/khi-solar-one-near-upington-achieves-a-technological-milestone (accessed March 30, 2016).
77. FTI Intelligence, op. cit. note 2.
78. GWEC, op. cit. note 1; WWEA, op. cit. note 1.
79. Page 88; Renewables 2017 Global Status Report; REN21: Paris, 2017.
80. Page 88; Renewables 2017 Global Status Report; REN21: Paris, 2017.
81. Page 94; Renewables 2017 Global Status Report; REN21: Paris, 2017.
82. Page 95; Renewables 2017 Global Status Report; REN21: Paris, 2017.
83. Page 94; Renewables 2017 Global Status Report; REN21: Paris, 2017.
84. Page 95; Renewables 2017 Global Status Report; REN21: Paris, 2017.
85. China added 23,370 MW for a total of 168,732 MW, from GWEC, op. cit. note 1; China added 23,328 MW for a total of 168,690 MW, from EurObserv'ER, op. cit. note 1, p. 3; and added 23,369 MW for a total of 168,730 MW, from FTI Consulting, op. cit. note 1, p. 51.
86. EurObserv'ER, op. cit. note 1, p. 5.
87. Additions of Yunnan (3.25 GW), Hebei (1.66 GW), and Jiangsu (1.49 GW) from China National Energy Board, cited in Xiao, op. cit. note 15.
88. FTI Consulting. op. cit. note 1, p. 21. India's Generation Based Incentive was set to expire and the higher Accelerated Depreciation for wind power was set to be halved (to 40%) in the first quarter of 2017, from idem.
89. Turkey added 1387.75 MW for a total of 6106.05 MW, from Turkish Wind Energy Association, Turkish Wind Energy Statistics Report (Ankara: January 2017), pp. 4, 5. http://www.tureb.com.tr/files/tureb_sayfa/duyurular/2017_duyurular/subat/turkiye_ruzgar_enerjisi_istatistik_raporu_ocak_2017.pdf; Turkey added 1387 MW in 2016 for a total of 6081 MW, from WindEurope, Wind in Power 2016 European Statistics, p. 9, https://windeurope.org/wp-content/uploads/files/about-wind/statistics/WindEurope- Annual-Statistics-2016.pdf (Brussels: February 9, 2017); added 1387 MW for a total of 6081 MW, from GWEC, op. cit. note 1; added 1382.8 MW for a total of 6101.1 MW, from FTI Consulting, op. cit. note 1, p. 50.
90. Pakistan added 282 MW for a total of 591 MW, followed by the Republic of Korea (201 MW, for a total of 1031 MW) and Japan (196 MW; 3234 MW), from GWEC, op. cit. note 1; Pakistan added 373 MW for a total of 709 MW, followed by Republic of Korea (201 MW; 1006 MW) and Japan (196 MW; 3223 MW), from FTI Consulting, op. cit. note 1, pp. 51, 54; Pakistan added 335 MW for a total of 591 MW, followed by the Republic of Korea (198 MW; 1031 MW) and Japan (196 MW; 3234 MW), from WWEA, op. cit. note 1.
91. Texas added 2.611 MW, followed by Oklahoma (1462 MW), Iowa (707 MW), Kansas (687 MW), and North Dakota (603 MW), and the top states for total capacity were Texas (20, 321 MW), Iowa (6917 MW) and Oklahoma (6645 MW), from AWEA, "U.S. Wind Industry Fourth Quarter 2016 Market Update". http://awea.files.cms-plus.com/FileDownloads/pdfs/4Q2016%20AWEA%20Market%20Report%20Public%20Version.pdf (Washington, DC: January 26, 2017).
92. AWEA, op. cit. note 28.
93. Canada added 702 MW for a total of 11,898 MW, from Canadian Wind Energy Association (CanWEA), "Installed capacity". http://canwea.ca/wind-energy/

installed-capacity/ (accessed February 17, 2017). For comparison, in 2015 Canada added 1506 MW for a total of 11,205 MW, from CanWEA, "Wind energy continues rapid growth in Canada in 2015", press release. http://canwea.ca/wind-energy-continues-rapid-growthin-canada-in-2015/ (Ottawa: January 12, 2016). Added 702 MW for a total of 11,870 MW, from FTI Consulting, op. cit. note 1, p. 52, and added 702 MW for a total of 11,900 MW, from GWEC, op. cit. note 1.

94. CanWEA. "Powering Canada's future". http://canwea.ca/wp-content/uploads/2017/01/Canada-Current-Installed-Capacity_e.pdf (Ottawa: December 2016); largest source of new generation from CanWEA, "Wind energy in Canada". http://canwea.ca/windenergy/installed-capacity (accessed March 25, 2017).

95. WindEurope, op. cit. note 23; shares of onshore and offshore based on data from GWEC, op. cit. note 1, p. 16.

96. WindEurope, op. cit. note 23, p. 7.

97. WindEurope, op. cit. note 23, pp. 7, 9.

98. China, the EU and the United States from IEA, World Energy Outlook 2015 (Paris: 2015), p. 346; India had annual turbine production capacity of about 10 GW in early 2017, from FTI Consulting, op. cit. note 1, p. 21. In addition, Brazil has 3–3.5 GW of manufacturing capacity, from Zhao, op. cit. note 40.

99. FTI Intelligence, op. cit. note 2; FTI Consulting, op. cit. note 91, pp. 6, 10.

100. FTI Consulting, op. cit. note 91, p. 12.

101. Siemens. "Siemens to build rotor blade factory for wind turbines in Morocco", press release. https://www.siemens.com/press/en/pressrelease/?press=/en/pressrelease/2016/windpower-renewables/pr2016030214wpen.htm&content[]=WP (Hamburg: March 10, 2016).

102. Weston, D.; Knight, S. *Windpower Monthly.* Senvion acquired Kenersys' (Germany) Indian factory and product portfolio, "Senvion announces job cuts to secure 'competitiveness'". http://www.windpowermonthly.com/article/1427097/senvion-announces-job-cuts-secure-competitiveness (accessed March 13, 2017); Weston, D. *Windpower Monthly.* "Innogy opens Ireland office". http://www.windpowermonthly.com/article/1422366/innogy-opens-irelandoffice (accessed January 27, 2017); *Renews Biz.* "Dong opens Taiwan base", http://renews.biz/104935/dong-opens-taiwan-base/ (accessed November 16, 2016).

103. Nordex; Acciona Windpower. "Nordex and Acciona Windpower join forces to create a major player in the wind industry", press release. http://www.nordex-online.com/en/news-press/news-detail.html?tx_ttnews%5Btt_news%5D=2671&tx_ttnews%5BbackPid%5D=1&cHash=4e2718b412 (Hamburg: October 4, 2015).

104. Reuters. Jose Elias Rodriguez, "Siemens, Gamesa to form world's largestwind farm business". http://in.reuters.com/article/gamesa-m-a-siemens-idINKCN0Z30OI (accessed June 17, 2016); *Windpower Monthly.* "Review of 2016—part one". http://www.windpowermonthly.com/article/1419176/review-2016-part-one (accessed December 23, 2016).

105. Reuters, "Offshore wind can match coal, gas for value by 2025-RWE, E.ON, GE, others". http://af.reuters.com/article/energyOilNews/idAFL8N18Y16U?sp=true (accessed June 6, 2016).

106. Catapult, op. cit. note 162.

107. World Energy Outlook 2016. Projections for 2015 and 2016 are from a linear extrapolation based on data for 2010-14 from IEA (Paris: 2016). http://www.worldenergyoutlook.org/publications/weo-2016/
108. US capacity data based on FERC, op. cit. note 31. However, note that EIA (EIA, *Electric Power Monthly*, February 2017, Table 6.1) shows a net reduction in US bio-power capacity for 2016, with a year-end total of 14.1 GW.
109. European Commission, "Directive 2009/28/EC of the European Parliament and of the Council of 23 April 2009 on the promotion of the use of energy from renewable sources and amending and subsequently repealing Directives 2001/77/EC and 2003/30/EC" (Brussels: 2009). http://eur-lex.europa.eu/legal-content/EN/ALL/?uri=CELEX:32009L0028
110. BMWi, op. cit. note 31.
111. IEA, op. cit. note 4.
112. Matsubara, op. cit. note 31. Capacity figure does not include co-firing capacity. Biopower expansion is fuelled mainly by forestry products including imported chips and pellets and palm kernel shells. The domestic supply chain of chips from forestry is so far limited and high-cost.
113. IEA, op. cit. note 4.
114. Government of India, MNRE, op. cit. note 31.
115. Brazil from EPE, op. cit. note 31, and from MME, op. cit. note 31.
116. Page 47; Renewables 2017 Global Status Report; REN21: Paris, 2017.
117. Page 92; Renewables 2017 Global Status Report; REN21: Paris, 2017.
118. Page 93; Renewables 2017 Global Status Report; REN21: Paris, 2017.
119. EIA, *Monthly Energy Review*, April 2017, Table 10.3. https://www.eia.gov/totalenergy/data/monthly/#renewable
120. Fuel ethanol data from F.O. Licht, op. cit. note 48.
121. IEA Bioenergy Task 39, *The Potential of Biofuels in China* (Paris: 2016). http://task39.sites.olt.ubc.ca/files/2013/05/The-Potentialof-biofuels-in-China-IEA-Bioenergy-Task-39-September-2016.pdf
122. Fuel ethanol data from F.O. Licht, op. cit. note 48.
123. Licht, F. O. "Biodiesel: World Production, by Country", op. cit. note 49.
124. Licht, F. O. "Biodiesel: World Production, by Country", op. cit. note 49.
125. Licht, F. O. "Biodiesel: World Production, by Country", op. cit. note 49.
126. German data from BMWI, op. cit. note 31, p. 9; other data from Licht, F. O., "Biodiesel: World Production, by Country", op. cit. note 49. Note that F. O. Licht estimates German biodiesel production at 3.0 billion liters.
127. Sapp, M. Biofuels Digest. "New rules to ensure Indonesia achieves 20% blending target". http://www.biofuelsdigest.com/bdigest/2016/10/25/new-rules-to-ensureindonesia-achieves-20-biodieselblending/ (October 25, 2016); Licht F. O. "Biodiesel: World Production, by Country", op. cit. note 49.
128. Licht, F. O. "Biodiesel: World Production, by Country", op. cit. note 49; USDA, FAS, GAIN, *China Biofuels Annual* (Washington, DC: February 7, 2017). https://gain.fas.usda.gov/Recent%20GAIN%20Publications/Biofuels%20Annual_Beijing_China%20-%20Peoples%20Republic%20of_1-18-2017.pdf
129. Licht, F. O. "Biodiesel: World Production, by Country", op. cit. note 49.
130. Based on data in US Environmental Protection Agency, "RIN Generation and Renewable Fuel Volume Production by Fuel Type from January 2017". https://www.

epa.gov/fuels-registrationreporting-and-compliance-help/spreadsheet-rin-generation-andrenewable-fuel-0, posted February 2017.

131. Bioenergy Insight. "Dong Energy makes Studstrup plant run on wood pellets instead of coal". http://www.bioenergy-news.com/display_news/11180/dong_energy_makes_studstrup_plant_run_on_wood_pellets_instead_of_coal/ (accessed October 13, 2016).

132. Bioenergy Insights. "Japan prepares for biomass power plant surge and increases imports of wood chips". http://www.bioenergy-news.com/display_news/11938/japan_prepares_for_biomass_power_plant_surge_and_increases_imports_of_wood_chips/ (accessed February 27, 2017).

133. EIA. "Monthly Densified Biomass Fuel Report". https://www.eia.gov/biofuels/biomass/ (accessed April 28, 2017).

134. Fletcher, op. cit. note 26.

135. *Biomass Magazine*. "Pellet plants—operational". http://biomassmagazine.com/plants/listplants/pellet/US/Operational/ (accessed January 26, 2017).

136. OFGEM. "Biomass sustainability", https://www.ofgem.gov.uk/environmental-programmes/ro/applicants/biomasssustainability (accessed May 1, 2017); Rasmussen, S. L., Danish Energy Association, "The Danish Industry Agreement for Sustainable Biomass", undated presentation, https://ens.dk/sites/ens.dk/files/Bioenergi/the_danish_industry_agreement.pdf

137. Bioenergy Insights. "Bioenergy torrefaction: rich rewards", September/October 2016, p. 26. http://www.biomasstorrefaction.org/wp-content/uploads/2016/09/A-flexible-biomasstorrefaction-plant-has-recently-been-unveiled-in-Canada.pdf

138. Scandinavian Biopower Oy, "Scandinavian Biopower to invest in a biocoal plant in Mikkeli Finland—construction works to start late 2017", press release. http://www.biomasstorrefaction.org/wp-content/uploads/2016/12/20161129-Bio-coal-plant-RELEASE-FINALVERSION-DEF.pdf (Mikkeli, Finland: November 29, 2016).

139. EIA. "Petroleum and other liquids, fuel exports by destination, fuel ethanol". https://www.eia.gov/dnav/pet/pet_move_expc_a_EPOOXE_EEX_mbbl_a.htm (accessed March 14, 2017).

140. EIA. "Monthly biodiesel production report", December 2016. https://www.eia.gov/biofuels/biodiesel/production/

141. USDA, FAS, GAIN, op. cit. note 70.

142. Heida, L. *Biofuels Digest*. "Biofuels Nigeria signs deal for 16.5 million biodiesel plant in Kogi State". http://www.biofuelsdigest.com/bdigest/2017/02/20/biofuels-nigeria-ltd-signs-deal-for-16-5-million-biodieselplant-in-kogi-state/

143. DuPont. "DuPont and New Tianlong Industry Co., Ltd. sign historic deal to bring cellulosic ethanol technology to China", press release (Changchu, China: July 16, 2015), http://www.dupont.com/corporate-functions/media-center/press-releases/dupont-NTLsign-historic-deal-cellulosic-ethanol-tech-china.html

144. Lane, J. *Biofuels Digest*. "Sugar, sugar: Toray, Mitsui set out to build monster cellulosic sugar plant in Asia". http://www.biofuelsdigest.com/bdigest/2017/01/16/sugar-sugartoray-mitsui-set-out-to-build-monster-cellulosic-sugar-plant-in-asia/ (accessed January 16, 2016).

145. *Washington Post*, IRENA. op. cit. note 124; *Chelsea Harvey*, "United Airlines is flying on biofuels. Here's why that's a really big idea". https://www.washingtonpost.com/

news/energyenvironment/wp/2016/03/11/united-airlines-is-flying-on-biofuelsheres-why-thats-a-really-big-deal/?utm_term=.1c3b08c4bf1a (accessed March 11, 2016).

146. Bioenergy Insight. "BP buys Clean Energy Fuels' biomethane arm". http://www.bioenergy-news.com/display_news/11952/bp_buys_clean_energy_fuels_biomethane_arm/ (accessed March 2, 2017).

147. *Bioenergy Insight,* "First African grid connected biogas powered electricity plant comes on line in Kenya". http://www.bioenergy-news.com/display_news/11654/first_african_gridconnected_biogaspowered_electricity_plant_comes_online_in_kenya/(accessed January 11, 2017).

# CHAPTER 14

# Adoption of PCM-Based Novel Technology: Role of Awareness and Communication in Society

SAURABH MISHRA

*Rajiv Gandhi Institute of Petroleum Technology, Jais, Amethi, India*

*E-mail: smishra@rgipt.ac.in*

## ABSTRACT

Now has dawned the era of tailored energy access demand. Phase change materials have the ability to dissolve the challenge; assuaging through its wide and varied applications, but phase change materials (PCMs being still on a growing scale have relatively small market as compared to other renewable energy sources. Various reports and studies have forecasted a steep growth in advanced phase change material market, but in spite of these staggering market growth projections, awareness among various stakeholders remains one of the biggest challenges. Greater difficulty of the companies in the business of PCM is the lack of awareness among various stakeholders. Awareness has been compromised a lot on the pretext of acquiring bits and pieces of related information through general advertising and discourse at large. Reward and awareness have a linked approach, which is nowadays basking the societal consumption pattern aloud. Governmental approach in converging stakeholder's opinion in the favor of renewable energy resources: innovative approaches and demands have a lot to depend on this approach.

## 14.1   INTRODUCTION

Renewable energy resources have emerged as strategically pertinent options before the world's scientific and commercial entities. In wake of growing thirst for cleaner and greener energy resources, global community through serious endeavors has been able to make inroads in developing various alternative energy resources. Energy storage material market has witnessed a growth in recent times, with the world populace expectation refusing to just being satiated by the fulfillment of the energy hunger. Now has dawned the era of tailored energy access demand. Phase change materials have the ability to dissolve the challenge; assuaging through its wide and varied applications ranging widely as-construction, electronics, textiles, heating, ventilation and air conditioning (HVAC), packaging, medical aids, and refrigeration only to mention a few. Thermal energy storage systems are adept in facilitating reduction in energy demand at peak consumption time. Several developers in Germany, Slovenia, Japan, Russia and the Netherlands are working on new materials and techniques for all TES systems(IRENA, 2013). The global advanced phase change material (APCM) market is classified broadly, as per the product type, application and the geography; on the basis of matter, inorganic APCM has dominated the growth sector in 2016. Also in global APCM market, as per application base categorization, Bio-based PCMs are expected to surge ahead although the buildings and construction sector is projected to be the leader (Credence Research Global, 2017).

Looking at the APCM market growth and development scenario, CNESA (Chinese Energy Storage Alliance) suggests, in 2016, development in the global energy storage industry sped up, reaching an annual compound growth rate of over 86%; suggesting actual commercialization of the energy storage industry around the world (CNESA, 2017). A new market research report published in July 2017 (Credence Research Global) claims that the APCM market is expected to reach US$ 832.6 Mn by 2025, expanding at a CAGR of 14.8% during the forecast period 2017-2025. India Energy Storage Alliance (IESA) estimates India's energy storage market to be approximately over 70 GW and 200 GWh by 2022, which is one of the highest in the world (Fraunhofer, 2017). However, in spite of these staggering market growth projections, awareness among various stakeholders remains one of the biggest challenges that can hamper the report projections (Serenapeter, 2017).

Higher market penetration is on cards for the PCM-based products as the market gets driven by stringent regulations being imposed and improvised by the various regulatory bodies across the world, making the need for versatile renewable products like PCMs even more sought after and practically viable. PCMs being still on a growing scale have relatively small market as compared to other renewable energy sources. The News Magazine (2015) referred to the point citing the greater difficulty of the companies in the business of PCM being lack of awareness among various stakeholders. To fight this, companies are trying hard to make people acquainted with the advantages of the PCM-based products and their varied applications.

## 14.2 COMMUNICATION AND AWARENESS AS KEY FACTORS

Technological development and advancement are the basis of mitigating modern day challenges born due to unruly harness made by human race in search of comfort. Equally important is that these solutions gain visibility and acceptability by various stakeholders at respective levels. Policy formulation and their advocacy are pertinent assets while seeking transition support.

Studies have agreed that awareness and education are important determinants in affecting a behavioral and environmental attitude change among people (Jackson 2004; 2005; Gardner and Stern, 1996). Reward and awareness have a linked approach, which is nowadays basking the societal consumption pattern aloud. Governmental approach in converging stakeholder's opinion is in the favor of renewable energy resources: innovations approaches and demands have a lot to depend on this approach. Rewarding can be done in an objective manner with special focus on identified stakeholders; varied actors (Bresciani et al., 2016) can further ordain it.

Surveys around the world have shown that people at large recognize the ecological danger looming large and tender their prompt commitment to work toward its deference. As is reflected in a study conducted by the U.K. government, Department for Environment, Food and Rural Affairs (2007), 93% know something about climate change, 50% claim to understand the concept of carbon footprint, 62% claim to have become more environmentally active, and 78% have the strong desire to "do the right thing" around climate change. But it has been noted that people tend to overstate their claims on matters of energy, environment, and climate change at times even effecting unequable commitments (Vermeir and Verbeke, 2006; Lane and Potter, 2007; McDonald et al., 2009). This also the probable cause that people buying choices are ungoverned by their basic claims of empathy on these issues (Veleva, 2010). Heterogeneous approaches are banded in consumer's understanding, habit, approach, and consistencies of market for products related to energy and environmental change (Heinzle and Wustenhagen 2011; Finisterro do Paço et al., 2009).

Awareness seems compromised a lot on the pretext acquiring bits and pieces of related information through general advertising and discourse at large. In addition, another cause in today's world is the information/ content liberation through internet. Although many would only believe that this information liberation has contributed a great deal in making common masses aware about many contemporary issues. However, interestingly, this in recent times has started to come up with its negatives also. In medical profession the problem has been described as IDIOT acronym, which means: Internet Derived Information Obstruction Treatment. If we have a cautious look at the problem it surfaces that the observation is not a singular point. In other areas of awareness and transition applications, the phenomenon may find its space.

This unaccounted and uncontrolled awareness biz may yield emotions, which are positive as well as negative among the end users. Wherein they may be, either completely put off with the idea and hypothesis or may completely develop trust consecutively. Making other stakeholders believe that the task of awareness and education has accomplished. Herein at this juncture of misconception the entire effort lacks the shape, it should had fruitfully taken.

The consumers through there raw embodiment have even more to be told than before. Fairly speaking, it would now require a two-step process:

i)   Making consumers aware/realize that their knowledge is half-baked.
ii)  Educating consumers to the right context and knowledge base.

This debunking process is of paramount importance to educate and make aware the end users. Empathy and positive emotions would reap as a resultant of the process. These positive emotions if harvested can serve, as much needed foundations for hoisting the subjects to the next level of technology-transition ready state. The theory of positive emotions propagates that positive emotions in people help amplify people's awareness and create openness, which further help make them experiment with new ideas and alternative actions, closing the gap from emotions to behavior in general (Fredrickson, 2014).

Government recognition to awareness among stakeholders has been much on the footing of fear and consequences (Frijda, 1986; Frijda et al., 1989; Tooby and Cosmides, 1990; Lazarus, 1991; Levenson, 1994; Oatley and Jenkins, 1996). The basic parameters of awareness thus are floated upon the negative emotions rather than positive emotions, which, on contrary, act to narrow the mindset to frivolous routine response, which commits boastingly but fails to act. Hence, positivity in emotions can not only work to amplify mindset, but also help acquire resources in particular, like cognitive (love of learning, capacity to understand, and assimilate complex material), social, psychological, and physical resources to accrue on awareness (Young, 2017).

Technology-transition has come through in a big way as an important tool to help the societies grow and gain sustainability with ease. Technology awareness and technology transition have been known to be separated by infamous divide. It has much to do with the phenomenon of "bounded awareness", which spells that people do not "see" accessible and perceivable information while making decisions. Contrarily, their focus is more adept at other equally accessible information, thus keeping out a chunk of potentially useful information (Chugh and Bazerman, 2007), which of course is the part of their awareness, but is bounded as of now. This failure in riveting attention on viable section of acquired awareness is "focusing failure", which results due to misalignment between the information needed for a good decision and the information included in awareness (Chugh and Bazerman, 2007). A similar attribute has been expressed by Schkade and Kahneman (1998) as "focusing illusion"—a tendency to make decisions based on focus only on subset of all available information where often important set of information is underweighted by unattended information.

Thus, mindfulness-based interventions are the need of the hour to facilitate the awareness aspect with transition effect in the societal arena. Ruminative explanation of mindfulness has been put up in various studies: Brown and Ryan's (2004) expressed it as "open or receptive attention to and awareness of ongoing events and experience." While Ellen Langer (1989) defined four vital characteristics of mindfulness as: (1) continuous creation of new categories for structuring new perceptions; (2) more openness to new information; (3) awareness of more than one perspective in problem solving; and (4) greater sensitivity to one's environment. Interventions based on these pertinent characteristics wash off the focusing failure and promote the awareness, which is in turn translatable to action, compelled by the acquired knowledge to adopt the transformation by imbibed willingness.

## 14.3   SOCIAL SCENARIO

Current social energy fabric can safely start to boast as having a fair share of clean energy. Although the journey is yet not complete, but it has safely crossed over the test of questions on viability and sustainability. In its various avatars, clean energy options have mandated the overhauling change that the world is put up; adopting the clean energy for sustainable future. World energy outlook, 2017 (Fig. 14.1) clearly marks a change in the global average annual net capacity share of renewable sources moving forward from about 120 GW in 2010–2016 to 160 Gw in 2017–2040.

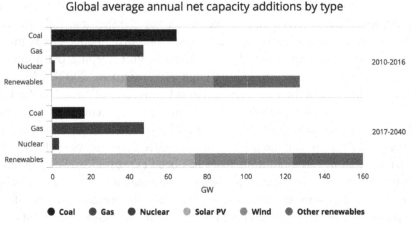

World Energy Outlook 2017, IEA

**FIGURE 14.1**   Increased contribution of renewable sources in global average annual net capacity.
*Source:* World Energy Outlook, 2017.

Energy democracy has been the new way that societies of today have scoped the liberalization of energy for all the citizens of the world. Countries like India and China have high primary energy demand targets setup

(Fig. 14.2) with focus on renewable sources through their policies and goals in a termed manner. This big demand needs the support of renewable sources in catching up with already worsening environmental situation in highly populated countries of the world. Developmental pollution has started to do more harm than good to the common masses. Advanced phase change materials with their ability to support the wide energy storage market can aid toward providing a cleaner and more viable source of energy security around the world. Such benefits have caused the social, economic, and environmental justice concerns to rise among the social fabric and translate the conventional fuel demands to renewable energy alternatives (Burke and Stephens, 2018).

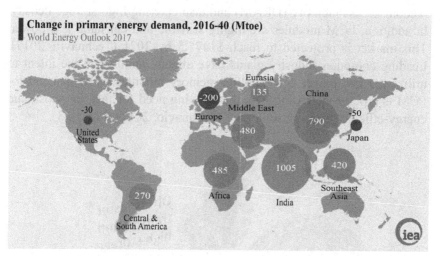

**FIGURE 14.2** High primary energy demand targets by China and India.
*Source:* World Energy Outlook, 2017.

Democratization of energy needs to be better delivered through modulation of various contextual renewable energy resources available such that these suit the ownership of end users of the market (Soutar and Mitchell, 2018). Energy poverty is another way that is put up to delegate the challenge of energy sufficiency across the world population. The dependence on conventional sources of energy has threadbare their limitations with regards to nature, availability, and rising costs. Especially in power-related sector where energy storage and allied usage has a high exhaustion rate, APCMs have the understandable capacity to dictate mitigation through sensible and latent heat utilization of materials as and when required, providing smart energy management.

Neoliberalism stands up as contrast to the concept of liberalization of clean energy. Today's national and international scenario stands witness to the parameters of economic growth and the energy consumption patterns. This relation between growth and energy liberalization can only be bettered out through an emancipatory energy transition. The need of the hour is not only to cultivate measures, which are sustainable, green, and renewable sources, but also posing ahead to tailor them such that they not only help create energy, but also spend less.

Phase change materials recent entry into global memory market has seized a favorable impact: PCM being four times faster than dynamic random access memory (DRAM) and also catering high storage density. In addition, PCM modules are highly scalable and consume less power. This market is projected to reach $1475.5 by 2021 (Technavio, 2017). Leading computer industry giants have already announced the intent to utilize the low power consumption memory modules. "The demand for PCM is rapidly increasing with the growing need for fast, low-cost, and energy-efficient memory solutions" (Technavio, 2017) (Fig. 14.3).

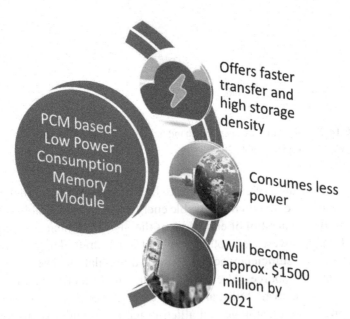

**FIGURE 14.3**   PCM-based global phase change memory market from 2017 to 2021.

## 14.4 CONCLUSION

Fast-paced population resurgence and high-energy demand have already burdened the conventional as well as unconventional energy resources. APCM bear the potential to predominantly redefine the renewable energy mix. A diverse portfolio of energy sources in tandem with good management and system design can help enhance security around the world (Edenhofer et al., 2011). Yet APCMs are faced up with unclear and poorly defined energy policies, inadequate focus on utility building and market growth. There is an urgent need to address regulatory aspect of adoption of renewable energy technology defining the standard and further scope of technological risks and failures. It has been noted that various regions and countries have made effort to frame policy structures, but these are mostly lacking the holistic vision and thus are unsuitable for implementation due to their immature nature. As per the International Renewable Energy Agency (IRENA) renewable energy initiatives have expressed that existing policy and regulatory environments are unsupportive and often act as barriers in rapid enhanced clean energy providence. In addition, the role of policymakers is of neglect toward the marginal contributors of the energy mix (IRENA, 2013).

APCMs with applications in HVAC, transportation, textiles, health, and energy storage have a widespread application arena. This can help in a potential growth in employment in the respective areas. It is indeed an asset to be recognized that renewable energy technologies can also be helpful in creating newer job opportunities. According to renewable energy study in 2008, around 2.3 million jobs worldwide were being provided by renewable energy sector, and this has gone on to improve the health health, education, gender equality, and environmental safety (Edenhofer et al., 2011). In addition, the cost of introducing any new technology is at the higher end in the beginning of endeavor, which is a reason where a lot of impetus get lost either due to lack at initial level of funding or for further improvement/advancement in the developed technologies. Fortunately, the cost technologies, like phase change materials, to deliver distributed renewable energy have declined sharply in recent years (Tawney et al., 2014).

The role policy and regulations in building a formidable communication with the various level of developers, stakeholders, and the end users is utmost pertinent. Defined regulatory structure and policy clarity would also help attract better finance opportunities in the area; also, it would help build awareness among the investors and the lenders and help them understand about distributed renewable energy opportunities (Alex and Athena, 2015). Democratic procedures have been at large responsible for the delay in deployment of renewable energy technologies (Peterson et al., 2015). Listening to voices of enterprises across geographical areas can provide valuable insights into which enabling policies and regulatory measures might facilitate increased investment in distributed renewable energy (Alex and Athena, 2015). Espouse transitioning to low-carbon energy systems like APCMs dependency on fundamental changes in technologies, policies, and institutions is galore and should be recognized with so as to better help communicate and proliferate the cause and action.

## KEYWORDS

- **awareness**
- **democratization**
- **mindfulness**
- **energy transition**
- **sensible energy**

## REFERENCES

1. Alex, D.; Athena, B. Clean Energy Access in Developing Countries: Perspectives on Policy and Regulation. World Resources Institute. 2015. https://www.wri.org/publication/clean-energy-access-developing-countries (accessed Jul 25, 2017).
2. Fredrickson, B. L. The Broaden-and-Build Theory of Positive Emotions. *Royal Soc. Publish.* 2014. https://www.ncbi.nlm.nih.gov/pmc/articles/PMC1693418/ (accessed Aug 03, 2017).
3. Brown, K. W.; Ryan, R. M. Perils and Promise in Defining and Measuring Mindfulness: Observations from Experience. *Clin. Psychol.: Sci. Prac.* **2004,** *11* (3), 242–248.
4. Bresciani, C.; Colorni, A.; Lia, F.; Nocerino, A. L. R. Behavioral Change and Social Innovation Through Reward: An Integrated Engagement System for Personal Mobility, Urban Logistics and Housing Efficiency. *Transportation Research Procedia.* **2016,** *14*, 353-361.
5. Chugh, D.; Bazerman, M. Bounded Awareness: What You Fail to See can Hurt You. *Rotman Magazine, Spring,* 2007, pp 21–25.

6. CNESA Energy Storage Industry White Paper. 2017. https://static1.squarespace.com /static/55826ab6e4b0a6d2b0f53e3d/t/5965bf4e78d171a8390f0cfb/1499840365633/ CNESA+White+Paper+2017+wm.pdf (accessed Aug 03, 2017).

7. Credence Research Global. Advanced Phase Change Material (APCM) Market - Growth, Future Prospects, and Competitive Analysis, 2017–2025. 2017. https:// www.credenceresearch.com/report/advanced-phase-change-material-apcm-market (accessed Aug 05, 2017).

8. DEFRA. A Framework for Pro-Environmental Behaviours, Department of Environment, Food and Rural Affairs. 2007. http://www.defra.gov.uk/evidence/series/index. htm (accessed Aug 05, 2017).

9. Edenhofer, O.; Pichs-Madruga, R.; Sokona, Y.; Seyboth, K.; Matschoss, P.; Kadner, S.; von Stechow, C. *Renewable Energy Sources and Climate Change Mitigation*; Cambridge University Press: Cambridge, UK, 2011.

10. Finisterra do Paço, A. M.; Barata Rapos, M. L.; Filho, W.L. Identifying the Green Consumer: A Segmentation Study. *J Target. Meas. Analysis Market.* **2009**, *17* (1),17–25.

11. Fraunhofer Introduction: Electrical Storage Systems ESS and Indian Scenario. 2017, Bangalore. https://www.fraunhofer.in/content/dam/indien/en/.../Report_ESS%20-% 20v2.pdf (accessed Aug 05, 2017).

12. Frijda, N. H. *The Emotions*; Cambridge University Press: Cambridge, UK, 1986.

13. Frijda, N. H.; Kuipers, P.; Schure, E. Relations Among Emotion, Appraisal, and Emotional Action Readiness. *J. Personality Social Psychol.* **1989**, *57*, 212–228.

14. Gardner, G. T.; Stern, P. C. *Environ. Problems Human Behav.*; Allyn & Bacon: USA, 1996.

15. Heinzle, S. L.; Wüstenhagen, R. Dynamic Adjustment of Eco-Labelling Schemes and Consumer Choice – the Revision of the EU Energy Label as a Missed Opportunity? *Business Strat. Environ.* **2012**, *21* (1), 60–70.

16. International Renewable Energy Agency (IRENA). IOREC 2012 International Off-Grid Renewable Energy Conference. 2013. http://www.irena.org/Document Downloads/Publications/IOREC_Key Findings and Recommendations.pdf (accessed Aug 04, 2017).

17. IRENA. Thermal Energy Storage Technology Brief. 2013. www.irena.org/Publications.

18. Soutar, I.; Mitchell, C. Towards Pragmatic Narratives of Societal Engagement in the UK Energy System. *Energy Res. Social Sci.* **2018**, *35*, 132–139.

19. Jackson, T. Negotiating Sustainable Consumption: A review of the Consumption Debate and its Policy Implications. *Energy Environ.* **2004**, *15*, 1027–1051.

20. Jackson, T. Motivating Sustainable Consumption. Sustain. *Dev. Res. Network* **2005**, *29*, 30.

21. Young, J. H. *Mindfulness Based Strategic Awareness Training*; Wiley Blackwell: U.K., 2017; p 19.

22. Lane, B.; Potter, S. The Adoption of Cleaner Vehicles in the UK: Exploring the Consumer Attitude-Action Gap. *J. Clean. Prod.* **2007**, *15*, 1085–1092.

23. Langer, E. J. *Mindfulness*; Reading: Addison-Wesley/Addison Wesley Longman: MA, USA, 1989.

24. Lazarus, R. S. *Emotion and Adaptation*; Oxford University Press: New York, 1991.

25. Levenson, R. W. Human Emotions: A Functional View. In *The Nature of Emotion: Fundamental Questions*; Ekman, P., Davidson, R., Ed.; Oxford University Press: New York, 1994; pp 123–126.:.

26. McDonald, S.; Oates, C.; Thyne, M. Comparing Sustainable Consumption Patterns Across Product Sectors. *Int. J. Consum. Stud.* **2009,** *33,* 137–145M.

27. Burke, J.; Stephens, J. C. Political Power and Renewable Energy Futures: A Critical Review. Energy Res. Soc. Sci. **2008,** *35,* 78–93.

28. Oatley, K.; Jenkins, J. M. *Understanding Emotions*; Blackwell: Cambridge, MA, 1996.

29. Technavio Phase Change Memory Market - Trends and Forecasts. 2017. https://www.businesswire.com/news/home/20170426006249/en/Phase-Change-Memory-Market---Trends-Forecasts (accessed Aug 05, 2017).

30. Phase Change Materials Market-Segmented by Product Type, End-user, Encapsulation Technology, and Geography-Growth, Trends and Forecast 2018-2023. 2018. Mordorintelligence. https://www.mordorintelligence.com/industry-reports/global-phase-change-material-market-industry (accessed Aug 03, 2017).

31. Schkade, D. A.; Kahneman. D. Does Living in California Make People Happy? A Focusing Illusion in Judgments of Life Satisfaction. *Psychol. Sci.* **1998,** *9* (5), 340–346.

32. Serenapeter Advanced Phase Change Material (APCM) Market Is Projected To Reach US$832.6mn By 2025 – Credence Research". Industry News. 2017. http://www.seotraininginpune.co.in/industry-news/advanced-phase-change-material-apcm-market-is-projected-to-reach-us832-6mn-by-2025-credence-research/ (accessed Aug 03, 2017).

33. Tawney, L.; Jairaj, B.; Lu, X. The Real Story Behind Falling Renewable Energy Investments, October 22, 2014. http://www.wri.org/blog/ real-story-behind-falling-renewable-energyinvestments (accessed Aug 05, 2017).

34. Tooby, J.; Cosmides, L. The Past Explains the Present: Emotional Adaptations and the Structure of Ancestral Environments. *Ethol. Sociobiol.* **1990,** *11,* 375–424.

35. Peterson, T. R.; Stephens, J. C.; Wilson, E. J. Public Perception of and Engagement with Emerging Low-Carbon Energy Technologies: A Literature Review *MRS Energy Sustain.*, **2015,** *2,* DOI: 10.1557/mre.2015.12.

36. Tanti, T.The Key Trends that will Shape Renewable Energy in 2018 and Beyond at World Economic Forum Annual Meeting, Jan 23–26, 2018; Davos-Klosters: Switzerland, 2018.

37. Veleva, V. R. Managing Corporate Citizenship: A New Tool for Companies. *Corp. Soc. Responsib. Environ. Manage.* **2010,** *17* (1), 40–51.

38. Vermeir, I.; Verbeke, W. Sustainable Food Consumption: Exploring the Consumer "Attidude-Behaviour Alintention" gap. *J. Agric. Environ. Ethics* **2008,** *19* 169–194.

39. Doshi, Y. Advanced Phase Change Materials Market by Type (Organic, Inorganic and Bio-based), Application (Building & Construction, Energy Storage, HVAC, Shipping & Transportation, Electronics, Textiles, and Others)-Global Opportunity Analysis and Industry Forecast, 2014-2022. Allied Market Research, 2017.

# Index

## T

## U

Printed in the United States
by Baker & Taylor Publisher Services